Instrumental Methods in Metal Ion Speciation

CHROMATOGRAPHIC SCIENCE SERIES

A Series of Textbooks and Reference Books

Editor: JACK CAZES

1. Dynamics of Chromatography: Principles and Theory, *J. Calvin Giddings*
2. Gas Chromatographic Analysis of Drugs and Pesticides, *Benjamin J. Gudzinowicz*
3. Principles of Adsorption Chromatography: The Separation of Nonionic Organic Compounds, *Lloyd R. Snyder*
4. Multicomponent Chromatography: Theory of Interference, *Friedrich Helfferich and Gerhard Klein*
5. Quantitative Analysis by Gas Chromatography, Josef Novák
6. High-Speed Liquid Chromatography, *Peter M. Rajcsanyi and Elisabeth Rajcsanyi*
7. Fundamentals of Integrated GC-MS (in three parts), *Benjamin J. Gudzinowicz, Michael J. Gudzinowicz, and Horace F. Martin*
8. Liquid Chromatography of Polymers and Related Materials, *Jack Cazes*
9. GLC and HPLC Determination of Therapeutic Agents (in three parts), Part 1 *edited by Kiyoshi Tsuji and Walter Morozowich*, Parts 2 and 3 *edited by Kiyoshi Tsuji*
10. Biological/Biomedical Applications of Liquid Chromatography, *edited by Gerald L. Hawk*
11. Chromatography in Petroleum Analysis, *edited by Klaus H. Altgelt and T. H. Gouw*
12. Biological/Biomedical Applications of Liquid Chromatography II, *edited by Gerald L. Hawk*
13. Liquid Chromatography of Polymers and Related Materials II, *edited by Jack Cazes and Xavier Delamare*
14. Introduction to Analytical Gas Chromatography: History, Principles, and Practice, *John A. Perry*
15. Applications of Glass Capillary Gas Chromatography, *edited by Walter G. Jennings*
16. Steroid Analysis by HPLC: Recent Applications, *edited by Marie P. Kautsky*
17. Thin-Layer Chromatography: Techniques and Applications, *Bernard Fried and Joseph Sherma*
18. Biological/Biomedical Applications of Liquid Chromatography III, *edited by Gerald L. Hawk*
19. Liquid Chromatography of Polymers and Related Materials III, *edited by Jack Cazes*
20. Biological/Biomedical Applications of Liquid Chromatography, *edited by Gerald L. Hawk*

21. Chromatographic Separation and Extraction with Foamed Plastics and Rubbers, *G. J. Moody and J. D. R. Thomas*
22. Analytical Pyrolysis: A Comprehensive Guide, *William J. Irwin*
23. Liquid Chromatography Detectors, *edited by Thomas M. Vickrey*
24. High-Performance Liquid Chromatography in Forensic Chemistry, *edited by Ira S. Lurie and John D. Wittwer, Jr.*
25. Steric Exclusion Liquid Chromatography of Polymers, *edited by Josef Janca*
26. HPLC Analysis of Biological Compounds: A Laboratory Guide, *William S. Hancock and James T. Sparrow*
27. Affinity Chromatography: Template Chromatography of Nucleic Acids and Proteins, *Herbert Schott*
28. HPLC in Nucleic Acid Research: Methods and Applications, *edited by Phyllis R. Brown*
29. Pyrolysis and GC in Polymer Analysis, *edited by S. A. Liebman and E. J. Levy*
30. Modern Chromatographic Analysis of the Vitamins, *edited by André P. De Leenheer, Willy E. Lambert, and Marcel G. M. De Ruyter*
31. Ion-Pair Chromatography, *edited by Milton T. W. Hearn*
32. Therapeutic Drug Monitoring and Toxicology by Liquid Chromatography, *edited by Steven H. Y. Wong*
33. Affinity Chromatography: Practical and Theoretical Aspects, *Peter Mohr and Klaus Pommerening*
34. Reaction Detection in Liquid Chromatography, *edited by Ira S. Krull*
35. Thin-Layer Chromatography: Techniques and Applications, Second Edition, Revised and Expanded, *Bernard Fried and Joseph Sherma*
36. Quantitative Thin-Layer Chromatography and Its Industrial Applications, *edited by Laszlo R. Treiber*
37. Ion Chromatography, *edited by James G. Tarter*
38. Chromatographic Theory and Basic Principles, *edited by Jan Åke Jönsson*
39. Field-Flow Fractionation: Analysis of Macromolecules and Particles, *Josef Janca*
40. Chromatographic Chiral Separations, *edited by Morris Zief and Laura J. Crane*
41. Quantitative Analysis by Gas Chromatography, Second Edition, Revised and Expanded, *Josef Novák*
42. Flow Perturbation Gas Chromatography, *N. A. Katsanos*
43. Ion-Exchange Chromatography of Proteins, *Shuichi Yamamoto, Kazuhiro Naka-nishi, and Ryuichi Matsuno*
44. Countercurrent Chromatography: Theory and Practice, *edited by N. Bhushan Man-dava and Yoichiro Ito*
45. Microbore Column Chromatography: A Unified Approach to Chromatography, *edited by Frank J. Yang*
46. Preparative-Scale Chromatography, *edited by Eli Grushka*
47. Packings and Stationary Phases in Chromatographic Techniques, *edited by Klaus K. Unger*
48. Detection-Oriented Derivatization Techniques in Liquid Chromatography, *edited by Henk Lingeman and Willy J. M. Underberg*
49. Chromatographic Analysis of Pharmaceuticals, *edited by John A. Adamovics*

50. Multidimensional Chromatography: Techniques and Applications, *edited by Hernan Cortes*
51. HPLC of Biological Macromolecules: Methods and Applications, *edited by Karen M. Gooding and Fred E. Regnier*
52. Modern Thin-Layer Chromatography, *edited by Nelu Grinberg*
53. Chromatographic Analysis of Alkaloids, *Milan Popl, Jan Fähnrich, and Vlastimil Tatar*
54. HPLC in Clinical Chemistry, *I. N. Papadoyannis*
55. Handbook of Thin-Layer Chromatography, *edited by Joseph Sherma and Bernard Fried*
56. Gas–Liquid–Solid Chromatography, *V. G. Berezkin*
57. Complexation Chromatography, *edited by D. Cagniant*
58. Liquid Chromatography–Mass Spectrometry, *W. M. A. Niessen and Jan van der Greef*
59. Trace Analysis with Microcolumn Liquid Chromatography, *Milos Krejcl*
60. Modern Chromatographic Analysis of Vitamins: Second Edition, *edited by André P. De Leenheer, Willy E. Lambert, and Hans J. Nelis*
61. Preparative and Production Scale Chromatography, *edited by G. Ganetsos and P. E. Barker*
62. Diode Array Detection in HPLC, *edited by Ludwig Huber and Stephan A. George*
63. Handbook of Affinity Chromatography, *edited by Toni Kline*
64. Capillary Electrophoresis Technology, *edited by Norberto A. Guzman*
65. Lipid Chromatographic Analysis, *edited by Takayuki Shibamoto*
66. Thin-Layer Chromatography: Techniques and Applications: Third Edition, Revised and Expanded, *Bernard Fried and Joseph Sherma*
67. Liquid Chromatography for the Analyst, *Raymond P. W. Scott*
68. Centrifugal Partition Chromatography, *edited by Alain P. Foucault*
69. Handbook of Size Exclusion Chromatography, *edited by Chi-San Wu*
70. Techniques and Practice of Chromatography, *Raymond P. W. Scott*
71. Handbook of Thin-Layer Chromatography: Second Edition, Revised and Expanded, *edited by Joseph Sherma and Bernard Fried*
72. Liquid Chromatography of Oligomers, *Constantin V. Uglea*
73. Chromatographic Detectors: Design, Function, and Operation, *Raymond P. W. Scott*
74. Chromatographic Analysis of Pharmaceuticals: Second Edition, Revised and Expanded, *edited by John A. Adamovics*
75. Supercritical Fluid Chromatography with Packed Columns: Techniques and Applications, *edited by Klaus Anton and Claire Berger*
76. Introduction to Analytical Gas Chromatography: Second Edition, Revised and Expanded, *Raymond P. W. Scott*
77. Chromatographic Analysis of Environmental and Food Toxicants, *edited by Takayuki Shibamoto*
78. Handbook of HPLC, *edited by Elena Katz, Roy Eksteen, Peter Schoenmakers, and Neil Miller*
79. Liquid Chromatography–Mass Spectrometry: Second Edition, Revised and Expanded, *Wilfried Niessen*
80. Capillary Electrophoresis of Proteins, *Tim Wehr, Roberto Rodríguez-Díaz, and Mingde Zhu*

81. Thin-Layer Chromatography: Fourth Edition, Revised and Expanded, *Bernard Fried and Joseph Sherma*
82. Countercurrent Chromatography, *edited by Jean-Michel Menet and Didier Thiébaut*
83. Micellar Liquid Chromatography, *Alain Berthod and Celia García-Alvarez-Coque*
84. Modern Chromatographic Analysis of Vitamins: Third Edition, Revised and Expanded, edited by *André P. De Leenheer, Willy E. Lambert, and Jan F. Van Bocxlaer*
85. Quantitative Chromatographic Analysis, *Thomas E. Beesley, Benjamin Buglio, and Raymond P. W. Scott*
86. Current Practice of Gas Chromatography–Mass Spectrometry, *edited by W. M. A. Niessen*
87. HPLC of Biological Macromolecules: Second Edition, Revised and Expanded, *edited by Karen M. Gooding and Fred E. Regnier*
88. Scale-Up and Optimization in Preparative Chromatography: Principles and Bio-pharmaceutical Applications, *edited by Anurag S. Rathore and Ajoy Velayudhan*
89. Handbook of Thin-Layer Chromatography: Third Edition, Revised and Expanded, *edited by Joseph Sherma and Bernard Fried*
90. Chiral Separations by Liquid Chromatography and Related Technologies, *Hassan Y. Aboul-Enein and Imran Ali*
91. Handbook of Size Exclusion Chromatography and Related Techniques: Second Edition, *edited by Chi-San Wu*
92. Handbook of Affinity Chromatography: Second Edition, *edited by David S. Hage*
93. Chromatographic Analysis of the Environment: Third Edition, *edited by Leo M. L. Nollet*
94. Microfluidic Lab-on-a-Chip for Chemical and Biological Analysis and Discovery, *Paul C.H. Li*
95. Preparative Layer Chromatography, *edited by Teresa Kowalska and Joseph Sherma*
96. Instrumental Methods in Metal Ion Speciation, *Imran Ali and Hassan Y. Aboul-Enein*

Instrumental Methods in Metal Ion Speciation

Imran Ali
National Institute of Hydrology
Roorkee, India

Hassan Y. Aboul-Enein
King Faisal Specialist Hospital and Research Centre
Riyadh, Saudi Arabia

CRC Press is an imprint of the
Taylor & Francis Group, an **informa** business
A TAYLOR & FRANCIS BOOK

Published in 2006 by
CRC Press
Taylor & Francis Group
6000 Broken Sound Parkway NW, Suite 300
Boca Raton, FL 33487-2742

First issued in paperback 2019

ISBN 13: 978-0-367-45380-0 (pbk)
ISBN 13: 978-0-8493-3736-9 (hbk)

Library of Congress Card Number 2005054894

Library of Congress Cataloging-in-Publication Data

Ali, Imran.
Instrumental methods in metal ion speciation : chromatography, capillary electrophoresis, spectroscopy, electrochemistry / by Imran Ali and Hassan Y. Aboul-Enein.
p. cm.
Includes bibliographical references and index.
ISBN 0-8493-3736-4 (alk. paper)
1. Metal ions--Toxicology. 2. Metal ions--Identification. 3. Metal ions--Classification. I. Aboul-Enein, Hassan Y. II. Title.

RA1231.M52A45 2005
615.9'253--dc22 2005054894

Visit the Taylor & Francis Web site at
http://www.taylorandfrancis.com

and the CRC Press Web site at
http://www.crcpress.com

Dedications

Dedicated to the memory of my late parents:
Basheer Ahmed and Mehmudan Begum

Imran Ali

To my loving family — Nagla, Youssef, Faisal, and Basil
— for your support and enlightenment

Hassan Y. Aboul-Enein

Preface

Transition metal ions, lanthanides, and actinides are widely distributed in environmental and biological materials. These metal ions exist in various oxidation states, having different physical and chemical properties leading to different toxicities. Many reports are available in the literature on the analysis of total concentrations of metal ions, but the data presented may not be reliable and may be incorrect due to the different toxicities of various oxidation states a particular metal ion can have, which confuses the toxicity concept. Therefore, to predict the exact toxicities of metal ion species, the determination of the concentrations of all species is required and essential. Knowledge of metal ion speciation is essential for environmental and industrial chemists, scientists from various disciplines, and regulatory authorities. Therefore, the separation and identification of metal ion species (speciation) is important and demanding in this developing world.

This book deals with the distribution, toxicities, and state-of-the-art techniques in metal ion speciation by gas chromatography, liquid chromatography (high-performance liquid chromatography [HPLC], ion chromatography [IC], ion-pair chromatography [IPC], micellar electrokinetic chromatography [MEKC], size exclusion chromatography [SEC], chiral chromatography [CC], capillary electrochromatography [CEC], and supercritical fluid chromatography [SFC]), capillary electrophoresis (CE), and electrochemical and radiochemical methods. This book describes metal ion speciation in detail with a main emphasis on experimental methodologies. Moreover, we attempt to explain the optimization techniques for metal ion speciation that will be helpful in designing future experiments in this area. The importance and uniqueness of this book lies in the fact that, unlike other books on this subject, it focuses only on metal ion speciation (analysis of different species of a particular metal ion). We hope that it will be a useful reference for scientists, researchers, academicians, and graduate students working in the field of metal ion speciation. This book will also fulfill the requirements of regulatory authorities for formulating the regulations and legislations to control the distribution of different metal ion species into our environment and other living entities.

Imran Ali
Roorkee, India
drimran_ali@yahoo.com

Hassan Y. Aboul-Enein
Riyadh, Saudi Arabia

About the Book

This book describes distribution, toxicities, and techniques of metal ion speciation. The techniques discussed are gas and liquid chromatography, capillary electrophoresis, electrochemical and radiochemical methods. The different liquid chromatographic approaches discussed are high-performance liquid chromatography (HPLC), ion chromatography (IC), ion-pair chromatography (IPC), micellar electrokinetic chromatography (MEKC), size exclusion chromatography (SEC), chiral chromatography (CC), capillary electrochromatography (CEC), and supercritical fluid chromatography (SFC). This book is divided into 12 chapters: the first chapter is an introduction to the principles of metal ion speciation followed by Chapter 2 through Chapter 11 that discuss the distribution, toxicity, sample preparation, and metal ion speciation into the environmental and biological samples using the above-mentioned techniques. Chapter 12 describes the perspectives of metal ion speciation and legislation. The optimization of the experimental parameters of the analytical techniques is discussed; hence, this book may be considered as an applied text in the area of metal ion speciation. This book is unique in the area of metal ion speciation, as it deals exclusively with metal ion speciation. By fulfilling the needs of the scientific community for an information resource focusing on metal ion speciation, it will become a necessary reference source for scientists, academicians, researchers, and others in such diverse positions as regulators, material industries specialists, clinicians and nutritional experts, agriculturalists, and environmentalists.

Acknowledgments

I express my deepest respect and gratitude and love to my wife Seema Imran who has helped and supported me in my efforts to carry out this work. My loving and sweet thanks also go to my dearest son Al-Arsh Basheer Baichain who has given me continual freshness and inspiration during the completion of this difficult job. I would also like to acknowledge my other family members and relatives who have helped me directly and indirectly during this period.

I also sincerely thank Professor Vinod K. Gupta, Chemistry Department, Indian Institute of Technology, Roorkee, India, who helped me to complete this book. Moreover, his moral support, which I received continuously, was monumental and will be forever treasured. Finally, the administration of National Institute of Hydrology, Roorkee, India, is also acknowledged for permitting me to write this book.

Imran Ali

I would like to express my thanks to the administration of King Faisal Specialist Hospital and Research Centre for their support of this work. Special thanks are extended to the editorial staff of Taylor & Francis publishers for their assistance in publishing this book. I am particularly grateful to my wife, Nagla El-Mojadaddy, for her forbearance and support throughout the preparation of this book. It is to her that I extend my deepest gratitude.

Hassan Y. Aboul-Enein

The Authors

Hassan Y. Aboul-Enein is a Principal Scientist and the Head of the Pharmaceutical Analysis and Drug Development Laboratory, Centre for Clinical Research, at King Faisal Specialist Hospital and Research Centre, Riyadh, Saudi Arabia. He is the author or coauthor of over 600 refereed journal articles, 30 book chapters, and 270 conference presentations. He is the author or coauthor of seven books, including *Chiral Separation by Liquid Chromatography and Related Technologies* (Marcel Dekker, Inc., New York), *The Impact of Stereochemistry on Drug Development and Use* (John Wiley & Sons, New York), *Chiral Pollutants: Distribution, Toxicity and Analysis by Chromatography and Capillary Electrophoresis* (John Wiley & Sons, Chichester, UK), among others. He is a member of the editorial boards of several journals, including *Analytica Chimica Acta*, *Chirality*, *Biomedical Chromatography*, *Analytical Letters*, *Journal of Liquid Chromatography and Related Technologies*, *Biopolymers*, *Biosensors and Bioelectronics*, to mention a few. Professor Aboul-Enein is a member of the World Health Organization's (WHO) advisory panel on international pharmacopeia and pharmaceutical preparations and a Fellow of the Royal Society of Chemistry (United Kingdom). He earned his B.Sc. degree (1964) in pharmacy and pharmaceutical chemistry from Cairo University, Cairo, Egypt, and his M.Sc. (1969) and Ph.D. (1971) degrees in pharmaceutical and medicinal chemistry from the University of Mississippi, U.S. Professor Aboul-Enein's current research interests are in the field of pharmaceutical and biomedical analysis and drug development with special emphasis on chiral chromatography, ion-selective electrodes, and other separation techniques.

Dr. Imran Ali is presently a scientist at the National Institute of Hydrology, Roorkee, India. He earned his M.Sc. (1986) and Ph.D. (1990) degrees from the University of Roorkee, India (now the Indian Institute of Technology, Roorkee, India). His research areas of interest are chiral analysis of biological and environmental active chiral compounds and metal ion speciation using chromatographic and capillary electrophoresis techniques. He also has expertise in water quality and wastewater treatment methodologies. Dr. Ali is author and coauthor of more than 100 journal articles and book and encyclopedia chapters. He is the coauthor of two books: *Chiral Separations by Liquid Chromatography and Related Technologies* (Marcel Dekker, Inc., New York) and *Chiral Pollutants: Distribution, Toxicity and Analysis by Chromatography and Capillary Electrophoresis* (John Wiley & Sons, Chichester, UK). Dr. Ali was awarded a Khosla Research Award in 1987 by the Indian Institute of Technology, Roorkee, India, for his work on the chiral resolution of amino acids. He is a life member of the Indian Science Congress Association, India.

The Authors

Contents

1 Introduction

1.1 IMPORTANCE OF THE ENVIRONMENT

"A sound mind exists in a sound body" is a well-known proverb in English literature, and the same is true for our environment. A neat and clean environment is essential for the proper growth, health, and persistence of human beings and other organisms. The conservation and protection of the environment are essential in the present industrialized and developing world. Overgrowth of the population in some countries is one of the most pressing problems of our age. The pollution problem of the environment has now reached a level that serves as a potential threat not only to the health of our environment but also to the entire population. The quality of our environment is deteriorating continuously due to the accumulation of undesirable constituents. The main sources of the contamination are industrialization, domestic activities, agricultural activities, atomic explosions, and other environmental and global changes. These activities and changes, if improperly controlled, can destroy the quality of our environment. Broadly, the environment is divided into three parts: the atmosphere, including the air sphere around the earth; the lithosphere, which consists of the earth; and the hydrosphere, which consists of all water bodies, including ocean, surface, and ground water. Hydrosphere and the atmosphere components of the environment are directly and readily available for contamination by pollutants. Therefore, the quality of these environmental constituents is deteriorating continuously, which is problematic and deserving of serious consideration. Again, the notorious pollutants find their way easily through water bodies, while the atmosphere is being contaminated by gases and volatile organic pollutants. Consequently, ground and surface waters at some places in the world are not suitable for drinking due to the presence of toxic pollutants. Similarly, the quality of the atmosphere in some cities and industrial areas of the world is not safe for human health. The importance of conservation and improvement of the environment is critical and urgent.[1–3] Environmental awareness has grown dramatically in recent decades, and several nations are taking the lead in implementing laws related to the environment. In view of this, environmental authorities are asking for accurate data and information on pollution levels and for the improvement of measures to control the contamination of the environment.

1.2 ENVIRONMENTAL POLLUTANTS

Any undesirable and toxic chemical, commodity, organism, or other object present in the environment may be considered an environmental pollutant. The pollutant may be present in solid, liquid, or gas form. In all forms, the presence of toxic pollutants is serious and harmful to human beings and other useful organisms. Broadly, environmental pollutants may be categorized into chemical and biological classes. Chemical pollutants are organic and inorganic compounds; biological contaminants are toxic microbes. The presence of inorganic pollutants, especially toxic metal ions, is a serious issue, as metal ions may often be carcinogenic in nature.[4–12] The route of these toxic metal ions to the human body is through water and other foodstuffs. Therefore, the monitoring of metal ions in water bodies and foodstuffs is essential and important. Prior to supplying water for drinking, bathing, agriculture, and other purposes, it is essential to determine the concentrations of these pollutants, if present. In addition, some toxic metal ions are also present in the atmosphere and indirectly affect our health; hence, the monitoring of air samples, especially of polluted areas, i.e., near roads, railway tracks, industries, and so forth, is critical. Trace elements date back to the early 20th century as many elements were recognized to exist at very low concentrations. During the following 60 years, all efforts have been made to focus on total trace element concentrations.

1.3 METAL IONS SPECIATION

Among the various metal ions, transition metals and some lanthanides and actinides are harmful and toxic. It is also interesting to observe that the transition metals, lanthanides, and actinides exist in different oxidation states. This is due to the presence of empty d orbitals in transition metal ions and f orbitals in lanthanides and actinides, respectively. Therefore, these metal ions occur in the environment in different oxidation states and form various species, for example, arsenic (As) as As(III), As(V), and organic arsenic species.[13] The different oxidation states of some toxic metal ions are given in Table 1.1. The different oxidation states of a particular metal ion possess different physical and chemical properties. Mainly, these oxidation states differ in their redox potential, complexation, and hydration properties. It is very important to observe and note that these different oxidation states have different toxicities. Basically, speciation is a concept used frequently in biological science, and it was adopted by those in analytical chemistry, expressing the idea that the specific chemical forms of an element should be considered individually. The separation and identification of the different oxidation states of a particular metal ion is known as speciation. The concept of speciation dates back to 1954 when Goldberg[14] introduced the concept of speciation to improve the understanding of the biogeochemical cycling of trace elements in seawater. Since then, this development has been growing exponentially to the point that research on trace element analysis is being con-

TABLE 1.1
Different Oxidation States of Some Toxic Metal Ions

Metal Ion	Oxidation States
As	As(III) and As(V)
Cd	Cd(I) and Cd(II)
Cu	Cu(I) and Cu(II)
Cr	Cr(III) and Cr(VI)
Co	Co(II) and Co(III)
Hg	Hg(I) and Hg(II)
Fe	Fe(II) and Fe(III)
Mn	Mn(I) and Mn(II)
Ni	Ni(II) and Ni(III)
Pb	Pb(II) and Pb(IV)
Sb	Sb(III) and Sb(V)
Se	Se(IV) and Se(VI)
Ti	Ti(II), Ti(III), and Ti(IV)
Zn	Zn(I) and Zn(II)

ducted. Today, research is almost exclusively focused on trace element species. Kinetic and thermodynamic information together with analytical data make it possible to differentiate between oxidized versus reduced, complexed or chelated versus free metal ions in dissolved form. Of course, the analysis of the total concentration of metal ions is required and essential. There are many reports available in the literature on the analysis of metal ions present in our environment, but the data presented are not reliable. This unreliability is due to the different toxicities of various oxidation states of particular metal ions, which confuses the toxicity concept.[15–16] Due to these facts, the knowledge of metal ion speciation is essential for environmental and industrial chemists and scientists of other analytical laboratories. In view of these facts, attempts have been made to explain the term *speciation* in the following sections.

Templeton et al.[17] described the speciation term as per the International Union for Pure and Applied Chemistry (IUPAC). Authors defined elemental speciation as follows:

1. Chemical species or chemical element: specific form of an element defined as to isotopic composition, electronic or oxidation state, or complex or molecular structure
2. Speciation analysis: separation and identification of one or more individual chemical species in a sample
3. Speciation of an element: distribution of an element among defined chemical species in a system; when elemental speciation is not feasible, the term *fractionation* will be used (as defined next)

4. Fractionation: classification of an analyte or a group of analytes from a certain sample according to physicochemical properties, such as size, solubility, bonding, reactivity, and so forth

It is often not possible to determine the individual concentrations of the different chemical species that add up to the total concentration of an element in a given matrix. Often, chemical species present in a given sample are not stable enough to be determined as such. During the procedure, the partitioning of the element among its species may be changed. For example, this can be caused by a change in pH necessitated by the analytical procedure, or by intrinsic properties of measurement methods that affect the equilibrium between species.

1.4 METAL ION SPECIATION AND THE ENVIRONMENT

Originally, the concentration of elemental species of anthropogenic origin was zero, but today, they are present in the environment, as they continue to be distributed in a manner that affects the life cycle. At the same time, it is also highly plausible that a certain background level of these anthropogenic metal ions can be tolerated without any adverse effect. In order to assess the impact of low background levels of element species, one has to develop separation and detection techniques that surpass the performance of the existing speciation methodology. Total trace element concentration may be static, and the species may be highly dynamic; hence, they change continuously with respect to changes in the environmental parameters such as pH, concentration of ligands for complex formation, physiological state of a cell, and state of health of a living body. The thermodynamic and kinetic stability of elemental species in the environment must also be taken into account. Unstable species in the environment are predominant, and this state requires some special attention. Some metal ions are stable in living systems, such as cobalt in vitamin B_{12} and iron in hemoglobin. The fate of the trace element in life is of great importance. Metal ions enter into the environment through natural or anthropogenic activities. These metal ions complex with suitable ligands. Natural methylation of metal ions under specific conditions is prevalent. The complex species (organic metal ion species), sometimes, are more or less toxic, as is the case with methylated mercury, or less toxic as in the case of arsenic.

As regard to the toxicities of a metal ion species, sometimes one oxidation state may be more toxic than the other. Various metallic ions are found in the environment in a variety of forms that are differentiated not only by their physical and chemical forms, but also by their diverse toxic activities with respect to living organisms. Changes in the degree of oxidation of an element also have an important affect on the degree of bioavailability and toxicity.[16] In an oxidized environment, As, Sb, and Se appear mostly as oxyanions.[15] The valency state of a metal ion controls the behavior of a metal ion in the aqueous system. For example, the toxicity of As(III) and Sb(III) is higher than that of their pentavalent

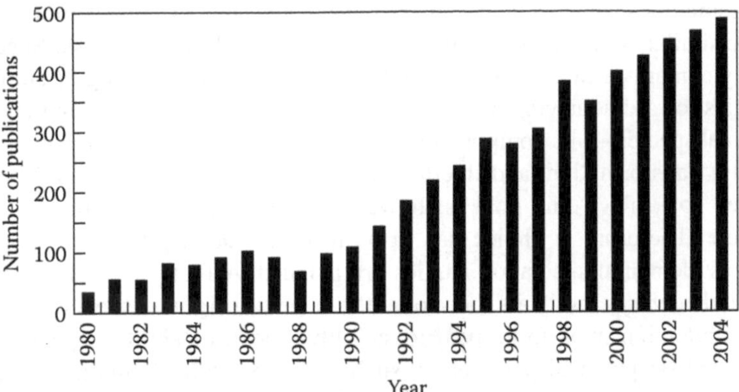

FIGURE 1.1 Growth in the number of research papers in speciation over a 25-year (1980–2004) period.

species.[18,19] The toxicity of the water-soluble species of arsenic varies in the order of arsenite > arsenate > monomethylarsenate (MMA) > dimethylarsinate (DMA).[20] Metals occur in natural waters in different physicochemical forms. Among these, simple hydrated metal ions are considered to be the most toxic forms, while strong complexes and species associated with colloidal particles are usually assumed to be nontoxic. Organometallic compounds are more toxic than their corresponding inorganic species, with the exception of arsenic. Therefore, the measurement of the total concentration of a metal ion cannot provide information about the actual physicochemical forms of a metal ion, required for understanding its toxicity, biotransformation, and so forth. Thus, in order to obtain information on toxicity and biotransformation of a metal ion in the environment, speciation of the individual element is necessary. In view of this, the elemental speciation analysis has now become part of the everyday life of environmental and analytical chemists and is also important to industry, academia, and government and legislative bodies. Therefore, diverse groups of people ranging from the regulators to those in the materials industry, clinicians, nutritional experts, agricultural chemists, and environmental scientists are now demanding data on speciation rather than the total concentrations of the elements. Therefore, the number of research papers is increasing continuously every year. The growth of the speciation of the elements over a 25-year period is shown in Figure 1.1. Taking all above discussed points into consideration, the determination of the concentrations of different oxidation states of a metal ion (speciation) is essential and urgent need of today.

1.5 METHODS OF METAL ION SPECIATION

The speciation of a metal ion in the environment is a difficult job, as the concentrations of heavy metals in the environment are generally very low. Moreover, sometimes, the physical and chemical properties of two or more species may be similar, and this makes the speciation again a Herculean task. Speciation strategy contains a careful plan or method or the art of developing or employing strategies toward achieving a goal. Scientists hope to learn everything about the elemental species — its composition, mass, biological and environmental cycles, stability, transformation, and interactions with inert or living matter. The work involved to achieve this goal is, however, challenging, if not impossible, to complete. Therefore, a choice has to be made to identify the most important issues as elemental speciation studies are pursued. A first group of compounds to be studied very closely are those of anthropogenic origin. In speciation studies, a lot of attention must to be paid to stability. Species stability depends on the matrix and on physical parameters, such as temperature, humidity, ultraviolet (UV) light, organic matter, and so forth. The isolation and purification of the species, the study of the possible transformation through the procedure, and the study of their characteristics and interactions are also important. New analytical procedures have to be devised, including appropriate quantification and calibration methodologies.

Various advanced techniques were developed to overcome this problem. Different analytical methods used for metal ion speciation include spectroscopic, polarographic, chromatographic, and electrochemical methods. Various technologies used for metal ion speciation are shown in Figure 1.2. Among these techniques, chromatographic and electrophoretic methods have been used frequently due to their ease of operation and reproducible results. The spectroscopic methods of speciation involve the use of various spectroscopic techniques, such as UV, visible, flame atomic absorption, atomic emission, hydride generation, graphite furnace, among others.[21–23] Besides the spectroscopic methods of speciation, polarography and voltammetry have also been used for the speciation of metal ions.[24–26]

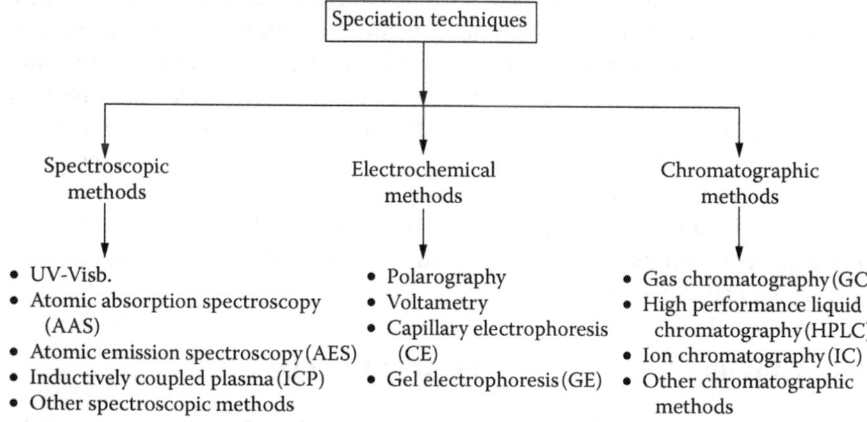

FIGURE 1.2 Different techniques for metal ions speciation.

1.5.1 Chromatographic Methods

The chromatographic methods involve the use of gas or liquid separately as the mobile phases. Therefore, the former kind of chromatography is called *gas chromatography* (GC), while the latter is termed *liquid chromatography* (LC). Due to some problems associated with gas chromatography, it could not be accepted as the method of choice for the speciation of metal ions. The major disadvantage of GC is its requirement of the conversion of inorganic metal ions into organic and volatile metal ion derivatives, which is carried out by a derivatization process. To its advantage, however, LC is the only and the best remaining technology for the speciation of a wide variety of metal ions. The main advantages of LC are its ability to differentiate metallic species in the environmental samples directly. During the course of time, various types of liquid chromatographic approaches were developed and used in this concern. The most important liquid chromatographic methods are high-performance liquid chromatography (HPLC) and ion chromatography (IC). However, sub- and supercritical fluid chromatography (SFC), capillary electrochromatography (CEC), and thin-layer chromatography (TLC) modalities can be used for this purpose.

Among the various liquid chromatographic techniques mentioned above, IC remains the best due to several advantages it has in comparison to others. High speed, sensitivity, and reproducible results make IC the method of choice in almost all laboratories. About 80% metal ion speciation has been carried out using the IC mode of liquid chromatography. Many types of columns are available in IC, which make this technique popular in metal ion speciation. A variety of mobile phases, including normal, reversed, and new polar organic phases, are used in IC. The composition of the mobile phases may be modified by the addition of various aqueous and nonaqueous solvents. Ion chromatography is a form of liquid chromatography that uses ion-exchange resins to separate atomic or molecular ions based on their interaction with the resin. Its greatest utility is in the analysis of ions for which there are no other rapid analytical methods. Most ion-exchange separations are carried out with metallic pumps and non-metallic columns. The column packing for ion chromatography consists of ion-exchange resins bonded to inert polymeric particles. The optimization of metal ion speciation is carried out by a number of experimental parameters.

The use of supercritical fluids as the mobile phases for chromatographic separation was first reported more than 30 years ago, but most of the growth in SFC has occurred recently. A supercritical fluid exists when both temperature and pressure of the system exceed the critical values, that is, critical temperature (Tc) and critical pressure (Pc). The critical fluids have their physical properties in between liquid and gas. Like gas, supercritical fluids are highly compressible, and properties of the fluid, including density and viscosity, can be maintained by changing the pressure and temperature conditions. In chromatographic systems, the solute diffusion coefficients are often of higher magnitude in supercritical fluids than in traditional liquids. On the other hand, the viscosities are lower than those of liquid.[27] At temperature below Tc and pressure above Pc, the fluid becomes a liquid. On the other hand, at temperature above Tc and pressure below Pc, the fluid behaves as a gas. Therefore, supercritical fluids can be used as part of a mixture of liquid and gas.[28] The commonly used supercritical fluids (SFs) are carbon dioxide, nitrous oxide, and trifluoromethane.[27–29] Its compatibility with most detectors, low critical temperature and pressure, low toxicity and environmental burden, and low cost make carbon dioxide the supercritical fluid of choice. The main drawback of supercritical carbon dioxide as a mobile phase is in its inability to elute more polar organic derivatives of metal ions and other compounds. This process can arise by the addition of organic modifiers to the relatively apolar carbon dioxide.

Basically, capillary electrochromatography (CEC) is a hybrid technique that works on the basic principles of capillary electrophoresis and chromatography.[30] This mode of chromatography is used either on packed or tubular capillaries/columns. The packed column in CEC was first introduced by Pretorius et al.[31] in 1974; the open tubular CEC was presented by Tsuda et al.[32] in 1982. In 1984, Terabe et al.[33] introduced another modification in liquid chromatography — micellar electrokinetic capillary chromatography (MEKCC). Of course, this mode also depends on the working principles of capillary electrophoresis and chromatography but also involves the formation of micelles. The high speed, sensitivity, lower limit of detection, and reproducible results make CEC and MECC the methods of choice in analytical science. However, these methods could not be used frequently for speciation, as these techniques are not fully developed, and research is underway on this subject.

The history of the development of thin-layer chromatography (TLC) is very old. As such, TLC could not be used for metal ion speciation, but it has been used for the separation of a simple mixture of metal ions. The derivatization of metal ions and, subsequently, separation on silica gel or reversed-phase RP-TLC plates represents a method of metal ion separation and identification. However, it can be used easily for this purpose.

In view of the importance of chromatography in metal ion speciation and to familiarize readers about chromatographic techniques, it is necessary to explain the chromatographic terms and symbols by which chromatographic speciation can be explained. Some of the important terms and equations of the chromatographic separations are discussed below. Chromatographic separations are characterized by retention (k), separation (α), and resolution factors (Rs). The values of these parameters can be calculated by the following standard equations[34]:

$$k = (t_r - t_0)/t_0 \tag{1.1}$$

$$\alpha = k_1/k_2 \tag{1.2}$$

$$Rs = 2\Delta t_r/(w_1 + w_2) \tag{1.3}$$

where t_r and t_0 are the retention time of the chromatogram of metal species and dead time (solvent front) of the column in minutes, respectively. Δt, w_1, and w_2

are the difference of the retention times of the two peaks of the separated metal ion species, base width of peak 1, and base width of peak 2, respectively. If the individual values of α and *Rs* are one or greater, the separation is said to be complete. If the individual values of these parameters are lower than one, the separation is partial or incomplete.

The number of the theoretical plate (*N*) characterizes the quality of a column. The larger the value of *N*, the more complicated the sample mixture can be that is to be separated with the column. The value of *N* can be calculated from the following equations:

$$N = 16(tr/w)^2 \tag{1.4}$$

or

$$N = 5.54[(tr)/w_{1/2}]^2 \tag{1.5}$$

where t_r, w, and $w_{1/2}$ are the retention time (min) of the peak, peak width at base, and peak width at half height of the peak, respectively. The height equivalent to a theoretical plate (HETP [*h*]) is a section of a column in which the mobile and the stationary phases are in equilibrium. Because a large number of theoretical plates are desired, *h* should be as small as possible. Naturally, there are no real plates in a column. The concept of the theoretical plate is a variable, and its value depends on particle size, flow velocity, mobile phase (viscosity), and especially on the quality of the packing. The value of *h* can be calculated from the following equation:

$$h = L/N \tag{1.6}$$

where *L* is the length of the column used.

1.5.2 Capillary Electrophoretic Method

In the last decade, a newly developed technique called capillary electrophoresis (CE) has emerged as one of the tools for the speciation of metal ions. The CE has certain advantages in comparison to the other techniques. The advantages of CE include inexpensiveness, simplicity, high speed of analysis, greater separation efficiency, unique selectivity, and high degree of matrix independence, all of which make it ideal for the speciation of metal ions.[35–37] Among the electrophoretic methods of metal ion speciation, various forms of capillary electrophoresis, such as capillary zone electrophoresis (CZE), capillary isotachphoresis (CIF), capillary gel electrophoresis (CGE), capillary isoelectric focusing (CIEF), affinity capillary electrophoresis (ACE), and separation on microchips have been used. However, in contrast to others, the CZE mode was used frequently for this purpose.[38] But, it is necessary to mention here that CE could not achieve a status of routine analysis in metal ion speciation. This is because of some drawbacks associated with these techniques. The limited applications of these methods are due to a lack of advancement in instrument and stationary phases (detection methods and availability of stationary phases used in the capillaries).

Again, it is necessary to explain some fundamental aspects of CE so that the reader can use this technique properly. The mechanism of the speciation in CE is based on the difference in the electrophoretic mobility of the separated metal ion species. Under CE conditions, the migration of the separated metal ion species is controlled by the sum of intrinsic electrophoretic mobility (μ_{ep}) and electroosmotic mobility (μ_{eo}), due to the action of electroosmotic flow (EOF). The observed mobility (μ_{obs}) of metal ions is related to μ_{eo} and μ_{ep} by the following equation:

$$\mu_{obs} = E\ (\mu_{eo} + \mu_{ep}) \tag{1.7}$$

where *E* is the applied voltage (kV).

The simplest way to characterize the separation of two components, resolution factor (*Rs*), is to divide the difference in the retention times by the average peak width as follows:

$$Rs = 2\ (t_2 - t_1)/(w_1 + w_2) \tag{1.8}$$

where t_1, t_2, w_1, and w_2 are retention times of peak 1 and peak 2 and widths of peak 1 and peak 2, respectively.

The value of the separation factor may be correlated with μ_{app} and μ_{ave} by the following equation:

$$Rs = (1/4)(\Delta\mu_{app}/\mu_{ave})N^{1/2} \tag{1.9}$$

where μ_{app} is the apparent mobility of the two separated metal ion species, and μ_{ave} is the average mobility of the separated metal ion species. The use of Equation 1.9 is to permit independent assessment of the two factors that affect separation, selectivity, and efficiency. The selectivity is reflected in the mobility of the analytes, while the efficiency of the separation process is indicated by *N*. Another expression for *N* is derived from the following equation:

$$N = 5.54\ (L/w_{1/2})^2 \tag{1.10}$$

where *L* and $w_{1/2}$ are capillary length and the peak width at half height, respectively. At this point, it is important to point out that it is misleading to discuss theoretical plates in CE; it is simply a carryover from chromatographic theory. The theoretical plate in CE is merely a convenient concept to describe the shape of the analyte peaks and to assess the factors that affect separation.

HETP may be considered as the function of the capillary occupied by the analyte, and it is more practical to measure separation efficiency in comparison to *N*. σ^2_{tot} is affected not only by diffusion but also by differences in the mobility values, Joule heating of the capillary, and interaction of the analytes with the capillary wall; hence, σ^2_{tot} can be represented as follows:

$$\sigma^2_{tot} = \sigma^2_{diff} + \sigma^2_T + \sigma^2_{int} + \sigma^2_{wall} + \sigma^2_{Electos} + \sigma^2_{Electmig} + \sigma^2_{Sorp} + \sigma^2_{Oth} \quad (1.11)$$

where $^2{}_{tot,}$ $^2{}_{diff}$, $^2{}_T$, $^2{}_{int}$, and $^2{}_{wall}$ are square roots of standard deviations of total, diffusion, Joule heat, injection, wall, electroosmosis, electromigration, sorption, and other phenomena, respectively.

1.5.3 Spectroscopic and Other Methods

Spectroscopic methods of the speciation involve the use of UV-visible, flame atomic absorption, atomic emission, hydride generation, graphite furnace, and so forth. Some of these techniques require the conversion of one form into another. Some metallic species form different colors with certain ligands, and they can be speciated by UV-visible spectroscopy. For example, arsenic can be speciated by molybdenum blue and silver diethyldithiocarbamate methods. Even methods such as atomic emission and atomic absorption spectrometry coupled with flow-injection analyzer system (FIAS) give results that are more precise and exact when only one form of an element is present. The atomic absorption spectrometry with hydride generation is one of the most widely used methods for the speciation of some metal ions. For example, the speciation of arsenic is carried out by determining As(III) and the total arsenic, which is determined by the reduction of As(V) to As(III). The prior oxidation of arsenite to arsenate by KI was reported.[39–43] In addition, a method was recently developed for determining trace amounts of arsenic(III) and total arsenic with L-cysteine as prereductant using flow injection hydride generation coupled with a nondispersive atomic absorption spectrometry (AAS) device made in-house.[44] Inductively coupled plasma (ICP) spectroscopy involves the use of plasma formed at high temperature. This technique can be used for speciation after hyphenation with a hydride generation unit. A lower limit of detection in ICP can be achieved by coupling it with a mass spectrometer.

Among electrochemical methods, potentiometric, polarographic, and cyclic, pulse, and stripping voltammetric techniques are used for metal ion determination.[45] With direct-current polarography, it is possible to detect different oxidation states of many metal ions. In addition to the above, some reports are available on metal ion speciation by other techniques, such as x-ray, x-ray fluorescence (XRF), extended x-ray absorption fine structure (EX-AFS), x-ray absorption near-edge structure (XA-NES) spectroscopy, isotopic dilution, sensors, and so forth.

1.6 DETECTION IN CHROMATOGRAPHY AND CAPILLARY ELECTROPHORESIS

Even though many detectors in chromatography and electrophoresis are used for metal ion speciation, the sensitivity of the detection of metal ion species remains a challenging problem, and the most commonly used method is indirect UV detection. The main advantage of indirect UV detection is its universal application. The complexation of metal ion species with some suitable ligands also increases the sensitivity of the detection. In addition, UV visualization agents (probes) have been used to increase the sensitivity of detection in the UV mode. Some other detectors, such as conductivity, amperometry, and potentiometry were also used for this purpose. And, chromatography and capillary electrophoresis have been coupled with AAS, atomic emission spectroscopy (AES), atomic fluorescence spectrometry (AFS), inductively coupled plasma (ICP), mass spectrometry (MS), inductively coupled plasma–mass spectrometry (ICP-MS), and time-of-flight mass spectrometry (TOF-MS) instruments for the detection of metal ion species.[46–52] MS, nuclear magnetic resonance spectroscopy (NMR), infrared (IR), and other techniques have been combined with chromatography and capillary electrophoresis to determine the structural information of the separated organo-metal ion species. The combination of these methods with chromatographic and electrophoretic modalities greatly improves the sensitivity, efficiency, and detection limits. Many combination methods were reported in the literature for gas chromatography, liquid chromatography, and electrophoresis techniques. These methods will be discussed in detail in subsequent chapters.

Most successful combinations employ element-specific detectors. AAS has been found to be a suitable detector in comparison to AES, AFS, and ICP-MS and is being used in many laboratories of the world.[53,54] Hydride generation atomic absorption spectrometry (HG-AAS) using the flow injection analyzer system (FIAS) system has been used for metal ion speciation. The limitation of this technique is its poor reproducibility and applicability only to arsenic, mercury, selenium, and lead metal ions. On the other hand, electrothermal atomic absorption spectrometry (ET-AAS) has high sensitivity and the capability to work with a wide range of metal ions, including Al,[55] As,[56] Cd,[57] Cr,[58] Fe,[59] Se,[60] Sn,[61] Sb,[62] and Zn.[63] Tsalev et al.[64] developed a permanent modification in an ET-AAS detector. This is a very promising development in chemical matrix modification in view of increasing sample throughput with fast programs, reducing reagent blanks, conducting preliminary elimination of unwanted modifier components, and having compatibility with online and in situ enrichment. Plasma Emission Atomic Spectroscopy (EAS) can detect both metals and nonmetals and different sources of plasma, such as microwave-induced plasma (MIP), direct-current plasma (DCP), and inductively coupled plasma (ICP), were reported in the literature. The best is ICP, because the continuous HPLC flow quenches the discharge of MIP, while DCP appears to be better than MIP as it provides a more stable plasma, especially with the introduction of mixed organic and aqueous phases. Briefly, HG-AAS, ET-AAS, ICP, and ICP-MS instruments have been coupled with various analytical techniques for metal ion speciation. However, the home-made combination is required commercially for better sensitivity, accuracy, reproducibility, and detection limits. It is pertinent to mention here that detailed discussion on the detection of different metal ion species in different chromatographic and electrophoretic techniques is presented in dedicated chapters in this text.

In view of the importance of metal ion speciation, attempts are made to explain the art of metal ion speciation by chromatography, electrophoresis, spectroscopy, and other methods. The present book comprises 12 chapters, including an introduction followed by 11 chapters. Ten chapters discuss the sources and distribution, toxicity, biodegradation, and speciation of metal ions by chromatography, capillary electrophoresis, spectroscopy, and other methods. The last chapter describes the legal aspects and future perspectives of metal ions speciation. In this book, the art of metal ion speciation is described in a well-defined, systematic, and scientific way that includes discussions on the optimization methods and mechanisms of the speciation.

REFERENCES

1. Franklin LB, *Wastewater Engineering: Treatment, Disposal and Reuse*, McGraw Hill, New York (1991).
2. Droste RL, *Theory and Practice of Water and Wastewater Treatment*, John Wiley & Sons, New York (1997).
3. John DZ, *Hand Book of Drinking Water Quality: Standards and Controls*, Van Nostrand Reinhold, New York (1990).
4. Toxic Substance Control Act, US EPA, III: 344 (1984).
5. Moore JW, Ramamoorthy S, *Heavy Metals in Natural Waters: Applied Monitoring and Impact Assessment*, Springer, New York (1984).
6. IARC Monographs, Suppl. 7, IARC, Lyon, Vol. 54: 40 (1987).
7. Dich J, Zahm SH, Hanberg A, Adami HO, *Cancer Causes & Control*, 8: 420 (1997).
8. Brusick D, *Toxicology & Industrial Health*, 9: 223 (1993).
9. Kaniansky D, Kremova E, Madajova V, Maser M, *Electrophoresis*, 18: 260 (1997).
10. Hutson DH, Roberts TR, *Environmental Fate of Pesticides*, Vol. 7, John Wiley & Sons, New York (1990).
11. Ali I, Jain CK, *Current Sci.*, 75: 1011 (1998).
12. Jain CK, Ali I, Int. J. Environ. *Anal. Chem.*, 68: 83 (1997).
13. Caruso JA, Sutton KL, Ackley KL (Eds.), *Elemental Speciation: Comprehensive Analytical Chemistry*, Vol. XXXIII: Elsevier, Amsterdam (2000).
14. Goldberg ED, *J. Geol.*, 62: 249 (1954).
15. Cutter GA, *Marine Chem.*, 40: 65 (1992).
16. Stoeppler M, *Hazardous Metals in the Environment*, Elsevier, Amsterdam (1992).
17. Templeton DM, Ariese F, Cornelis R, Danielsson LG, Muntau H, Van Leeuwen HP, Lobinski R, *Pure Appl. Chem.*, 72: 1453 (2000).
18. Berman E, *Toxic Metals and Their Analysis*, Heyden & Sons, London (1980).
19. Gesamp, (IMO/FAO/UNESCO/WMO/WHOIAEA/UN/UNEP Joint Group of experts on the scientific aspects of marine pollution) review of potentially harmful substances — Arsenic, mercury and selenium, Rept. Stud. GESAMP 28 (1986).
20. Stugeron RE, Siu KWM, Willie SN, Bermann S, *Analyst*, 114: 1393 (1989).
21. Batley GE, Low GKC, *Trace Element Speciation: Analytical Methods and Problems*, CRC Press, London (1989).
22. Krull IS, *Trace Metal Analysis and Speciation*, Elsevier, Amsterdam (1991).
23. Jain CK, Ali I, *Wat. Res.*, 34: 4304 (2000).
24. Pretty JR, Evans EH, Blubaugh EA, Shen WL, Caruso JA, Davidson TM, *J. Anal. At. Spectrom.*, 5: 437 (1990).
25. Pretty JR, Blubaugh EA, Evans EH, Caruso JA, Davidson TM, *J. Anal. At. Spectrom.*, 7: 1131 (1992).

26. Pretty JR, Blubaugh EA, Caruso JA, *Anal. Chem.*, 65: 3396 (1993).
27. Weast RC (Ed.), *Handbook of Chemistry and Physics*, 54th ed., CRC Press, Cleveland (1973).
28. Berger, TA, in *Packed Column SFC*, Smith RM (Ed.), The Royal Society of Chemistry, Chromatography Monographs, Cambridge (1995).
29. Schoenmakers PJ, in *Packed Column SFC*, Smith RM (Ed.), The Royal Society of Chemistry, Chromatography Monographs, Cambridge (1988).
30. Cronin JR, Pizarello S, *Science*, 275: 95 (1997).
31. Pretorius V, Hopkins BJ, Schieke JD, *J. Chromatogr.*, 99: 23 (1974).
32. Tsuda T, Nomura K, Nagakawa G, *J. Chromatogr.*, 248: 241 (1982).
33. Terabe S, Otsuka K, Ichikawa K, Tsuchiya A, Ando T, *Anal. Chem.*, 56: 111 (1984).
34. Meyer VR (Ed.), *Practical High Performance Liquid Chromatography*, John Wiley & Sons, New York (1993).
35. Gareil P, *Analusis,* 18: 447 (1990).
36. Tessier A, Turner DR (Eds.), *Metal Speciation and Bioavailability in Aquatic Systems*, John Wiley & Sons, New York (1996).
37. Baraj B, Niencheski LF, Soares JA, Martinez M, *Fresenius' J. Anal. Chem.*, 367: 12 (2000).
38. Chankvetadze B, *Capillary Electrophoresis in Chiral Analysis*, John Wiley & Sons, New York (1997).
39. Bogdanova VI, *Microkim. Acta*, II: 317 (1984).
40. Matsubara C, Yamamoto Y, Takamura K, *Analyst* 112: 1257 (1987).
41. Nasu T, Kan R, *Analyst*, 113: 1683 (1988).
42. Tamari Y, Yamamoto N, Tsuji H, Kasuka Y, *Anal. Sci.*, 5: 481 (1989).
43. Palanivelu K, Balasubramaniam N, Ramakrishna TV, *Talanta*, 39: 555 (1992).
44. Yin X, Hoffmann E, Ludke C, Fresenius' *J. Anal. Chem.*, 355: 324 (1996).
45. Rajeshwar K, Ibanez JG, *Environmental Electrochemistry: Fundamentals and Application is Pollution Abatement*, Academic Press, New York (1997).
46. Yeng ES, Kuhr WG, *Anal. Chem.*, 63: 275A (1991).
47. Nielen MWF, *J. Chromatogr.*, 588: 321 (1992).
48. Wang T, Hartwick RA, *J. Chromatogr.*, 607: 119 (1992).
49. Poppe H, *Anal. Chem.*, 64: 1908 (1992).
50. Beckers JL, *J. Chromatogr. A*, 679: 153 (1994).
51. Kok WT, *Chromatographia*, 51: 52 (2000).
52. Leach AM, Heisterkamp M, Adams FC, Hieftje GM, *J. Anal. At. Spectrom.*, 15: 151 (2000).
53. Donard OFX, in *Environmental Analysis: Applications and Quality Assurance*, Barcelo D (Ed.), Elsevier, New York, Chap. 6 (1993).
54. Ellis LA, Roberts DJ, *J. Chromatogr. A*, 774: 3 (1997).
55. Bantan T, Milacic R, Mitrovic B, Pihlar B, *Fresenius' J. Anal. Chem.*, 365: 545 (1999).
56. Burguera M, Burguera JL, *J. Anal. At. Spectrom.*, 8: 229 (1993).
57. Fernandez FM, Tudino MB, Troccoli OE, *J. Anal. At. Spectrom.*, 15: 687 (2000).
58. Vassileva E, Hadjivanov K, Stoychev T, Daiev C, *Analyst*, 125: 693 (2000).
59. Bermejo P, Pena E, Dominguez R, Bermejo A, Fraga JM, Cocho JA, *Talanta*, 50: 1211 (2000).
60. Yan XP, Sperling M, Welz B, *Anal. Chem.*, 71: 4353 (1999).

61. Bermejo-Barrera P, Soto-Ferreiro RM, Dominguez-Gonzalez R, Bermejo-Barrera A, *Ann. Chim.*, 86: 495 (1996).
62. Farkasovska I, Zavadska M, Zemberyova M, *Chem. Listy.*, 93: 173 (1999).
63. Das AK, Chakarborty R, Fresenius' *J. Anal. Chem.*, 357: 1 (1997).
64. Tsalev L, Slaveykova VI, Lampugnani L, D'Ulivo A, Georgieva R, *Spectrochim. Acta Part B*, 55: 473 (2000).

2 Metal Ion Species: Sources and Distribution

Basically, all metal ions are present in the earth's crust in different abundances and oxidation states. There are different ores containing various metal ions. Among the most important ores are bauxite (aluminum), stibnite (antimony), realgar, orpiment and arseno pyrites (arsenic), chromite (chromium), magnatite (iron), pyrolusite and braunite (manganese), galena (lead), cinnabar (mercury), cassiterite (tin), patronite, and vamidinite and carnotite (vanadium). All metal ions find their ways from these ores to different environmental components. Some get dissolved into aquifers and contaminant groundwater, while the others are extracted by humans and used for different purposes. But the side effects of these activities resulted into the pollution of surface water, air, vegetables, and other foodstuffs. In addition, some metal ions are present in the surfaces of the soil and contaminate the surface water, thereby also contaminating the food cycle through plants and some animals. The formation of the earth and the development of civilization have left metal ions widely distributed in our ecosystem. Many metal ions have been found in different planets of our universe. The heavy metal ions occur in different oxidation states. The change in degree of oxidation of an element also has an important effect on the degree of bioavailability and toxicity.[1] Metals occur into natural waters in different physicochemical forms. Among them, simple hydrated metal ions are considered to be the most toxic forms, while strong complexes and species associated with colloidal particles are usually assumed to be nontoxic.

Some metal ions, at low concentrations, are essential for human, animal, and plant growth, but higher concentrations of metal ions are toxic and produce diseases and lethal effects. In biological systems, metal ions disturb various enzymatic functions by interacting with them. Therefore, to understand the toxicity of metal ions, which will be discussed in Chapter 3, the knowledge of the biotransformation of metal ions in different sources and species is very important. In view of these points, attempts are made to discuss the sources of contamination, distribution, and biotransformation of metal ions species in our environment and biological systems. It is very important to add here that efforts were made to describe the distribution of metal ion species, but the speciation subject is in its development, and therefore, in some instances, only the distribution of single metal ions species is given. The present chapter describes these issues in a systematic way.

2.1 SOURCES OF CONTAMINATION

The contamination of our environment due to toxic metal ions is a very serious and challenging issue. It is essential to know the sources of the pollution of metal ions to control and minimize pollution. Knowledge of pollution sources is also helpful in explaining the transportation behaviors of metal ions. The sources of contamination can be classified into point and nonpoint categories. The point sources include the activities of industries and domestic survival (home lawns, ornaments, pests control, etc.), while the nonpoint sources are of geological origin, from agriculture, forestry, and related activities.

There are many types of industries that use metal ions as raw materials, consequently releasing some of them as their by-products. Moreover, metallurgical industries are the major sources of metal ion contamination. The major industries using and releasing metal ions are classified in Table 2.1.[2] The industries dealing with arsenic as the raw material can be considered the sources of arsenic contamination at low concentrations. Arsenic has been detected in the atmosphere near industries that utilize the material. Matschullat[3] reported the presence of arsenic in the atmosphere. The author reported the emission of arsenic from copper smelting, cobalt combustion, glass production, and steel production processes. Chromium metal ions, that is, Cr(III) and Cr(VI), were reported as being

TABLE 2.1
Major Industries Employing and Releasing Heavy Metals

Sl. No.	Industries	Cd	Co	Cr	Cu	Fe	Mn	Pb	Zn
1.	Pulp, papermill; building paper, board mills			X	X			X	X
2.	Organic chemicals, petrochemicals	X		X		X		X	X
3.	Alkalis, chlorine, inorganic chemicals	X		X		X		X	X
4.	Fertilizers	X		X	X	X	X	X	X
5.	Petroleum refining	X		X	X	X		X	X
6.	Steel works	X		X	X	X		X	X
7.	Nonferrous metal works	X		X	X			X	X
8.	Motor vehicle, aircraft plating	X		X	X				
9.	Flat glass, cement, asbestos			X					
10.	Textiles			X					
11.	Leather tanning			X					
12.	Steam generation			X					X

found in the air near stainless steel welding industries.[4] Mercury is known to contaminate the environment through lamps, batteries, thermometers, and amalgams in the manufacturing of chlorine, pesticides, fungicides, catalysts, paints, pigments, and so forth.[5–7] Mercury was also reported to be found in the atmosphere[8] and exists in a variety of forms that have different biological and environmental behaviors. The electroplating industries are the major sources of chromium and nickel contamination. The interconversion of these forms controls environmental mobility. Two possible biogeochemical cycles of transport and distribution were described.[9,10] A global cycle of transport involving the atmospheric circulation of elemental mercury vapor from its sources to the ocean, and a local one that is dependent on the methylation of inorganic mercury resulting from industrial waste are well known. Tributyltin is the synthetic and powerful biocidal substance produced since the 1960s and primarily applied as an antifouling agent for vessels and fish net paints used in marine activities. Tributyltin-producing industries all over the world may be considered the pollution sources of tin metal ion.[11] Beside various industries, metal testing and other research laboratories also contribute to the contamination of the environment. Most significant laboratories polluting our environment include those testing and developing pesticides, metalloproteins, and other chemicals.

The hazardous effects take place when toxic metal ions enter the drinking water supply. Hence, the contamination of groundwater, which becomes the drinking water supply throughout the world, by geological formations (nonpoint sources) is of great significance. The presence of metal ions in groundwater is generally associated with geochemical environments, such as basin-fill deposits of alluvial-lacustrine origin, volcanic deposits, inputs from geothermal sources, and mining wastes.[12,13] The occurrence of metal ions in groundwater depends on the local geology, hydrology, and geochemical characteristics of the aquifer materials. Furthermore, organic content in the sediments as well as land-use patterns may play an important role in controlling the mobility of metal ions in alluvial aquifers. The other important processes responsible for metal ion pollution are dispersion, diffusion, advection, turbulence, gustiness, and plumes. At present, groundwater in many areas of the world, especially at low depths, was reported to be contaminated with toxic metal ions such as arsenic, mercury, lead, and antimony. An example of this type of contamination is the pollution of groundwater due to arsenic. Arsenic contamination in groundwater due to geological origin is a worldwide problem, and arsenic was detected in the groundwater of many countries, including Bangladesh, India, the United States, China, Chile, Taiwan, Mexico, Argentina, Poland, Canada, Hungary, and Japan.[14]

Some metal ions are used in the preparation of pesticides, herbicides, and fungicides. Arsenic, mercury, lead, and copper are the main metal ions used for the preparation of these insecticides. These insecticides are used in agriculture, forestry, and domestic activities and, hence, contaminate the environment. Uncontrolled anthropogenic activities, such as the smelting of metal ores, and volcanic activity contribute to environmental pollution with metal ions. Ionic alkyllead compounds are products of the degradation of tetraalkyllead compounds, which are mainly used as gasoline additives. Despite the fact that the use of leaded gasoline has decreased, their concentrations in the urban environment remain high.[15,16] As lead is the most significant toxin of the heavy metals, industrial decisions to add lead to paints, dyes, and gasoline, have contributed greatly to lead contamination. Household activities, such as the use of pesticides, paints, and pigments, may also be considered significant sources of metal ions pollution. The rusting of iron, particularly on rainy days, is another source of iron contamination. Copper is a common contaminant in drinking water, released by the piping material used for household connections from street water mains to in-house water taps. The corrosion caused by calcium and alkalinity results in a layer of Cu_2O with smaller amounts of CuO, $Cu(OH)_2$, and $CuCO_3$. The ratio of Cu(I) and Cu(II) is interesting as it reveals details of the corrosion process. Briefly, metal ions find their way from the earth crust into the environment by many natural and man-made routes.

2.2 DISTRIBUTION OF METAL IONS

Metal occurs in nature mostly in combined forms but can also, at times, be found in its free state. The active metal ions are found in the form of their salts, while metals, which have little or no affinity for oxygen and water, are found in their free state. Generally, chromium, vanadium manganese, and iron (most electropositive) metal ions are found in the form of their chlorides or nitrates, while the low electropositive metals such as arsenic, lead, zinc, and mercury are found in the form of sulfides. In contrast, the weakest metal ions, such as platinum and gold, are found as native but rare forms. The heavy metal ions are distributed worldwide and are found in almost every component of the universe. The distribution of metal ions is discussed herein in water, sediments, soils, air, and aquatic and terrestrial biota.

2.2.1 Distribution in Water

Water is the main component of the ecosystem; unfortunately, it is the most contaminated by toxic metal ions. All forms of water resources — ground, surface, and waste — are contaminated due to heavy toxic metal ions. The oxidation of different mineral species causes metal ions to become soluble, and they enter the surrounding environment through drainage water. Knowledge of the geographical distribution of metal ions species in the natural water system is important for environmental consideration of the geochemical and biological cycling of the element. Furthermore, this knowledge provides insight into the geochemical process responsible for elevated metal ion concentrations in different geological environments.

As discussed above, many toxic metal ions are found in groundwater. At present, arsenic contamination in the groundwaters of the United States, China, Chile, Bangladesh, Taiwan, Mexico, Argentina, Poland, Canada, Hungary, Japan, and India is a challenge for scientists.[17–20] Falter and Schöler[21] reported the presence of methyl-, ethyl-, methoxyethyl-, ethoxyethyl-, phenyl, and inorganic mercury species in drinking water from Lithuania. Pozdniakova et al.[22] reported Fe(II) and Fe(III) species in groundwater from Kirkville, MO, USA. Similarly, Xu and Ma[23] reported Fe(II) in groundwater. Boyle and Jonasson[24] reported high mercury contents in groundwater near rocks containing high levels of mercury metal ion in Canada. The poisoning of groundwater due to mercury, cadmium, lead, and chromium was reported in various parts of the world. Timerbaev et al.[25] reviewed the speciation of metal ions in tap water (groundwater) in the Netherlands and reported the presence of arsenic, strontium, cadmium, chromium, copper, and iron and various species of these metal ions.

In addition to groundwater, surface water is also crucial, as most major cities depend on the surface water supply. The quality of the surface water is also deteriorating continuously due to the effluent discharges of many industries and domestic cities. As per Timerbaev et al.,[25] many metallic species were reported in surface water, that is, in rivers, seas, ponds, and lakes. Houba et al.[26] determined the presence of cadmium, zinc, copper, and lead in the Vesdre River, Belgium. Sakai et al.[27] and Saeki et al.[28] selected the Toyhira River and other lakes in Japan for their studies, and they determined that copper, lead, zinc, manganese, chromium, and cadmium heavy metal ions were in these natural water resources. The monitoring of toxic metal ions was carried out by Latimer and coworkers[29] in the Pawtuxet River, Rhode Island. They detected cadmium, chromium, copper, lead, and nickel in the river. Cadmium, cobalt, copper, iron, manganese, lead, and zinc were analyzed in the water of a most turbid river, the Huanghe, China.[30] In Pakistan, the sampling at 17 points on the river Ravi was carried out for the detection of heavy metal ions. Levels of zinc, iron, manganese, cadmium, lead, and nickel were determined in the river water.[31] The various heavy metal ions were analyzed in the streams of the mountain peaks of central Maryland.[32] Barak and Kress[33] found copper, cadmium, and zinc metal ions in the river Kishon, Israel. Levels of various toxic heavy metal ions were determined in 12 Chilean rivers.[34] In France, the Seine river water was analyzed for toxic metal ions, and the presence of cadmium, copper, lead, and zinc was reported.[35] Zinc was detected in the Przmsza river in Poland.[36] In India, various workers determined the presence of toxic metal ions in Indian rivers. Jain[37] determined the presence of iron, zinc, and copper in the Kali River, India. The status and the trend of river water pollution due to toxic metal ions in India was discussed and described by Prebha and Selvapathy.[38] The scientists at the Central Pollution Control Board, New Delhi, analyzed some water samples of the Yamuna River in July 1995 and March 1996. The metal ions reported were cadmium, chromium, copper, iron, nickel, lead, and zinc.[39] The Ganga (a 25-km stretch only) at Moradabad, India, was selected for the study of toxic metal ions.[40] Levels of copper, chromium, zinc, iron, lead, and nickel were determined in the river water. Priyadarshani[41] detected zinc, copper, nickel, cadmium, lead, manganese, mercury, cobalt, and iron in the water of the Safi River. Recently, Ali and Jain[42] determined the presence of iron, cobalt, nickel, copper, and zinc in the Solani River at Roorkee, India. Förstner and Wittmann[2] reported the presence of many metal ions in the rivers of the United States, Europe, and Asian continents.

The presence of magnesium was reported in seawater.[43–45] Flandysz et al.[11] determined various concentrations of monobutyltin, dibutyltin, and tributyltin in seawater. Similarly, Botana et al.[46] described the presence of monobutyltin, dibutyltin, and tributyltin species in seawater. Regan et al.[47] reported the presence of cobalt, copper, iron, and zinc in pond water. Xu and Ma[23] reported Fe(II) in lake water. Zufiaurre et al.[15] described methyl-, ethyl-, and methylethyl lead in rainwater. Similarly, Nerin and Pons[48] reported the presence of trimethylbutyllead, triethylbutyllead, and tetraethylbutyllead in rainwater. The Greenland snow and ice cap was shown to contain universal archives of the large-scale atmospheric pollution of the Northern Hemisphere by lead and other metal ions.[49] The first evidence of global contamination by lead was first reported by Murozumi et al.[50] who detected lead in snow and fern samples covering the period from 1753 to 1965. This discovery was controversial but was later affirmed by the assessment of natural lead levels in old Greenland ice dated 5500 B.C.[51] Pleßow and Heinrichs[52] presented the speciation of various trace elements in acidic pore waters from waste rock dumps. Pantsar-Kallio and others[53,54] reported the speciation of chromium metal ions in wastewater. Similarly, Arar and Pfaff[55] reported the presence of hexavalent chromium in industrial wastewater. Carro and Mejuto[8] reviewed the presence of organomercury species in different kinds of waters. Briefly, all types of water resources have been contaminated due to the heavy toxic metal ions.

2.2.2 Distribution in Sediment

Because reports of surface water contamination due to heavy metal ions are common, the presence of heavy metal ions in the sediment is also a common finding. A dynamic equilibrium is established between sediment and water. Many reports were published on the presence of heavy metal ions in the sediment. Due to the transfer process between different environmental components and the natural ability of the sediments to accumulate metal ions, their concentrations are generally higher than those present in adjacent water and air. The enrichment of metal ions in sediments by anthropogenic pollution is controlled by geological, mineralogical, hydrological, and biological factors. Various metal ions were detected in the sediments of rivers, lakes, and sea basins. Thomas[56] reported the contamination of Ontario lake sediment by mercury. Similarly, Lai et al.[57] reported the presence of methylmercury in sediment samples from Ottawa, Canada; Whang and Whang[58] speciated butyltin(I) and butyltin(II) species in marine-spiked sediment samples from Taiwan; Falandysz et al.[11] reported mono-, di-, and tributyltin species in the sediment samples collected from the Gulf of Gdansk, Baltic Sea; Feng and Narasaki[59] reported tributyltin and triphenyltin in the marine sediment samples from Japan; Yang et al.[60] and Pritzl et al.[61] also detected the presence of tin metal ion in sediment samples collected from different places in the Czech Republic; Yang et al.[60] reported the speciation of trimethyltin, triethyltin, and triphenyltin ion species. Milde et al.[62] described the analysis of butyltin in the sediment sample collected from the Czech Republic; Ceulemans and Adams[63] reported monobutyltin, dibutyltin, and tributyltin species in harbor sediments; Cai et al.[64] described the presence of monobutyltin, dibutyltin, and tributyltin species in sediment samples; Guntians and coworkers[65] reported the speciation of organic tin, lead, and mercury metal ions in sediments; Goldsmidt [66] described the presence of several metal ions in sediment samples; Tack and Verloo[67] reviewed the presence of metal ions in sediment and soil samples; and Förstner and Wittmann[2] described the presence of arsenic, selenium, zinc, lead, cadmium, chromium, copper, cobalt, nickel, mercury, silver, and antimony metal ions in the sediment of lakes in North America and coastal areas of North America and the British islands.

2.2.3 Distribution in Soil

Agricultural and other types of soils are contaminated by heavy metal ions. The soils are contaminated by irrigation with groundwater containing heavy metal ions and by the use of metal-ion-containing fertilizers and pesticides. Some metal ions present in the air also contaminate the soil through rain. In addition, other man-made activities, such as industrial and domestic discharges, contribute to the contamination of soils. The humus layer of soil plays a crucial role by acting as a natural biogeochemical barrier that suppresses the percolation of metal ions with seepage water, and thus, strongly accumulating the elements.[66]

Many types of metal ions were reported in the soil in different parts of the world. Some reviews were published on the distribution of metal ions in the soil.[67,68] Gibson and Willet[69] speciated aluminum metal ion species in soil samples from Australia. Göttlein and Blasek[70] reported aluminum in soil samples from Germany. Mitrovic and Milacic[71] described the speciation of aluminum in forest soils; the authors reported the presence of aluminum fulvate and aluminum humate. Matschullat[3] reviewed the status of arsenic levels in different soils and reported variable concentrations. Under aerobic conditions, arsenic was found in the form of arsenate [As(V)], bound to clay minerals, to Fe and Mn oxides/hydroxides, and to organic substances. The bonding strength was found to be directly proportional to arsenic concentration. Prohaska et al.[72] reported As(III), As(V), Monomethylarsenate (MMA), and Dimethylarsinate (DMA) species in soil samples. Yang et al.[73] described arsenic species in soil samples. Krajnc et al.[74] characterized the complexes formed in soil between copper, chromium, and fulvic acids. Hempel et al.[75] reported mercury contamination of an area of land in Germany. Carro et al.[76] reported methylmercury in sediment samples from Spain. Michalke and coworkers[77] tried to determine platinum levels in platinum spiked soils. Hirner et al.[78] analyzed 13 soil samples and reported 24 compounds of nine metal ions. The major compounds reported were inorganic and organic in nature, containing lead, antimony, and tin metal ions. Recently, Wang and coworkers[79] reported speciation of rare earth elements (REEs) in soil samples collected from Zhangzhou, Fujian Province, China. The results showed that the distribution of REEs in different fractions of the pooled soil samples studied followed the order of soluble species (46.76%) > species bound to organic matter (22.08%) > species in the residue (16.77%) > species bound to iron manganese oxides (2.02%).

2.2.4 Distribution in Air

The atmosphere is not untouched by the contamination of metal ions due to man-made and natural activities. Metal ions in the atmosphere are spread over a large area depending on the temperature, humidity, wind speed, and other determining factors. The determination of levels of metal ions in ambient air, aerosols, and dust is a focus of many environmental studies. Metal ions may be absorbed onto dust particles and fly ash that then enter the atmosphere. The levels of metal ions in aerosols are of concern.[80–87] Tripathi et al.[88] estimated the presence of aluminum metal ion in the air. The results indicated that the average concentration of aluminum in air particulate samples was 5.3 μgm^3. The daily intake of aluminum by the adult population of Bombay, India, is 6.4 mg/day. Matschullat[3] reviewed the presence of arsenic in the atmosphere and reported 28,230 to 54,270 ton/year as arsenic flux released from different sources. Dabek-Zlotorzynska and coworkers[82] described the presence of manganese, strontium, cadmium, and some other metal ions in aerosol samples. Similarly, Mihalyi et al.[84] reported the presence of magnesium and some alkaline earth metal ions in aerosols. Fung et al.[87] determined magnesium, zinc, and alkaline earth elements on dust particles suspended in the atmosphere. Organolead compounds are the products of the degradation of tetraalkyllead in petrol-propelled vehicles; these compounds are mainly used as gasoline additives. Although the use of leaded gasoline has decreased, high concentrations of lead compounds are reported in the atmosphere.[16] Nerin and Pons[89] reported tetramethyllead, tetrabutyllead, trimethylethyllead, dimethyldiethyllead, and methytriethyllead species in air samples. Furthermore, the same group[90] reported trimethylbutyllead, dimethyl-ethylbutyllead, triethylbutyllead, methyldiethylbutyllead, dibutyldiethyllead, and tetrabutyllead compounds in dust particles collected near roadways in Spain. Some alkali and alkaline earth metal ions were also found in the atmosphere.[82,84,85,87] Some toxic metal ions present in rainwater and discussed above may be considered air pollution, as the rainwater dissolved these metal ions from the atmosphere. Förstner and Wittmann[2] reported aluminum, antimony, arsenic, cadmium, chromium, cobalt, copper, iron, lead, mercury, nickel, selenium, silver, vanadium, and zinc in air samples from Germany.

2.2.5 Distribution in Aquatic Biota

The information concerning metal ions at high concentrations in animals and plants is an indication of the degree of contamination of the environment. The role of heavy metals in aquatic organisms was reported in earlier literature with an emphasis on the toxicity of individual metal ions.[91] Both essential (copper, zinc, iron, cobalt) and nonessential elements (arsenic, cadmium, lead, mercury) are found in different aquatic organisms. Metal ions accumulation in aquatic organisms is controlled by biotic and abiotic factors. The biotic factors are physiological behavior, pattern of life cycle, seasonal variation in organisms, and species-specific and individual variability, while the abiotic factors include temperature and dissolved oxygen of water, hardness, pH, salinity, and the presence of several organic and inorganic compounds.

Knauer and Martin[92] reported cadmium, copper, manganese, zinc, and lead in phytoplankton (diatoms) in Monterey Bay, California. Similarly, Morris[93] reported manganese metal ion in phytoflagellate (*Phaeocystis*). Schulz-Baldes and Lewin[94] described lead metal ion in *Phaeodactylum tricornutum* (diatoms) and *Platymonas subcordiformis* (flagellate). Furthermore, the authors found an accumulation of lead even in waters containing low levels of lead metal ion. Preston et al.[95] reported the presence of copper, lead, nickel, manganese, iron, and zinc in *Fucus vesiculosus* in the coastal waters of Great Britain. Similarly, other researchers[96,97] reported high concentrations of copper, lead, and silver metal ions in *Fucus vesiculosus*. Martin and Broenkow[98] reported chromium, copper, iron, manganese, and zinc in freshwater algae. The various species of arsenic were determined in green algae, red algae, brown algae, bacteria, and other phytoplanktons.[99] Dietz[100] described the presence of mercury in two moss species, *Fontinalis antipyretica* and *Hygroamblystegium*. In 1972, Reay[101] reported the presence of different arsenic species in higher aquatic plants in New Zealand. Similar observations were made by Dietz.[100] The different concentrations of cadmium, zinc, lead, and copper were reported in roots, stems, and leaves of different aquatic plants.[2] Arsenic speciation studies reveal the presence of many arsenic species in crustacea,[99,102] mollusca[99,102] (in pisces,[99,102] porifera,[99,102] and echinodermata[99]), and coelentera[99,102] aquatic classes of animals.

Generally, zooplanktons feed on phytoplankton; hence, the presence of heavy metal ions in zooplanktons is very common. Several reports were published on the detection of metal ions in zooplanktons.[2] Knauer and Martin[92] reported a variation in the concentrations of mercury metal ion in zooplanktons (euphausiids, copepods, ctenophores, coelenterates, etc.). Furthermore, the authors described mercury variation depending on the seasonal changes. Skei et al.[103] reported 25.21 ppm mercury in some zooplanktons. Philips[104] described the variation of zinc concentrations in *Mytilus edulis* (a bivalve creature) in different water depths with higher concentrations at greater depths. Metal ions were reported in crustaceans up to >3000 ppm.[105] Johnels et al.[106] and Hasselrot and Göthberg[107] reported mercury metal ion in *Asellus aquaticus*. Similarly, many metal ions were studied in different species of fish. Müller and Prosi[108] reported cadmium, copper, and zinc in different organs of roaches fish (*Rutilus rutilus*). Papadopoulou et al.[109] detected zinc metal ion in mackerel (*Platichthyes flesus*), and a negative linear function was observed between the age and the zinc content of the fish. Similarly, MacKay et al.[110] found very high concentrations of mercury and selenium in black marlin (*Makaira indica*) in northeast Australian water. Furthermore, the same authors carried out the variation coefficients of arsenic, mercury, cadmium, copper, lead, selenium, and zinc in muscle and liver tissues of 42 black marlin fish. Freitas et al.[111] reported MeHgCl and $HgCl_2$ in freshwater perch (*Perca flavescens*). The authors also reported different concentrations of these mercury species. Pereiro et al.[112] speciated mercury metal ion in tuna and dogfish, and the authors described the presence of diethylmercury and ethylmethylmercury species. Similarly, Logar et al.[113] determined diethylmercury and ethylmethylmercury species in dogfish liver. Other metals including arsenic, copper, lead, and cadmium were reported in some freshwater fish.[2]

The concentrations of heavy metals (cadmium, copper, cobalt, iron, lead, manganese, mercury, nickel, and zinc) were determined in tissues and organs of loggerhead (*Caretta caretta*) and green turtles (*Chelonia mydas*) collected from Japanese coastal waters in order to elucidate body distribution of the metals and to develop a nonlethal monitoring technique using the carapace. A majority of metal burdens were present in the muscle, liver, bone, and carapace of sea turtles. The high copper concentrations exceeding 10 μg/g were observed in the liver of these two turtle species. The mean zinc concentrations in fat tissues of loggerhead and green turtles were 94.6 and 51.3 μg/g, respectively, which were about 10-fold higher than those reported in other marine animals. The concentrations of manganese, zinc, and mercury in the carapace were correlated with whole body burdens, indicating that the carapace is a useful nonlethal indicator for monitoring heavy metal levels in the body of the sea turtle.[114] Furthermore, the authors reported arsenic metal ion species in the liver, kidney, and muscles of three species of sea turtles — green turtles (*Chelonia mydas*), loggerhead turtles (*Caretta caretta*), and hawksbill turtles (*Eretmochelys imbricata*). The order of arsenic concentration was muscles > kidneys > liver. Arsenic concentrations in the hawksbill turtles mainly feeding on sponges were higher than the other two turtles primarily feeding on algae and mollusks. The authors reported arsenobenite in all tissues of turtles. The distributions of some metal ion species in different aquatic biota are summarized in Table 2.2.

2.2.6 Distribution in Terrestrial Biota

All terrestrial animals and plants depend on the water supply from ground and surface waters; hence, it is a common observation to find toxic metal ions in these organisms. The evolutionary studies of heavy metal tolerance contribute to our response to contaminated soils on several levels. Such types of studies provide evidence of the toxic effects of heavy metal contamination on nonadapted genotypes. The evolutionary studies also show that the mechanism for metal tolerance is uptake, not exclusion, such that metal-tolerant genotypes are also metal accumulators. This last insight, in conjunction with physiological research on metal-tolerant plants, led to a growing use of plants as part of a cleanup technology for dealing with contaminated sites.[122,123] The heavy metal tolerance was first reported by the Czech scientist S. Prat in 1934 and has since been studied extensively by a number of scientists in Europe and the United States. Bradshaw and coworkers,[124] in particular, conducted extensive experiments on the evolutionary properties of plants growing in contaminated sites, such as mine spoils. Their findings include that plants in contaminated sites are genetically adapted to be tolerant of heavy metals. Metal-tolerant plants do not compete well in noncontaminated sites. The selection is so strong that genetic adaptation to contaminated sites takes place even though there is potential for gene flow from nearby nontolerant populations,

TABLE 2.2
Distribution of Some Metal Ion Species In Different Aquatic Biota

Metal Ion Species	Aquatic Organisms	Ref.
Arsenic	Plants	101
	Crustacea, mollusca, porifera, echinodermata, and coelentera	99, 102
	Green, loggerhead, and hawksbill turtles	115
	Dogfish muscles	116
	Seaweeds	117
Chromium	Geoporphyrins and shales oil	118
Iron	Geoporphyrins and shales oil	118
Mercury	Perca flavescens	111
	Dogfish and tuna fish	112
	Dogfish	113
	Swordfish	119, 120
	Cod, oyster, carp, trout, salmon, whale	121
	Many aquatic animals (review)	25
	Many aquatic animals (review)	8
Tin	Geoporphyrins and shales oil	118
Zinc	Geoporphyrins and shales oil	118

even at relatively low levels of contamination, such as roadside lead pollution from auto exhaust in urban areas with the imposed selection for metal tolerance. This adaptation of plants to heavy metal contamination is of particular interest, because it is a characteristic that appears to have evolved in part in response to human disturbance.[124–126] Dwivedi and Dey[127] explored the possibility of the translocation of heavy metals into humans and animals. The authors studied 28 commonly used medicinal plants and estimated their heavy metal contents. The plant materials were collected from the same sources used by traditional healers and commercial drug manufacturers. The plants were identified, authenticated, and processed for the analysis of toxic metals. Lead and cadmium levels were estimated in leaves, stems, bark, roots, or seeds, depending on the medicinal value of the plant portion. The mean lead concentration in medical herbs ranged from 2.624 ppm (standard deviation = 0.426) to 32.757 ppm (standard deviation = 0.124), and the cadmium concentration ranged from 0.056 ppm (standard deviation = 0.002) to 0.419 ppm (standard deviation = 0.006). Interestingly, the heavy metal concentrations (i.e., lead and cadmium) were higher in leaves than in stems, bark, or roots, and the lowest values were recorded in seeds. No published reports on the permissible levels of toxic metals in commonly used medicinal plants of India have come to the authors' attention; therefore, it was difficult to determine the role of toxic metals in drug-induced health hazards. However, the presence of toxic metals in different plants led to the conclusion that prolonged consumption of such medicinal plants may be detrimental to health. Szpunar and Lobinski[128] reviewed the speciation of metal ions in environmental biota, that is, plant and animal tissues. The areas discussed by them include the speciation of selenium and arsenic and the analytical chemistry of metal complexes with phytochelatins and metallothioneins.

Bantan et al.[129] determined low molecular weight aluminum organic acid complexes (LMW-Al) in plant sap. The plant sap of *Sempervivum tectorum* with a high concentration of aluminum (9.3 μg cm^{-3}) and *Sansevieria trifasciata* with an appreciably lower aluminum concentration (0.065 μg cm^3) were analyzed. The speciation study in plant sap indicated that the species present in these samples existed as negatively charged LMW-Al complexes. The predominant LMW-Al negatively charged complexes in *Sempervivum tectorum* and *Sansevieria trifasciata* were found to be aluminum citrate and aluminum aconitate. It was shown that other LMW organic acids do not form complexes with aluminum in the samples studied. Lin et al.[130] studied the aluminum tolerance in wheat by a large volume solution culture and a small pot soil test. The significant correlations were observed between root and shoot tolerance indices in wheat genotypes by using the same or different screening methods. It was also shown that the root tolerance parameters, such as relative root length (RRL) and relative root dry weight (RRW), were more sensitive for differentiating aluminum tolerance in wheat genotypes, and this indicated higher standard deviation (SD), coefficient of variation (CV), and distribution of data. Although shoot tolerance parameters, such as relative shoot length (RSL) and relative shoot dry weight (RSW), were less sensitive for differentiating aluminum tolerance within wheat genotypes indicated by lower SD, CV, and distribution of data, shoot growth parameters were reliable indicators of aluminum tolerance, because they were sufficient to discriminate aluminum-tolerant and aluminum-sensitive genotypes in wheat. Shoot growth, especially shoot length, could be easily, rapidly, and nondestructively determined, and then used effectively to screen for aluminum tolerance in a large-scale screening or breeding program. Weber et al.[131] determined iron species in phytosiderophores of the mugineic acid family, that is, in the root washings of iron-deficient wheat and barley plants and to a xylem exudate of nondeficient maize.

Arsenic (As) uptake by turnip, growing without soil culture conditions, was studied by Carbonell-Barrachina and coworkers.[132] The reported species were arsenite, arsenate, monomethylarsonic acid (MMA), and dimethylarsinic acid (DMA), with concentrations ranging from 1 to 5 mg/L. Arsenic phytoavailability and phytotoxicity were primarily determined by arsenic speciation. Organic arsenicals, especially MMA, were clearly phytotoxic to this turnip cultivar. The plant arsenic concentrations significantly increased with increasing arsenic application rates. Organic arsenicals showed a higher upward translocation than their inorganic counterparts, contributing to the greater phytotoxicity and lower dry matter productions of these organic treatments. Both inner root and outer root skin arsenic concentrations were above the maximum limit set for arsenic content in food crops (1 mg kg^{-1}). Koch et al.[133] reported inorganic arsenic species in plants from Yellowknife, Northwest Territories, Canada. Bohari et al.[134] reported arsenic species (arsenite, arsenate, monomethylarsonate [MMA], and dimethylarsinate [DMA]) in plants. Arsenate appears to be predominant in soils, roots, and leaves; unidentified species (probably arsenosugars) play an important role (60%) in rice fruits. Carrot was found to be the most contaminated edible plant part, containing 1 mgkg[1], essentially as arsenate species. MMA was detected in all soils and some plant parts, especially in shallots at low levels, whereas DMA was found only in one soil sample and in hot pepper leaves. Arsenite is a minor component of all soils; it is also present in some plant parts at low levels. However, no evident relationships were found between arsenic speciation in the various plant parts, and much more detailed studies will be necessary to elucidate arsenic behavior in plants.

Romero-Puertas et al.[135] reported cadmium in plants in Spain. Nocito et al.[136] studied the effect of cadmium on the high-affinity sulfate transport of maize (*Zea mays*) roots and related it to the changes in the levels of sulfate and nonprotein thiols during cadmium-induced phytochelatin (PC) biosynthesis. An amount of 10 μM $CdCl_2$ in the nutrient solution induced a 100% increase in sulfate uptake by roots. Tie et al.[137] studied the distribution and chemical pattern of cadmium in the grasses planted in cadmium-contaminated soil in a western suburb of Shenyang, China. The results showed that cadmium was mainly distributed in the grass roots, and characteristics of cadmium distribution in various grasses were different. Some grasses had the ability to resist cadmium, some had a strong affinity to bioaccumulate cadmium. The results obtained from samples of different grasses extracted by different solvents showed that the amount of acid-soluble cadmium in grasses was about 63.3% of the total, and the amounts of water-soluble and organic cadmium were small, accounting for 0.7% and 0.4% of the total, respectively. Acid-soluble cadmium was more active than the others during transformation in the soil–plant system, while water-soluble cadmium depended on the change of time and temperature, and organic cadmium became more active with the increase of solvent polarity.

Kotrebai et al.[138] reported selenium in hyperaccumulative phytoremediation plants — *Astragalus praleongus* and *Brassica juncea* at 517 and 138 μgg[1] selenium, respectively; yeast 1200, 1922, and 2100, μgg[1] selenium in dry sample; ramp (*Allium tricoccum*) 48, 77, 230, 252, 405, and 524 μgg[1] selenium in dry sample; onion (*Allium cepa*) 96 and 140 μgg[1] selenium in dry sample; and garlic (*Allium sativum*) 68, 112, 135, 296, and 1355 μgg[1] selenium in dry sample. Zhang and Frankenberger[139] determined selenium species in plants during their study of the bioavailability and toxicity of selenium in contaminated soil/sediment. This study showed that considerable amounts of the accumulation of Se(VI) into the plant was metabolized into selenium-amino acids during growth of the plant. Montes-Bayon et al.[140] studied selenium at levels (μg/L) in various *Brassica* species. This is an example of selenium phytoremediation. *Brassica juncea* (Indian mustard) was used to accumulate selenium by growing it with sodium selenite as the selenium source under hydroponic conditions, resulting in selenium accumulation at parts per million (ppm) levels in various parts of the plants.

Jung et al.[141] investigated environmental contamination derived from metalliferous mining activities. In the study area, the Dalsung Cu-W mine in Korea, soils, various crop plants, stream waters, sediments, and particulates were sampled in and around the mine and analyzed for arsenic, antimony, and bismuth. Arsenic, antimony, and bismuth contents in plant samples varied depending upon their species and parts, with higher concentrations in spring onions, soybean leaves, and perilla leaves, and lower levels in red peppers, corn grains, and jujube grains. These results confirm that elemental concentrations in plant leaves are much higher than those in plant grains. Murillo et al.[142] reported the accumulation of chemical elements in two crops of sunflower and sorghum affected by the spill. Total concentrations of arsenic, antimony, cadmium, copper, lead, manganese, tellurium, and zinc in spill-affected soils were greater than in adjacent, unaffected soils. Leaves of spill-affected crop plants had higher nutrient (K, Ca, and Mg for sunflower and N and K for sorghum) concentrations than the controls, indicating a fertilizing affect caused by the sludge. Seeds of spill-affected sunflower plants accumulated more arsenic, cadmium, copper, and zinc than controls, but values were below toxic levels. Leaves of sorghum plants accumulated more arsenic, bismuth, cadmium, lead, manganese, tellurium, and zinc than controls, but these values were also below toxic levels for livestock consumption. In general, none of the heavy metals studied in both crops reached either phytotoxic or toxic levels for humans or livestock. Nevertheless, a continuous monitoring of heavy metal accumulation in soil and plants must be established in the spill-affected area. Steinkellner et al.[143] reported arsenic(III), cadmium(II), lead(II), and zinc(II) in the tradescantia pollen mother cells (Trad MCN) and meristematic root tip cells of *Allium cepa* and *Vicia faba*. Zayed et al.[144] reported the presence of chromium species in various vegetable crops absorbing and accumulating chromium(III) and CrO_4^{2} into roots. The crops tested included cabbage (*Brassica oleracea* L. var. *capitata* L.), cauliflower (*Brassica oleracea* L. var. *botrytis* L.), celery (*Apium graveolens* L. var. *dulce* [Mill.] Pers.), chive (*Allium schoenoprasum* L.), collard (*Brassica oleracea* L. var. *acephala* DC.), garden pea (*Pisum sativum* L.), kale (*Brassica oleracea* L. var. *acephala* DC.), lettuce (*Lactuca sativa* L.), onion (*Allium cepa* L.), spinach (*Spinacia oleracea* L.), and strawberry (*Fragaria ananassa* Duch.). The authors also studied the translocation of both chromium species from roots to shoots, indicating limited accumulation of chromium by roots that were a hundredfold higher than by shoots, regardless of the chromium species supplied. Higher chromium concentrations were detected in members of the *Brassicaceae* family, such as cauliflower, kale, and cabbage. Weber et al.[131] reported up to five different iron metal ion species in phytosiderophores of the mugineic acid family which were separated and detected as different phytosiderophore species and two amino acid species. One more species detected but not identified by the authors was probably a decomposition product of phytosiderophores. Song et al.[145] studied the effects of the ecotoxicity of cadmium, copper, lead, and zinc in the single form for four types of soils (red loam soils, meadow brown soils, chestnut soils, and dark brown soils), and their combined effect was determined with meadow brown soils. The results showed that with the same content of heavy metals, root elongation was strongly inhibited compared to seed germination, and the latter was more sensitive to heavy metals pollution. The inhibition rate of heavy metals pollution on the root elongation of wheat was significantly related to the contents of organic matter and Kjedahl. Under low concentrations of cadmium, copper, lead, and zinc, an elongation effect in root was observed. Conversly, high concentrations of these metal ions resulted in an inhibition effect in the single form, while a synergistic effect occurred in the combined form. Hui[146] determined lead metal ion concentrations and distributions in plants and animals found it dispersed throughout different parts of the organisms. The concentrations of lead in plants vary according to the species' abilities to inhibit lead uptake from the soil. Horn snails had a mean lead concentration of 1987 μg/L dry mass, which was a hundred times greater than that in the leaves of the plant species with the highest mean concentration of 18.1 μg/L dry mass at the same site. The authors also reported that Avian predators of gastropods might have received minimum exposure to lead due to calcium in their shells, but the incidental ingestion of soil in addition to the direct ingestion of shot pellets might provide significant exposure to the birds. Jedrzejczak[147] determined the total mercury concentrations in 573 samples of agricultural crops and foods of plant origin (i.e., cereals, fruits, and vegetables, and their products) commercially available on the Polish market.

The various species of different metal ions were reported in almost all kinds of animals. Much work remains to be carried out in the determination of metal ions in animals. However, das Neves et al.[148] studied the presence of chromium(VI) in mouse spleen. Dean et al.[149] reported arsenic in the tissues of chickens. Crews et al.[150] determined levels of cadmium in the kidney of pig. Westöö[151] reported the presence of methylmercury in chicken tissue and liver of pig. Wrobel et al.[152] speciated aluminum and silicon in human serum. Aluminum citrate was found to be the main low molecular mass aluminum species in serum, amounting to about 12 ± 5% of the total aluminum in an aluminum-loaded serum sample. The proposed speciation procedure permitted the simultaneous identification and determination of three aluminum species in metal-spiked serum. The results for silicon suggested that it seemed to be unspecifically adsorbed to several serum proteins, and its speciation was not affected. Soldado Cabezuelo et al.[153] determined levels of aluminum in human serum. Bantan et al.[154] reported the presence of aluminum complexes in human serum. The authors reported aluminum phosphate and aluminum citrate as the species.

Despite deadly toxic effects, arsenic was reported in different body tissues and excretory products of human beings. Jimenez de Blas et al.[155] determined arsenic in human urine. The reported species were monomethylarsonic acid (MMA) and dimethylarsinic acid (DMA). Six arsenic compounds including arsenocholine, arsenobetaine, dimethylarsinic acid, methylarsonic acid, arsenous acid, and arsenic acid were detected in human urine by Zheng et al.[156] Mester and Pawliszyn[157] determined two methylated arsenic species in human urine samples. The reported arsenic species were DMA and MMA. Le et al.[158] reported the presence of MMA, DMA, As(III), and As(V) in human urine. He et al.[159] speciated arsenic species (i.e., arsenite, arsenate, MMA, and DMA) in blood samples.

Merrifield et al.[160] reported copper speciation in the alpha and beta domains of recombinant human metallothionein. Pons et al.[161] reported ionic alkyllead in human urine (i.e., butylated with BuMgCl). The speciation of ionic alkylleads in the urine of a petrol station worker showed a value of 27.9 ng/L of $Me_3Pb(II)$ in urine. Organomercury species, methylmercury (MeHg), ethylmercury (EtHg), phenylmercury (PhHg), and inorganic mercury [Hg(II)] at nanogram per liter (ng/L) levels were described in human urine by Yin et al.[162] Wang et al.[163] reported the speciation of trace elements in proteins in human and bovine serum. The distribution of various elements, including cobalt, iron, manganese, molybdenum, vanadium, and lanthanides, in human or bovine serum samples was shown. Alkali metals and tellurium were present primarily as free metal ions and were not bound to proteins. Jiang et al.[164] speciated organotin compounds and other major trace metal elements in different human organs. Copper, zinc, and other metal ions were reported in bacteria and some other microorganisms.[165,166] Kresimon et al.[167] reported organogelenium, arsenic, selenium, tin, antimony, and lead species in human urine after fish consumption. The distribution of some metal ion species in human beings is given in Table 2.3.

TABLE 2.3
Distribution of Some Metal Ion Species in Humans

Metal Ion Species	Body Part/Excretion Products	Ref.
Antimony	Urine	167
Arsenic	Urine	155–158
	Blood	99, 159
	Kidney	99, 168
	Liver	99, 168
	Muscle tissue	99
	Bones	99
	Hair	99
	Finger- and toenails	99
Chromium	Urine	169
Cobalt	Urine and blood	163
Copper	Urine and blood	170, 171
Iron	Urine and blood	163
Lead	Urine	161, 162, 171
	Blood	172
Manganese	Urine and blood	163
Nickel	Urine	173
Molybdanum	Urine and blood	163
Selenium	Urine	167, 174
	Blood	175
Tin	Urine	167
Vanadium	Urine and blood	163, 118
Zinc	Blood	170, 176

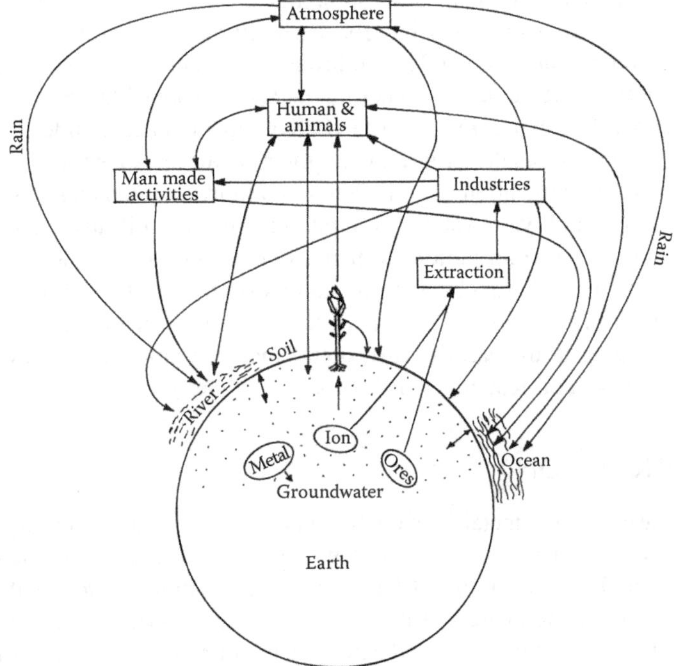

FIGURE 2.1 The distribution of metal ions in the environment.

Some metal ions were also reported in different kinds of food products. Basically, the type of metallic species and its concentrations in foodstuffs depend on the origin of the food products. For example, many metallic species were detected in seafood samples and food products grown on heavily contaminated soil and water resources. Aluminum is a ubiquitous element found in every food product. The other sources of aluminum are corn, yellow cheese, salt, herbs, spices, tea, and tap water. In households, aluminum-made wares are major sources of the element.[177] Larsen et al.[178] determined various arsenicals in seafood samples. Liang et al.[179] reported mercury species in seafood samples. Benramdane et al.[180] reviewed the presence of arsenic species in different foodstuffs. The distribution of metal ions in the environment is shown in Figure 2.1.

2.3 MECHANISMS OF METAL IONS CONTAMINATION

The mechanisms of the distribution and contamination of the environment by metal ions are simple to describe. Basically, the origin of metal ions is in the earth's crust, and they are in direct contact with groundwater. Metal ions are leached into groundwater from their ores in the earth's crust. The excessive withdrawal of groundwater creates spaces in aquifers that are filled by atmospheric air. The air present in these spaces oxidizes some metal ions in the ores that then contaminates groundwater. Sometimes, chemical reduction and bacteriological action are also responsible for the leaching of metal ions into groundwater, for example, arsenic is released through the reduction process and bacteriological action. Geological weathering is also responsible for groundwater contamination. The exposure of pyrite (FeS_2) and of other sulfide minerals to atmospheric oxygen results in one of the most acidic of all known weathering reactions. The contamination of soil occurs due to irrigation using contaminated ground, surface, and wastewater. The contamination of soil also occurs during rainy seasons. Major contributions to metal pollution of surface waters and soil are due to effluent discharges by many metal industries. The use of leaded gasoline and other man-made activities also lead to contamination of the environment. Briefly, beginning at the earth's crust, metal ions contaminate our environment by undergoing several reactions, processes, and cycles.

2.4 CONCLUSION

From the earth's crust, metal ions contaminate our environment through natural and man-made occurrences. Almost every component of the environment suffers from pollution due to metal ions. Moreover, contamination due to metal ions was also reported on some planets of the universe. As discussed, various metal ions are proven to be harmful, lethal, and carcinogenic in nature. In spite of this, some people of the world, especially those in developing and underdeveloped countries, are drinking water and eating foodstuffs containing various metal ions. Even small amounts of metal ion can alter the biological system, and, therefore, water and other foodstuffs with higher concentrations of metal ions than the permissible limits are dangerous to health. Different toxic metal ion species were reported in various body parts of human beings, making this a serious subject deserving great thought. Briefly, all of us, in the collaboration of governments, should try to avoid the activities responsible for metal ion contamination. To avoid groundwater contamination, its excessive withdrawal should be banned and it should also be recharged with contaminated free water. In addition, metal-ions-polluting industries should install treatment plants for metal ions removal in order to minimize water contamination.

REFERENCES

1. Stoeppler M, *Hazardous Metals in the Environment*, Elsevier, Amsterdam (1992).
2. Förstner U, Wittmann GTW (Eds.), *Metal Pollution in the Aquatic Environment*, Springer Verlag, Berlin (1983).
3. Matschullat J, *Sci. Total Environ.*, 249: 297 (2000).
4. Milacic R, Scancar J, Tusek J, *Anal. Bioanal. Chem.*, 372: 549 (2002).
5. Nriagu JO, *Nature*, 338: 47 (1989).
6. Hempel M, Chau YK, Dutka BJ, McInnis R, Kwan KK, Liu D, *Analyst*, 120: 721 (1995).

7. Antonovich VP, Bezlutskaya IV, *Anal. Chem.*, 51: 106 (1996).
8. Carro AM, Mejuto MC, *J. Chromatogr. A*, 882: 283 (2000).
9. Purves D, *Trace Elements Contamination of the Environment*, Elsevier, New York (1985).
10. Fitzgerald WF, Clarkson TW, *Environ. Health Prospect.*, 96: 159 (1991).
11. Flandysz J, Brzostowski A, Szpunar J, Rodriguez-Pereiro I, J. *Environ. Sci. Health*, A37: 353 (2002).
12. Welch AH, Lico MS, Hughes JL, *Ground Wat.*, 26: 333 (1988).
13. Korte NE, Fernando Q, *Critic Rev. Environ. Control*, 21: 1 (1991).
14. Jain CK, Ali I, *Water Res.*, 34: 4304 (2000).
15. Zufiaurre R, Pons B, Nerin C, *J. Chromatogr. A*, 779: 299 (1997).
16. Van Cleuvenbergen RJA, Dirx W, Quevauviller Ph, Adams FC, *Int. J. Environ. Anal. Chem.* 47: 21 (1992).
17. Robertson FN, *Environ. Geochem. & Health*, 11: 171 (1989).
18. Frost F, Frank D, Pierson K, Woodruff L, Raasina B, Davis R, Davies J, *Environ. Geochem. & Health*, 15: 209 (1993).
19. Das D, Chatterjee A, Mandal BK, Samanta G, Chakroborty D, Chanda B, *Analyst*, 120: 917 (1995).
20. Chatterjee A, Das, D, Mandal BK, Chowdhury TR, Samanta G, Chakraborty D, *Analyst*, 120: 643 (1995).
21. Falter R, Schöler HF, *Fresenius' J. Anal. Chem.*, 353: 34 (1995).
22. Pozdniakova S, Padarauskas G, Schwedt G, *Anal. Chim. Acta*, 351: 41 (1997).
23. Xu J, Ma Y, *J. Microl. Sep.*, 8: 137 (1996).
24. Boyle RW, Jonasson IR, *J. Geochem. Explor.*, 2: 251 (1973).
25. Timerbaev AR, Dabek-Zlotorzynska E, van den Hoop MAGT, *Analyst*, 124: 811 (1999).
26. Houba C, Remacle J, Dubois D, Thorez J, *Water Res.*, 17: 1281 (1983).
27. Sakai H, Kojima Y, Saito K, *Water Res.*, 20: 559 (1986).
28. Saeki K, Okazaki M, Matsumoto S, *Water Res.*, 27: 1243 (1993).
29. Latimer JS, Carey CG, Hoffman EJ, Quinn JG, *Water Resour. Bull.*, August issue, 791 (1988).
30. Zhang J, Huang WW, *Water Res.*, 27: 1 (1993).
31. Mohammed J, Hayat S, *Proc. Pak. Conf. Zool.*, 16: 283 (1996).
32. Katz BG, Bricker OP, Kennedy MM, *Am. J. Sci.*, 285: 931 (1985).
33. Barak H, Kress N, *Mar. Pollut. Bull.*, 34: 706 (1997).
34. Pizarro J, Vila I, Manuel C, *Verth-Int. Ver. Theor. Angew Limnol.*, 26: 948 (1998).
35. Alexandrine E, Jean-Marie M, Thevenot DR, *Water, Air, Soil Pollut.*, 108: 83 (1998).
36. Pistelok F, Galas W, *Poll. J. Environ. Stud.*, 8: 47 (1999).
37. Jain CK, *J. Hydrology*, 182: 105 (1996).
38. Prebha S, Selvapathy P, *Indian J. Environ. Pollut.*, 17: 641 (1997).
39. Agarwal A, *Homicides by Pesticides — What Pollution Does to Our Bodies*, Center for Science & Environment, Excellent Printing Press, New Delhi (1997).
40. Sharma SD, Pande KS, *Pollut. Res.*, 17: 201 (1998).
41. Priyadarshani N, *Indian Environ. Port.*, 18: 511 (1998).
42. Ali I, Jain CK, *Pollut. Res.*, 17: 321 (1998).
43. Tangen A, Lund W, Frederiksen RB, *J. Chromatogr. A*, 767: 311 (1997).
44. Schnierle P, Hauser PC, *J. Chromatogr. A.*, 779: 347 (1997).
45. Fukushi K, Hiiro K, *Fresenius' J. Anal. Chem.*, 356: 150 (1996).

46. Botana JC, Pereiro IR, Torrijos RC, *J. Chromatogr. A*, 963: 195 (2002).
47. Regan FB, Meaney MP, Lunte SM, *J. Chromatogr. B*, 657: 409 (1994).
48. Nerin C, Pons B, *Quimica Anal.*, 13: 209 (1994).
49. Adams FC, Heisterkamp M, Candelone JP, Laturnus F, van de Velde K, Boutron CF, *Analyst*, 123: 767 (1998).
50. Murozumi M, Chow TJ, Patterson CC, *Geochim. Cosmochim. Ac*ta, 33: 1247 (1969).
51. Ng A, Patterson CC, *Geochim. Cosmochim. Acta*, 45: 2109 (1981).
52. Pleßow A, Heinrichs H, *Aquatic Geochem.*, 6: 347 (2000).
53. Pantsar-Kallio M, Manninen PKG, *J. Chromatogr. A*, 750: 89 (1996).
54. Pantsar-Kallio M, Manninen PKG, *J. Chromatogr. A*, 779: 139 (1997).
55. Arar FJ, Pfaff JD, *J. Chromatogr. A*, 546: 335 (1991).
56. Thomas RL, Can. *J. Earth Sci.*, 9: 636 (1972).
57. Lai EPC, Zhang W, Trier X, Georgi A, Kowalski S, Kennedy S, Muslim T, Dabek-Zlotorzynska E, *Anal. Chim. Acta*, 364: 63 (1998).
58. Whang KS, Whang CW, *Electrophoresis*, 18: 241 (1997).
59. Feng YL, Narasaki H, *Anal. Bioanal. Chem.*, 372: 382 (2002).
60. Yang HJ, Jiang SJ, Yang YJ, Hwang CJ, *Anal. Chim. Acta*, 312: 141 (1995).
61. Pritzl G, Stuer-Lauridsen F, Carlsen L, Jensen AK, Thorsen TK, *Int. J. Environ. Anal. Chem.*, 62: 147 (1996).
62. Milde D, Plzak Z, Suchanek M, *Coll. Czech Chem. Commun.*, 62: 1403 (1992).
63. Ceulemans M, Adams FC, *Anal. Chim. Acta*, 317: 161 (1995).
64. Cai Y, Rapsomanikis S, Andreae MO, *Anal. Chim. Acta,* 274: 243 (1993).
65. Guntians MBD, Lobinski R, Adams FC, *J. Anal. At. Spectrom.*, 10: 111 (1995).
66. Goldschmidt VM, *J. Chem. Soc.* (London), 655 (1937).
67. Tack FMG, Verloo MG, *Int. J. Environ. Anal. Chem.*, 59: 225 (1995).
68. Das AK, Chakraborty R, Cervera ML, Delaguardia M, *Talanta*, 42: 1007 (1995).
69. Gibson JAE, Willet IR, *Commun. Soil Sci. Plant Anal.*, 22: 1303 (1991).
70. Göttlein A, Blasek R, *Soil Sci.*, 161: 705 (1996).
71. Mitrovic B, Milacic M, *Sci. Total Environ.*, 258: 183 (2000).
72. Prohaska T, Pfeffer M, Tulipan M, Stingeder G, Mentler A, Wenzel WW, *Fresenius' J. Anal. Chem.*, 364: 467 (1999).
73. Yang Q, Hartmann C, Smeyers-Verbeke J, Massart DL, *J. Chromatogr. A*, 717: 415 (1995).
74. Krajnc M, Stupar J, Milicev S, *Sci. Total Environ.*, 159: 23 (1995).
75. Hempel M, Wilken RD, Miess R, Hertutich J, Beyer K, *Water, Air, Soil Pollut.*, 80: 1089 (1995).
76. Carro AM, Lorenzo RA, Cela R, *LC-GC*, 16: 926 (1998).
77. Michalke B, Lusting S, Schramel P, *Electrophoresis*, 18: 196 (1997).
78. Hirner AV, Grüter UM, Kresimon J, *Fresenius' J. Anal. Chem.*, 368: 263 (2000).
79. Wang Q, Huang B, Guan Z, Yang L, Li B, *Fresenius' J. Anal. Chem.*, 370: 1041 (2001).
80. Dabek-Zlotorzynska E, Dlouhy JF, *J. Chromatogr. A.*, 671: 389 (1994).
81. Dabek-Zlotorzynska E, Dlouhy JF, *J. Chromatogr. A.*, 706: 527 (1995).
82. Dabek-Zlotorzynska E, Dlouhy JF, Houle N, Piechowski M, Ritchie S, *J. Chromatogr. A*, 706: 469 (1995).
83. Dasgupta PK, Kar S, *Anal. Chem.*, 67: 3853 (1995).
84. Mihalyi B, Molnar A, Meszaros E, *Ann. Chim.*, 86: 439 (1996).

85. Krivacsy Z, Molnar A, Tarjanyi E, Gelencser A, Kiss G, Hlavy J, *J. Chromatogr. A*, 781: 223 (1997).
86. Dabek-Zlotorzynska E, Piechowski E, Liu F, Kennedy S, Dlouhy JF, *J. Chromatogr. A*, 770: 349 (1997).
87. Fung YS, Lau KM, Tung HS, *Talanta*, 45: 619 (1998).
88. Tripathi RM, Mahapatra S, Raghunath R, Vinod Kumar A, Sadasivan S, *Sci. Total Environ.*, 299: 73 (2002).
89. Nerin C, Pons B, *Appl. Organomet. Chem.*, 8: 607 (1994).
90. Nerin C, Pons B, Zufiaurre R, *Analyst*, 120: 751 (1995).
91. Doudoroff P, Katz M, *Wastes*, 25: 802 (1953).
92. Knauer GA, Martin JH, *Limnol. Oceanorg.*, 18: 597 (1973).
93. Morris AW, *Nature*, 233: 427 (1971).
94. Schulz-Baldes M, Lewin AR, *Biol. Bull.*, 150: 118 (1976).
95. Preston A, *Proc. Royal Soc. Ser. B*, 180: 421 (1972).
96. Stenner RD, Nickless G, *Nature*, 247: 198 (1975).
97. Lande E, *Environ. Pollut.*, 12: 187 (1977).
98. Martin JH, Broenkow WW, *Science*, 190: 884 (1975).
99. Bowen HJM, *Environmental Chemistry of the Elements*, Academic Press, London (1979).
100. Dietz F, *Wasser/Abwasser*, 113: 269 (1972).
101. Reay PF, *J. Appl. Ecol.*, 9: 557 (1972).
102. Francesconi KA, Edmonds JS, Morita M, in Nriagu JO (Ed.), *Arsenic in the Environment Part I: Cycling and Characterisation*, John Wiley & Sons (1994).
103. Skei JM, Saunders M, Price NB, *Mar. Pollut. Bull.*, 7: 34 (1976).
104. Philips DJH, *Mar. Biol.*, 38: 59 (1976).
105. Stenner RD, Nickless G, *Mar. Pollut. Bull.*, 6: 89 (1975).
106. Johnels AG, Westermark T, Berg W, Persson PI, Sjöstrand B, *Oikos*, 18: 323 (1967).
107. Hasselrot TB, Göthberg A, *Proc. Int. Conf. Persist. Chem., Aquatic Ecosystem*, Ottawa, Canada, 37 (1974).
108. Müller G, Prosi F, *Naturforsh*, 33c: 7 (1978).
109. Papadopoulou C, Kanias GD, Elli MK, *Mar. Pollut. Bull.*, 9: 106 (1978).
110. MacKay NJ, Kazacos MN, Williams RJ, Leedowq MI, *Mar. Pollut. Bull.*, 6: 57 (1975).
111. Freitas ASW, de Qadri SU, Case BE, *Proc. Int. Conf. Transp. Persist. Chem. Aquatic Ecosys.*, Ottawa, Canada, 31 (1974).
112. Pereiro IR, Wasik A, Lobinski R, J. *Anal. At. Spectrom.*, 13: 743 (1998).
113. Logar M, Horvat M, Falnoga I, Stibilj V, *Fresenius' J. Anal. Chem.*, 366: 453 (2000).
114. Sakai H, Saeki K, Ichihashi H, Suganuma H, Tatsukawa STR, *Mar. Pollut. Bull.*, 40: 701 (2000).
115. Saeki K, Sakakibara H, Sakai H, Kunito T, Tanabe S, *Bio. Met.*, 13: 241 (2000).
116. Branch S, Ebdon L, O'Neill P, J. *Anal. At. Spectrom.*, 9: 33 (1994).
117. Gallagher PA, Shoemaker JA, Wei X, Brockhoff-Schwegel CA, Creed JT, *Fresenius' J. Anal. Chem.*, 369: 71 (2001).
118. Ebdon L, Evans EH, Pretoripus WG, Rowlands SJ, *J. Anal. At. Spectrom.*, 9: 939 (1994).
119. Panaro KW, Erickson D, Krull IS, *Analyst*, 112: 1097 (1987).

120. Holak W, *J. Assoc. Off. Anal. Chem.*, 72: 926 (1989).
121. Donais MC, Uden PC, Schantz MM, Wise SA, *Anal. Chem.*, 68: 3859 (1996).
122. Salt DE, *Bio. Technol.*, 13: 468 (1995).
123. Adler T, *Science News*, 150: 42 (1996).
124. Bradshaw AD, *Trans. P, Roy. Soc.* (London), 333: 289 (1991).
125. Antonovics J, *Adv. Ecol. Res.*, 7: 1 (1971).
126. Antonovics J, International Conference on Heavy Metals in the Environment, Toronto, Ontario, Canada, p169 (1975).
127. Dwivedi SK, Dey S, Arch. *Environ. Health*, 57: 229 (2002).
128. Szpunar J, Lobinski R, *Fresenius' J. Anal. Chem.*, 363: 550 (1999).
129. Bantan T, Milacic R, Mitrovic B, Pihlar B, *Fresenius' J. Anal. Chem.*, 363: 545 (1999).
130. Lin X, Zhang Y, Luo A, *Ying Yong Sheng Tai Xue Bao*, 13: 766 (2002).
131. Weber G, Neumann G, Romheld V, *Anal. Bioanal. Chem.*, 373: 767 (2002).
132. Carbonell-Barrachina AA, Burlo F, Valero D, Lopez E, Martinez-Romero D, Martinez-Sanchez F, *J. Agric. Food Chem.*, 47: 2288 (1999).
133. Koch I, Wang L, Ollson CA, Cullen WR, Reimer KJ, *Environ. Sci. Technol.*, 34: 22 (2000).
134. Bohari Y, Lobos G, Pinochet H, Pannier F, Astruc A, Potin-Gautier M, J. *Environ. Monit.*, 4: 596 (2002).
135. Romero-Puertas MC, McCarthy I, Sandalio LM, Palma JM, Corpas FJ, Gomez M, del Rio LA, *Free Radic. Res.*, S25 (1999).
136. Nocito FF, Pirovano L, Cocucci M, Sacchi GA, *Plant Physiol.*, 129: 1872 (2002).
137. Tie M, Liang Y, Zhang C, Li F, Dong H, Zang S, Ying Yong Y, *Ying Yong Sheng Tai Xue Bao*, 13: 175 (2002).
138. Kotrebai M, Birringer M, Tyson JF, Block E, Uden PC, *Analyst*, 125: 125 (2000).
139. Zhang Y, Frankenberger WT Jr., *Sci. Total Environ.*, 26: 269 (2001).
140. Montes-Bayon M, Yanes EG, Ponce D, Jayasimhulu K, Stalcup A, Shann J, Caruso JA, *Anal. Chem.*, 74: 107 (2002).
141. Jung MC, Thornton I, Chon HT, *Sci. Total Environ.*, 259: 81 (2002).
142. Murillo JM, Maranon T, Cabrera F, Lopez R, *Sci. Total Environ.*, 242: 281 (1999).
143. Steinkellner H, Mun-Sik K, Helma C, Ecker S, Ma TH, Horak O, Kundi M, Knasmuller S, *Environ. Mol. Mutagen*, 31: 183 (1998).
144. Zayed A, Mel Lytle C, Qian JH, Terry T, *Planta*, 206: 293 (1998).
145. Song Y, Zhou Q, Xu H, Ren L, Gong P, *Ying Yong Sheng Tai Xue Bao*, 13: 459 (2002).
146. Hui CA, *J. Toxicol. Environ. Health A*, 65: 1093 (2002).
147. Jedrzejczak R, *Food Addit. Contam.*, 19: 996 (2002).
148. das Neves RP, Santos TM, de Pereira ML, de Jesus JP, *Cytobios*, 106: 27 (2001).
149. Dean RJ, Ebdon L, Foulkes ME, Crews HM, Massey SS, *J. Anal. At. Spectrom.*, 4: 285 (1989).
150. Crews HM, Dean JR, Ebdon L, Massey RC, *Analyst*, 114: 895 (1989).
151. Westöö G, *Acta Chem. Scand.*, 21: 1790 (1967).
152. Wrobel K, Gonzalez EB, Wrobel K, Sanz-Medel A, *Analyst*, 120: 809 (1995).
153. Soldado Cabezuelo AB, Montes Bayon M, Blancon Gonzalez E, Garcia Alonso JI, Sanz-Medel A, *Analyst*, 123: 865 (1998).
154. Bantan T, Milacic R, Mitrovic B, Pihlar B, *J. Anal. At. Spectrom.*, 14: 1743 (1999).
155. Jimenez de Blas O, Vicente Gonzalez S, Seisdedos Rodriguez R, Hernandez Mendez J, *J. AOAC Int.*, 77: 441 (1994).

156. Zheng J, Kosmus W, Pichler-Semmelrock F, Kock M, J. *Trace Elem. Med. Biol.*, 13: 150 (1999).
157. Mester Z, Pawliszyn J, *J. Chromatogr. A*, 873: 129 (2000).
158. Le XC, Chris, Lu X, Ma M, Cullen WR, Aposhian VH, Zheng B, *Anal. Chem.*, 72: 5172 (2000).
159. He B, Jiang GB, Xu X, *Fresenius' J. Anal. Chem.*, 368: 803 (2000).
160. Merrifield ME, Huang Z, Kille P, Stillman MJ, *J. Inorg. Biochem.*, 15: 88 (2002).
161. Pons B, Carrera A, Nerin C, *J. Chromatogr. B*, 716: 139 (1998).
162. Yin X, Frech W, Hoffmann E, Lüdke C, Skole J, *Fresenius' J. Anal. Chem.*, 361: 761 (1998).
163. Wang J, Houk RS, Dreessen D, Wiederin DR, *J. Biol. Inorg. Chem.*, 4: 546 (1999).
164. Jiang GJ, Zhou Qun, He B, *Environ. Sci. & Technol.*, 34: 2697 (2000).
165. Kunito T, Saeki K, Oyazu H, Matsumoto S, *Ecotoxicol. & Environ. Safety*, 44: 174 (1999).
166. Kunito T, Saeki K, Goto S, Hyashi H, Oyazu H, Matsumoto S, *Bioresour. Technol.*, 79: 1325 (2001).
167. Kresimon J, Grüter UM, Hirner AV, *Fresenius' J. Anal. Chem.*, 371: 586 (2001).
168. Fergusson JE, *The Heavy Elements: Chemistry, Environmental Impact and Health Effects*, Pergamon Press, Oxford (1990).
169. Zoorob G, Tomlinson M, Wang J, Caruso J, *J. Anal. At. Spectrom.*, 10: 853 (1995).
170. Matz SG, Elder RC, Teppeman K, *J. Anal. At. Spectrom.*, 4: 767 (1989).
171. Shum SCK, Pang HM, Houk RS, *Anal. Chem.*, 64: 2444 (1992).
172. Gercken B, Barnes RM, *Anal. Chem.*, 63: 283 (1991).
173. Tomlinson MJ, Wang J, Caruso JA, *J. At. Anal. Spectrom.*, 9: 957 (1994).
174. Yang KL, Jiang SJ, *Anal. Chim. Acta*, 307: 109 (1995).
175. Olivas RM, Donard OXF, Gilon N, Potin-Gautlier M, *J. Anal. At. Spectrom.*, 11: 1171 (1996).
176. Kumar U, Dorsey JG, Caruso JA, *J. Chromatogr. Sci.*, 32: 282 (1994).
177. Ochmanski W, Barabasz W, *Przegl Lek*, 57: 665 (2000).
178. Larsen EH, Pritz G, Hansen SH, *J. Anal. At. Spectrom.*, 8: 1075 (1993).
179. Liang LN, Jiang GB, Liu JF, Hu JT, *Anal. Chim. Acta*, 477: 131 (2003).
180. Benramdane L, Bressolle F, Vallon JJ, *J. Chromatogr. Sci.*, 37: 330 (1999).

156. [illegible] M, [illegible] K, [illegible] M, [illegible] (1995).
157. [illegible] Y, [illegible] *Chromatogr. A*, 874, 129 (2000).
158. Li XG, Chen [illegible], Ma [illegible], Chen [illegible], [illegible] Rapid Commun. Chem. 72, 1822 (2000).
159. He S, Hang [illegible] [illegible] *J.* [illegible], 283, 1410 (2006).
160. [illegible] (2012).
161. [illegible] (1987).
162. [illegible] *Anal. Chem.* [illegible] (1985).
163. [illegible]
164. [illegible]
165. [illegible] (1999).
166. [illegible] 28 (2005).
167. Kristiansen I, Gutter M, [illegible] A, [illegible] *Anal. Chem.* 77, 2600 (2005).
168. [illegible] *Polymer Characterization* [illegible] *Techniques for Polymer* [illegible], Oxford (1991).
169. [illegible] M, [illegible] 31, 323 (1943).
170. [illegible] SC, [illegible] *Appl. Spectrosc.* [illegible] (1980).
171. [illegible] SCR, [illegible] *Polymer* [illegible] (1992).
172. [illegible] *Anal. Chem.*, 65, 237 (1993).
173. Tomlinson M, [illegible] *Macromolecules*, 2, 159 (1992).
174. [illegible] *Sci. Appl.* 30, 3488 (1985).
175. [illegible] *Appl. Spectrosc.* [illegible] (1994).
176. [illegible] *Chromatogr.* [illegible] (1994).
177. [illegible] *Polym. J.* [illegible] (2000).
178. [illegible] *J. Anal. Appl. Pyrolysis* [illegible] (1993).
179. [illegible] *Polym. Chem.* [illegible] 131 (2003).
180. [illegible] *Macromolecules*, 32, 389 (1999).

3 Metal Ion Species: Toxicities, Biotransformation, and Biodegradation

Metal ions are essential components of biological systems; nevertheless, even essential elements may have toxic or carcinogenic properties. The contamination of our environment due to heavy metals is a very serious and challenging problem. Heavy metals occur in the environment in different physicochemical forms. Among them, simple hydrated metal ions are considered to be the most toxic forms, while strong complexes and species associated with colloidal particles are usually assumed to be nontoxic. Organometallic compounds are more toxic than the corresponding inorganic species, with the exception of arsenic. If unrecognized or inappropriately treated, heavy metal toxicity can result in significant morbidity and mortality. Therefore, heavy metals are associated with various lethal disorders and diseases including carcinogenesis. Heavy metals are finding their ways into the human body through water, food, vegetables, air, and so forth. After ingestion, metal ions are absorbed through the gastrointestinal tract into the bloodstream and from there reach different parts of the body where they disturb biological functions. The general route of entry of metal ions into the human body is shown in Figure 3.1. The toxicities of metal ion species depend on the age and sex of the human. The biological activities of organic and inorganic metal ion species in living systems are given in Table 3.1.[1]

Many metallic ions are found in the environment as different species, which are differentiated not only by their physical and chemical forms but also by their diverse toxic activities with respect to living organisms (speciation). The speciation affects the bioavailability and toxicity of elements and, hence, is important in toxicology and nutrition. The exploitation of the speciation profiles in medical management is widely unexplored. Isotopic speciation in the body can also offer clues to the sources of exposure. The changes in the degree of oxidation of an element also have an important effect on the degree of bioavailability and toxicity.[2] The elements occur in the environment in different oxidation states and species, for example, As as As(III) and As(V); Sb as Sb(III) and Sb(V); and Se as Se(VI) and Se(IV). In an oxidized environment, As, Sb, and Se appear mostly as oxya-

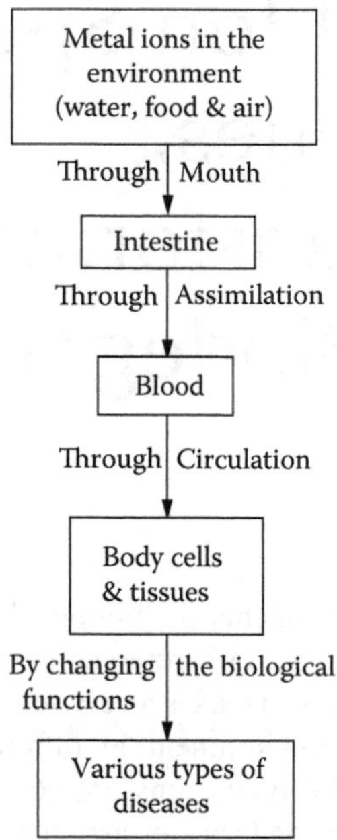

FIGURE 3.1 The general route of entry for metal ion toxicities into the human body.

TABLE 3.1
Comparison of the Activities of Organic and Inorganic Metal Ion Species in Biological Systems

Biological Action	Organic Species	Inorganic Species
Adsorption	Yes	Yes
Plasma protein binding	Albumin	Globulin
Target organs	Yes	Yes
Metabolized	Yes	Yes
Excretion (main route)	Bile	Urine
Induction	Enzyme (P-450)	Metallothionein
Oxidation states	Not important	Important
Biodegradation	Yes	No

Source: From Furst, A., in *Toxicology of Metals: Clinical and Experimental Research*, Ellis Horwood Ltd., Chichester, 1987, 17. With permission.

nions. The valency state of an element plays an important role in the behavior of that element in the aqueous system. For example, the toxicities of As(III) and Sb(III) are higher than those of their pentavalent species. Briefly, the toxicities of the different oxidation states (species) of a particular metal ion may be different, and one of the species may be more toxic. Therefore, knowledge of the toxicities of different species of a particular metal ion is very important in the environmental and toxicological points of view.

In international legislation concerning trace elements in food, most environmental and occupational health regulations are based on total element contents and are frequently given as maximum limits or guideline levels. However, only few regulations pay attention to the molecular species in which the elements are found and bound. International legislation concerning contaminants in food is presently being established by the Codex Alimentarius Commission, which is an independent United Nations organization under the joint Food and Agriculture Organization (FAO)/World Health Organization (WHO) Food Standards Programme. The development of the Codex General Standard for Contaminants and Toxins in Food provides the framework for future international legislation on metals as contaminants in food. For certain food additives, which include some essential minerals, speciation is an integral part of the set of specification criteria, because only certain defined chemical compounds are permitted as sources of the essential element. The development of more species-specific analytical and toxicological data and improved communication with legislators will be necessary before it will become possible to lay down species-specific regulations in all the cases where the specialized scientist will consider it reasonable.

The existing data on toxicity due to the total concentration of a particular metal ion is not sufficient to describe the exact toxicity, as different species may have different toxicities. Therefore, to obtain the exact toxicities of metal ions, there is a great need to find out the toxicities of the different forms of all metal ions. This chapter describes the bioactivities and toxicities of the different species of toxic heavy metal ions. In this chapter, we will discuss only the different species of the toxic heavy metals.

3.1 MODES OF BIOACTIVITY AND TOXICITIES OF METAL IONS

Metal ions are the main component of the biological system, as they are found in all organisms ranging from microbes to human beings. Some metal ions interact with certain microbes and alter their enzymatic activities.[3–5] Therefore, before discussing the toxicities of metal ions, it is important to discuss the bioactivities of metal ions. Some important activities of metal ions in biological systems are discussed herein.

The choice of cell line for *in vitro* biological tests, which assess the cytotoxicity of dental materials, remains controversial, yet this issue is important, because these tests are widely used to rate the biocompatibility of new and existing materials, and many different cell lines are commonly used. The quantification of responses of four cell lines (Balb/c 3T3, L929, ROS 17/2.8, and WI-38) to 14 metal ions that were released from dental materials and relation of these responses to the metabolic activity and population doubling times of these cells were discussed by Wataha et al.[6] The authors measured the activity of succinic dehydrogenase (SDH) to monitor the metabolic activity and cytotoxic response. The cell lines responded differently to most metal ions. In general, the Balb/c 3T3 line was most sensitive, and the WI-38 line was the least sensitive. However, there were many exceptions depending on the metal ions. The passage number of the cells also affected the cytotoxic response. It was concluded that the cytotoxicity of materials that released metal ions significantly depended on the cell line selected. Based on the findings that cell lines ranked the toxicities of the metal ions, it was logical to use these types of cell lines for *in vitro* tests to rank the cytotoxicities of materials.

The role of reactive oxygen species, with the subsequent oxidative deterioration of biological macromolecules in the toxicities associated with transition metal ions, is very important. Recent studies[7] showed that metals, including iron, copper, chromium, and vanadium, undergo redox cycling, while cadmium, mercury, and nickel, as well as lead, deplete glutathione and protein-bound sulfhydryl groups, resulting in the production of such reactive oxygen species as superoxide ion, hydrogen peroxide, and hydroxyl radical. As a consequence, enhanced lipid peroxidation, DNA damage, and altered calcium and sulfhydryl homeostasis occurred. Fenton-like reactions may be commonly associated with most membranous fractions, including mitochondria, microsomes, and peroxisomes. The phagocytic cells may be another important source of reactive oxygen species in response to metal ions. Furthermore, various studies suggested that the ability to generate reactive oxygen species by redox cycling quinones and related compounds may require metal ions. Recent studies suggested that metal ions may enhance the production of tumor necrosis factor alpha (TNF-α) and may activate protein kinase C, as well as induce the production of stress proteins. Thus, some mechanisms associated with the toxicities of metal ions are similar to the effects produced by many organic xenobiotics. Specific differences in the toxicities of metal ions may be related to the differences in solubilities, absorption, transport, chemical reactivity, and the complexes that are formed within the body. England et al.[8] reviewed the current studies that were conducted with transition metal ions and lead and that were concerned with the production of reactive oxygen species and oxidative tissue damage. Furthermore, these authors attempted to predict the toxicities of 24 metal ions for a given species, using physicochemical parameters associated with the ions. The authors chose indicators of toxicity for biological systems of increasing levels of complexity, starting with individual biological molecules and ascending to mice as being representative of higher-order animals. The numerical values for these indicators were normalized to a scale of 100 for Mg(II) (essentially nontoxic) and 0 for Cd(II) (very toxic). To predict toxicities to humans, extrapolations across biological species were carried out for each of the metal ions considered. The predicted values were then compared with threshold limit values (TLVs) from the literature. Both methods for predicting toxicities have their advantages and disadvantages, and both have limited success for metal ions. However, the second approach suggests that the TLV for Cu(II) should be lower than that currently recommended.

de la Fuente et al.[9] investigated the effects of cadmium, lead, and arsenic on the apoptosis of human immune cells. The peripheral blood mononuclear cells (MNCs) were incubated with increasing concentrations of these metals, and then cellular apoptosis was determined by flow cytometry and by DNA electrophoresis. The authors reported that arsenic induced a significant level of apoptosis at 15 μM after 48 h of incubation. Cadmium had a similar effect but at higher concentrations (65 μM). In addition, cadmium exerted a cytotoxic effect on MNCs that seemed to be independent of the induction of apoptosis. In contrast, concentrations of lead as high as 500 μM were nontoxic and did not induce a significant degree of apoptosis. Additional experiments showed that arsenic at concentrations as low as 1 μM had a significant proapoptotic effect when cells were cultured in the presence of this pollutant for more than 72 h. Cells that were not T cells were more susceptible than T lymphocytes to the effects of arsenic and cadmium. Interestingly, MNCs from children chronically exposed to arsenic showed a high basal rate of apoptosis and a diminished *in vitro* sensibility to this metalloid. The authors reported both arsenic and cadmium as able to induce apoptosis of lymphoid cells and suggested that this phenomenon might be contributed to their immunotoxic effect *in vivo*.

In addition to As(III), Cd(II), Ni(II), and Co(II) being disturbing agents for different types of DNA repair systems at low noncytotoxic concentrations, some metals exert high affinities for sulfhydryl (SH) groups. For example, zinc finger structures in the DNA-binding motifs of DNA repair proteins are potential targets for toxic metal ions. The bacterial formamidopyrimidine-DNA glycosylase (Fpg protein) involved in base excision repair was inhibited by Cd(II), Cu(II), and Hg(II) with increasing efficiencies, whereas Co(II), As(III), Pb(II), and Ni(II) had no effect. Furthermore, Cd(II) still disturbed enzyme function when bound to metallothionein. The strong inhibition was also observed in the presence of phenylselenyl chloride followed by selenocystine, while selenomethionine was not inhibitory. Regarding the mammalian XPA protein involved in the recognition of DNA lesions during nucleotide excision repair, its DNA-binding capacity was diminished by Cd(II), Cu(II), Ni(II), and Co(II), while Hg(II), Pb(II), and As(III) were ineffective. Finally, the hydrogen-peroxide-induced activation of the poly(ADP-ribose)polymerase (PARP) involved in DNA strand break detection and apoptosis was greatly reduced by Cd(II), Co(II), Ni(II), and As(III). Similarly, the disruption of correct p53 (suppressor gene) folding and DNA binding by Cd(II), Ni(II), and Co(II) was shown by other authors. Therefore, zinc-dependent proteins involved in DNA repair and cell-cycle control may represent sensitive targets for some toxic metals, such as Cd(II), Ni(II), Co(II), and Cu(II), as well as for some selenium compounds. Relevant mechanisms of inhibition appear to be the displacement of zinc by other transition metals as well as redox reactions leading to thiol/disulfide interchange.[10]

Wolterbeek and Verburg[11] carried out the prediction of metal toxicity by evaluating the relationships between general metal properties and toxic effects. For this, metal toxicity data were taken from 30 literature data sets, which varied largely in exposure times, organisms, effects, and effect levels. The authors selected general metal properties, including the electrochemical potential delta, the ionization potential, the ratio between atomic radius and atomic weight (AR/AW), and the electronegativity (Xm). The results suggested that toxicity prediction might be performed on the basis of these fixed metal properties without any adaptation to specific organisms, without any division of metals into classes, and without any grouping of toxicity tests. The results further indicated that metal properties contributed to the observed effects in relative importance, which depended on specific effects, effect levels, exposure times, selected organisms, and ambient conditions. The discussion strongly suggested that prediction should be by interpolation rather than by extrapolation of calibrated toxicity data. The concept was that unknown metal toxicities might be predicted on the basis of observed metal toxicities in calibration experiments. Considering the used metal properties, the calibration covered the largest number of metals by the simultaneous use of Ge(IV), Cs(I), Li(I), Mn(VII), Sc, and Bi in toxicity studies. Based on the data from the 30 studies, metal toxicities could be ordered in a relative way. This ordering indicated that the natural abundance of metals or metal ions in the earth's crust might be regarded as a general comparative measure of the metal toxicities. The problems encountered in toxicity interpretation and ordering of toxicities indicated that the control of the solution acidity, the metal solubility, and the metal oxidation state may be key problems to overcome in future metal ion toxicity studies.

Qu et al.[12] investigated the mechanism of cross-tolerance to nickel in arsenic-transformed cells. The chronic arsenite-exposed (CAsE) cells (TRL 1215 cells, which were continuously exposed to 0.5 μM arsenite for 20 or more weeks) and control TRL 1215 cells were both exposed to nickel for 24 h, and cell viability was determined by metabolic integrity. The LC_{50} for nickel was 608 ± 32 μM in CAsE cells as compared to 232 ± 16 μM in control cells, a 2.6-fold increase. CAsE and control cells were treated with 200 μM nickel for 4 h, and cellular free radical production was measured using electron spin resonance spectrometry. Hydroxyl radical generation was decreased in CAsE cells. Thiobarbituric acid reactive substances, indicative of lipid peroxidation, and 8-oxo-2′-deoxyguanosine, indicative of oxidative DNA damage, were reduced in CAsE cells. The flow cytometric analysis using Annexin/FITC revealed that nickel-induced apoptosis was reduced in CAsE cells. CAsE cells showed generalized resistance to oxidant-induced toxicity as evidenced by a marked reduction in sensitivity to hydrogen peroxide. Interestingly, intracellular reduced glutathione (GSH) levels were significantly increased in CAsE cells, and when GSH was depleted, CAsE cells lost their nickel resistance. The mechanisms of arsenic-induced cross-tolerance to cytotoxicity, genotoxicity, and apoptosis induced by nickel appears to be related to a generalized resistance to oxidant-induced injury, probably based, at least in part, on increased cellular GSH levels.

Seoane and Dulout[13] studied the aneugenic and clastogenic ability of cadmium chloride(II), cadmium sulfate(II), nickel chloride(II), nickel sulfate(II), chromium chloride(III), and potassium dichromate(IV) through a kinetochore-stained micronucleus test. The traditional genotoxicity assays evaluated DNA damage, gene mutations, and chromosome breakage. However, these tests were not adequate in detecting aneugenic agents that do not act directly on DNA. Staining kinetochores in the cytokinesis-blocked micronucleus assay was a useful way to discriminate between clastogens and aneuploidogens and might allow for rapid identification of aneuploidy-inducing environmental compounds. Human diploid fibroblasts (MRC-5) were employed. All compounds increased micronuclei frequency in a statistically significant way. However, increases in kinetochore-positive micronuclei frequencies were higher than in kinetochore-negative ones. The authors reported the genotoxic ability of the cadmium and chromium salts studied. Aneugenic as well as clastogenic ability could be observed with this assay. Nickel salts, as it was expected because of their known weak mutagenicity, showed lower genotoxic effects than the other metal salts studied. As the test employed only allowed for the detection of malsegregation, it is proposed that this mechanism was at least one by which the tested metal salts induced aneuploidy. On the other hand, visualization of kinetochores in all experiments suggested that the compounds studied did not act by damaging these structures.

3.2 TOXICITIES OF METAL IONS

Among the natural environment's chemical control indicators, heavy metals occupy a particularly important place, as they are the most toxic chemical elements. Some heavy metals as microelements are of vital importance for animal and plant development, however, when present in larger concentrations, they act as typical toxicants. One of the most dangerous characteristics of heavy metals is that they accumulate in living organisms even when their concentrations in the surrounding environment are comparatively low, and so over a longer period of time, these concentrations can reach dangerous levels. Out of 53 naturally occurring elements, 53 are heavy metals. Again out of 53 heavy metals, only 17 metals (As, Ag, Cd, Co, Cr, Cu, Fe, Hg, Mn, Mo, Ni, Pb, Sb, U, V, W, and Zn) are available to living cells and form soluble cations indicating biological effects. Human beings are inevitably exposed to carcinogens from environmental, occupational, medicinal, and other sources. The current approaches for evaluating carcinogenic risk rely on the information obtained from experiments with animal studies. These studies are carried out at high exposure levels, often at maximum tolerable doses, and the dose responses observed are extrapolated linearly to low exposure levels. The results from such direct extrapolations are unreliable because of the large differences in dose levels of exposure encountered in environmental or occupational settings compared with the high exposures in animal experiments. A good approach is to assess the internal dose of a toxic compound by measuring the extent of chemical interaction of the compounds with biological macromolecules, such as nucleic acids, peptides, and proteins. These types of measurements are preferable to those of external doses, because they take into account individual differences in genetics, absorption, metabolism, distribution, and excretion. Determination of the dose of active compounds that reaches the target site of action within the body enables one to estimate the risk associated with the exposure. Normally, the toxicity of metal ions is described by measuring median lethal dose (LD_{50}) values or median lethal concentration (LC_{50}). However, some authors reported toxicities using IC_{50} and EC_{50} terms.

All the metal ions described in the periodic table are not toxic; therefore, only toxic metal ions are discussed herein. Among various metal ions, most common heavy metals implicated in acute and chronic conditions include antimony, arsenic, chromium, lead, mercury, and selenium. However, the higher concentrations of other metal ions, such as aluminum, vanadium, copper, zinc, platinum, and rhodium, may be toxic. The concept of the existence of different species (speciation) occurs only in transition, lanthanides, and actinide metal ions (heavy metal ions) due to the presence of empty d (transition) and f (lanthanides and actinide) orbitals, respectively. Therefore, we will restrict our discussion to the toxicities of different species of these heavy metal ions. Considerable interest has grown in assessing the risk posed by metals, trace elements, and other metallic compounds to the environment and to human health. In recent years, it was realized that some parts of the globe contain naturally endemic areas related to trace element excess, resulting in chronic poisoning due to the contamination of drinking water, foodstuffs, volcanic eruptions, and so forth. An understanding of the potential environmental and health effects of metals is of critical importance in ensuring that the production, use, and disposal of metal ions is environmentally cautious, in minimizing exposure to toxic levels by using scientifically sound regulations, and in developing appropriate analytical techniques for their analyses even at low levels. Exposure to toxic metal ions occurs through percutaneous absorption, ingestion, and inhalation. Dermal toxicity results due to a reaction on the skin, for example, allergic dermatitis induced by nickel. Ingestion and inhalation may result in a variety of toxic pathology responses in human tissues and organs (i.e., skin, liver, heart, kidney, etc.). In view of these points, the toxicities of different forms of metal ions are discussed below.

3.2.1 Aluminum (Al)

Aluminum was generally considered nontoxic, and the complete metabolism of this metal is not known. However, at higher concentrations, aluminum may be toxic at both environmental and therapeutic levels, depending on ligand interactions. Dietary acids that normally occur in fruits and vegetables and commonly serve as taste enhancers are good ligands of the Al(III) ion. Malic acid is one of these and is predominant in food and beverages. The higher concentration of aluminum may produce mental status changes, learning disabilities, speech disturbances, and coarse tremors, and is known to produce abnormal brain functioning. It also inhibits cell division during the "S phase"; fosters bone disorders, including fractures; causes a microcytic hypoproliferative anemia; is a toxic agent in the etiology of Alzheimer's disease; and kills liver cells. The solubilized polymeric aluminium hydroxo complexes are highly toxic for fish.[14] It also causes blood to clot at 3 to 4 ppm, causing strokes and heart attacks. The European Economic Community (EEC) recommended a guideline of 0.2 mg/L of aluminum in drinking water.

Venturini-Soriano and Berthon[15] investigated the influence of aluminum bioavailability through speciation calculations based on Al(III)–malate complex formation constants especially determined for physiological conditions. According to the results obtained, malate appears to be extremely effective in maintaining the solubility of $Al(OH)_3$ over the whole pH range of the small intestine under normal dietary conditions. In addition, two neutral Al(III)–malate complexes were formed with percentages that were at maximum from very low malate levels. When aluminum was administered therapeutically as its trihydroxide, the amount of metal neutralized by malate peaks as its solubility pH range regresses to its original limits in the absence of malate. The enhancing effect of malate toward aluminum absorption was, therefore, virtually independent of the aluminum levels in the gastrointestinal tract. The presence of phosphate in the gastrointestinal juice was expected to limit the potential influence of malate on aluminum absorption. Under normal dietary conditions, phosphate effectively reduces the fraction of aluminum neutralized by malate without nullifying it. Aluminum phosphate was predicted to precipitate when aluminum levels were raised, as with the administration of aluminum hydroxide, but a significant amount of neutral aluminum malate remained in solution. Even therapeutic aluminum phosphate was not totally safe in the presence of malate, even at low malate concentrations. As plasma simulations predict that no compensatory effect in favor of aluminum excretion may be expected from malate, simultaneous ingestion of malic acid with any therapeutic aluminum salt should be avoided.

Illmer et al.[16] described the effect of aluminum concentration on the growth of microbes in soil. They reported the growth of bacterial cells as inversely proportional to the concentrations of aluminum. Illmer and Schinner[17] used *Pseudomonas* and *Azobactor* species to describe the toxicity of aluminum metal ion. The authors studied the effect of aluminum concentration on the motility and swarming of the above-cited bacterial species. The motility and swarming were found to be reduced by increasing the aluminum concentration. Furthermore, the same authors[18] used *Pseudomonas* bacterium species to describe the toxicity of aluminum in terms of Weibull function. Illmer and Erlebach[19] studied the effect of aluminum concentration on the cell size and intracellular water of *Arthrobactor* species PI/1-95. The authors reported the decrease in cell growth and water content by increasing aluminum concentration. Illmer and Mutschlechner[20] studied the effect of pH on the toxicity of aluminum metal ion on *Arthrobactor* species. The authors reported EC100 indicating a complete inhibition of microbial growth with 185 μM aluminum for *Arthrobactor* 1 and 11 m*M* of aluminum *Arthrobactor* 2.

The various complexes of aluminum may cause diseases in humans, especially by hampering many metabolic processes including the turnover of calcium, phosphorus, and iron in the body. The different salts of aluminum may bind to DNA and RNA, inhibit such enzymes as hexokinase, acid, and alkaline phosphatases, phosphodiesterase, and phosphooxydase. Aluminum salts are especially harmful to the nervous and hematopoietic systems and to the skeleton. Aluminum reaches an organism through food, water, and cosmetics, and through the use of aluminumware and containers. The toxicity comes from the substitution of Mg and Fe ions causing disturbances in intracellular signaling, excretory functions, and cellular growth. The neurotoxic action of aluminum probably comes from the substitution of Mg ions in ATP, which influences the function of every ATP-using enzyme. There are observations in experimental models proving that aluminum salts are responsible for Alzheimer's disease development. The toxicity of aluminum to the skeletal system results in diminished resistance, thus increasing the tendency of bones to break, which is a result of lower collagen synthesis and a slowing down of mineralization. Low erythropoietin production, inhibition of haem-synthesizing enzymes, and binding of aluminum to transferrin are effects shown in anemia. The carcinogenic effects of aluminum were not proved or denied, but high concentrations of aluminum were found in many neoplastic cells.[21]

3.2.2 Arsenic (As)

Arsenic is known to be one of the oldest poisons used by man. The toxicity of the water-soluble species of arsenic varies in the order of arsenite [As(III)] > arsenate [As(V)] » monomethylarsonate (MMA) > dimethylarsinate (DMA). The results of clinical findings for arsenic poisoning from drinking arsenic-contaminated water show the presence of almost all the stages of arsenic clinical manifestation.[22] After the intake of arsenic into the human body, approximately 50% of the arsenic is excreted by the urine and a small portion is excreted through the feces, skin, hair, nails, and lungs. Arsenic in urine, feces, skin, hair, nails, and lungs has been used as an indicator of the arsenic hazard to the general population.[23–25] The people drinking arsenic-contaminated water generally showed arsenical skin lesions, which are a late manifestation of arsenic toxicity. Long-term exposure to arsenic-contaminated water may lead to various diseases such as conjunctivitis, hyperkeratosis, hyperpigmentation, cardiovascular diseases, disturbances in the peripheral vascular and nervous systems, cancer of the skin and gangrene, leukomelanosis, nonpitting swelling, hepatomegaly, and splenomegaly.[26–29] Effects on the lungs, uterus, genitourinary tract, and other parts of the body were detected in the advance stages of arsenic toxicity. In addition, high concentrations of arsenic may result in an increase in stillbirths and spontaneous abortions.[30] The other side effects of arsenic contamination are sensory disturbances, visual disturbances, fatigue, loss of energy, shock, coma, convulsions, muscular paralyses, blindness, atrophy, and kidney damage.

Arsenocosis may be divided into four categories: preclinical, clinical, complication, and malignancy. The preclinical stage is determined by the presence of arsenic in urine and other body tissues without presentation of any symptoms of arsenocosis. The clinical stage is attributed to the presence of clinical symptoms and the detection of arsenic in hair, nails, and skin scales. The stage of compli-

TABLE 3.2
LD_{50} Values of Some Arsenicals in Rats

Arsenicals	LD_{50} (mg/kg)
Potassium arsenite	14
Calcium arsenate	20
Monomethylarsonate (MMA)	700–1800
Dimethylarsinate (DMA)	700–2600
Arsenobetaine	>10,000

Sources: Andreae, M.O., in *Organometallic Compounds in the Environment: Principles and Reactions*, John Wiley & Sons, New York, 1986; Penrose, W.R., *CRC Crit. Rev. Environ. Control*, 4, 465, 1974; Kaise, T., and Fukui, S., *Appl. Organomet. Chem.*, 6, 155, 1992. With permission.

cation is associated with the clinical symptoms of affected lungs, liver, muscles, and eyes. The stage of malignancy is determined by affected skin, lungs, bladder, and other organs.[23] Not all persons exposed to an arsenic hazard will show arsenocosis due to genetic factors.[31] The LD_{50} values of arsenicals in rats are given in Table 3.2. The permissible limit for arsenic assigned by the WHO is 0.01 mg/L.

Methylation was considered to be the primary detoxification pathway of inorganic arsenic. Inorganic arsenic is methylated by many but not all animal species to monomethylarsonic acid [(MMA(V)], monomethylarsonous acid [(MMA(III)], and dimethylarsinic acid [(DMA(V)]. As(V) derivatives were assumed to produce low toxicity, but the relative toxicity of MMA(III) remains unknown. The toxicities of arsenate, arsenite, MMA(V), MMA(III), and DMA(V) were determined and found to cause a change of the human hepatocytes. Leakage of lactate dehydrogenase (LDH) and intracellular potassium (K^+) and mitochondrial metabolism of the tetrazolium salt (XTT) were used to assess cytotoxicity due to arsenic exposure. The mean LC_{50} based on LDH assays in phosphate media was 6 μM for MMA(III) and 68 μM for arsenite. Using the assay for K^+ leakage in phosphate media, the mean LC_{50} was 6.3 μM for MMA(III) and 19.8 μM for arsenite. The mean LC_{50} based on the XTT assay in phosphate media was 13.6 μM for MMA(III) and 164 μM for arsenite. The results of the three cytotoxicity assays (LDH, K^+, and XTT) revealed the following order of toxicity in chang human hepatocytes: MMA(III) > arsenite > arsenate > MMA(V) = DMA(V). The data demonstrated that MMA(III), an intermediate in inorganic arsenic methylation, was highly toxic and again raised the question as to whether methylation of inorganic arsenic is a detoxification process.[35]

Omura et al.[36] studied the testicular toxicities of two compound semiconductor materials, gallium arsenide (GaAs) and indium arsenide (InAs), and arsenic oxide (As_2O_3) in rats by repetitive intratracheal instillation of these substances in suspension twice a week for a total of 16 times. A single instillation dose was

7.7 mg/kg in the GaAs and the InAs groups and 1.3 mg/kg in the As_2O_3 group. A significant decrease in sperm count and a significant increase in the proportion of morphologically abnormal sperm were found in the epididymis in the GaAs group. Particularly, abnormal sperm with a straight head increased markedly in this group. In the GaAs treated rats, there was a 40-fold increase in the degenerating late elongated spermatids at the postspermiation stages, stages IX, XI, and XI. From these results, it is indicated that GaAs disturbed the spermatid head transformation at the late spermiogenic phases and caused spermiation failure. InAs caused a sperm count decrease in the epididymis, though its testicular toxicity was relatively weak compared with that of GaAs. As_2O_3, a probable dissolution arsenic product of GaAs and InAs *in vivo*, did not show any testicular toxicities in this study. It seems likely that, along with arsenics, gallium and indium play a role in the testicular toxicities of GaAs and InAs.

3.2.3 Cadmium (Cd)

Cadmium is supposed to be a toxic metal ion that produces nausea and vomiting at a 15 mg/L concentration. It is a severely toxic metal ion but is not lethal. The kidneys are the critical target of cadmium contamination. The disorders associated with the kidneys in cadmium contamination are renal dysfunction, hypertension, and anemia. The other side effects of cadmium are extreme restlessness and irritability, headache, chest pain, increased salivation, choking, abdominal pain, diarrhea, tenesmus, throat dryness, cough, and pneumonitis. Exposure to cadmium chloride aerosol in rats was reported to result in tumor formation (carcinogenesis). The different cadmium compounds found to be carcinogenic in nature are CdS, CdO, $CdSO_4$, and $CdCl_2$. The WHO guideline for cadmium is 0.003 mg/L.

Sandrint and Maier[37] studied the lowering of pH of a microbiological medium from 7 to 4, which decreased cadmium toxicity during naphthalene biodegradation by a *Burkholderia* sp. The authors studied the cadmium speciation and accumulation in the system to explain this effect. With respect to cadmium accumulation, cells contained in media adjusted to pH 4 accumulated only 2.76 ± 0.76 mg Cd/g cells, whereas cells in media adjusted to pH 7 accumulated 8.52 ± 0.71 mg Cd/g cells. These data suggested that cadmium toxicity was correlated with increased cadmium accumulation rather than the formation of CdOH as pH is increased. At low pH, the decrease in cadmium accumulation might be caused by increased competition between hydrogen and cadmium ions for binding sites on the cell surface or by increased metal efflux pump activity due to an increase in the proton gradient that drives the efflux pump.

3.2.4 Cobalt (Co)

Cobalt is an essential element for growth in humans, as it is the main constituent of vitamin B_{12} and, hence, it is a less toxic element. However, at higher concentrations, it may be hazardous to humans. In a concentration greater than 1 mg/kg of body weight, it may be considered toxic. Because the concentration of cobalt as related to its potential toxicity in water is negligible, the authorities did not formulate guidelines for cobalt metal ion. However, USSR agencies have fixed 1.0 mg/L as the permissible concentration of cobalt.[38]

Lison et al.[39] carried out a study to integrate recent understandings of the mechanisms of genotoxicity and carcinogenicity of the different cobalt compounds. The authors reported two different mechanisms of genotoxicity, DNA breakage induced by cobalt metal and especially hard metal particles. The inhibition of DNA repair by cobalt(II) ions contributed to the carcinogenic potential of cobalt compounds. There was evidence that soluble cobalt(II) cations exerted genotoxic and carcinogenic activity *in vitro* and *in vivo* in experimental systems, but evidence in humans is lacking. The experimental data indicated some evidence of a genotoxic potential for cobalt metal *in vitro* in human lymphocytes, but there was no evidence of a carcinogenic potential. There was evidence that hard metal particles exerted genotoxic and carcinogenic activity *in vitro* and in human studies, respectively. There was insufficient information for cobalt oxides and other compounds. Although many areas of uncertainty remain, an assessment of the carcinogenicity of cobalt and its compounds requires a clear distinction between the different compounds of the element and needs to take into account the different mechanisms involved.

3.2.5 Chromium (Cr)

Trivalent chromium [(Cr(III)] is essential for animal and human health and growth at trace levels, with a safe and relatively innocuous level of 0.20 mg/day, whereas hexavalent chromium [Cr(VI)] is a potent carcinogen and is extremely toxic to animals and humans. Recently, Mulyani et al.[40] reported Cr(III) oxidation to Cr(VI) in living systems leading to carcinogenesis at higher concentration of Cr(III). Thus, accumulated chromium in food plants may represent potential health hazards to animals and humans if the element is accumulated in the hexavalent form or in high concentrations. The notorious effects of Cr(VI) are on liver and kidneys and respiratory organs with hemorrhagic effects and dermatitis and ulceration of the skin in chronic and subchronic exposure. One-day exposure is safe at 1.4 mg/L for a child and 5.0 mg/L for an adult. In an animal study for Cr(VI), a 60-days exposure was tested, and 14.4 mg/L was found to be the no observed adverse effect level (NOAEL). With an uncertainty factor of 100, the level at ten days of assessment is 1.4 mg/L (child) and 5 mg/L (adult). A NOAEL value of 2.41 mg/kg/day was obtained with rats supplied with 25 mg/L for Cr(VI). It is worthy to mention here that with an uncertainty factor of 500, the resulting adjusted acceptable daily intake was 0.17 mg/L. In view of these points, chromium is classified by the International Agency for research on Cancer (IARC) in group 1 (carcinogenicity in humans and animals). A toxic dose for man is reported at about 0.5 g/kg/day for potassium dichromate. The carcinogenesis due to Cr(VI) was reported in humans and rodents. The WHO guideline for chromium metal ion is 0.05 mg/L, as Cr(VI) and total concentration of chromium metal.

Kim et al.[41] studied the reaction kinetics of hexavalent chromium with ferrous ions to determine the influence of reduction on the toxicity of chromium to aquatic organisms. The changes in chemical forms of the chromate in the presence of ferrous ions were examined in a bioassay system using *Daphnia magna* as a test organism. This study demonstrated that the reaction kinetics of chromate with ferrous ions showed a significant decrease of chromate concentration with the second-order rate coefficient (k) for the reduction of Cr(VI) being determined as 55.2 M^1s^1. The concentration of Cr(VI) remaining in the solution decreased as the ratio of ferrous ion to chromate increased, revealing a nonstoichiometric reaction due to oxygenation and the moderately alkaline pH of the solutions. The toxicity test indicated that the bioavailability of chromate to *D. magna* was reduced in the presence of Fe(II) and that it decreased further with increasing Fe(II) concentrations. However, the toxic effect of chromate to aquatic organisms was not controlled kinetically in the presence of ferrous ions. It was also found that LC_{50} of chromate to *D. magna* decreased about 1.5-fold as the test period increased from 24 to 48 h in the presence of Fe(II).

Dayan and Paine[42] reported in clinical and laboratory-based studies that Cr(VI) is responsible for most toxic actions, although much of the underlying molecular damage may be due to its intracellular reduction to the even more highly reactive and short-lived chemical species Cr(III) and Cr(V). Exposure to Cr(VI) can result in various point mutations in DNA and to chromosomal damage, as well as to oxidative changes in proteins and to adduct formation. The relative importance of these effects of chromium ions and of the free oxidizing radicals that may be generated in the body, causing tumors and allergic sensitization, remain to be demonstrated. The biochemical studies of the damaging effects to DNA and of the pathogenesis of the allergic reactions to chromium ions have not kept up with advances in the understanding of the molecular basis of the effects of other carcinogens and allergens.

Zayed et al.[43] conducted a study to determine the extent to which various vegetable crops absorb and accumulate Cr(III) and Cr(VI) ions into roots and shoots and to ascertain the different chemical forms of chromium in these tissues. Two greenhouse hydroponic experiments were performed using a recirculating nutrient culture technique that allowed all plants to be equally supplied with chromium at all times. In the first experiment, 1 mg/L chromium was supplied to 11 vegetable plant species as Cr(III) or Cr(VI), and the accumulation of chromium in roots and shoots was compared. The crops tested included cabbage (*Brassica oleracea* L. var. *capitata* L.), cauliflower (*Brassica oleracea* L. var. *botrytis* L.), celery (*Apium graveolens* L. var. *dulce* [Mill.] Pers.), chive (*Allium schoenoprasum* L.), collard (*Brassica oleracea* L. var. *acephala* DC.), garden pea (*Pisum sativum* L.), kale (*Brassica oleracea* L. var. *acephala* DC.), lettuce (*Lactuca sativa* L.), onion (*Allium cepa* L.), spinach (*Spinacia oleracea* L.), and strawberry (*Fragaria* × *ananassa* Duch.). In the second experiment, x-ray absorption spectroscopy (XAS) analysis on chromium in plant tissues was performed in roots and shoots of various vegetable plants treated with Cr(VI) at either 2 mg/L chromium for 7 days or 10 mg/L chromium for 2, 4, or 7 days, respectively. The crops used in this experiment included beet (*Beta vulgaris* L. var. *crassa* [Alef.] J. Helm), broccoli (*Brassica oleracea* L. var. *Italica* Plenck), cantaloupe (*Cucumis melo* L. gp. Cantalupensis), cucumber (*Cucumis sativus* L.), lettuce, radish (*Raphanus sativus* L.), spinach, tomato (*Lycopersicon lycopersicum* [L.] Karsten), and turnip (*Brassica rapa* L. var. *rapifera* Bailey). The XAS speciation analysis indicated that Cr(VI) was converted in the root to Cr(III) by all plants tested. Translocation of both chromium forms from roots to shoots was extremely limited, and accumulation of chromium by roots was 100-fold higher than that by shoots, regardless of the chromium species supplied. The highest chromium concentrations were detected in members of the Brassicaceae family, such as in cauliflower, kale, and cabbage. Based on these findings and observations by other researchers, a hypothesis for the differential accumulation and identical translocation patterns of the two chromium ions can be proposed.

3.2.6 Copper (Cu)

Copper is suggested to be an essential element for growth, as it is required in many enzymatic reactions. It is a nontoxic element at lower concentrations; however, in high concentrations, certain adverse effects such as nausea, Wilson's disease, and so forth, were reported. Limited data are available on the toxicity of copper, and toxicity occurs in the forms of Cu(I), Cu(II), and Cu-organic species, that is, as pesticides and fungicides. Depression and schizophrenia have also been linked to high levels of copper. Copper and its compounds have been found toxic for rodents. Elemental copper, cupric chloride, and sulfate have been found to have similar toxicities for mice. The oral toxicity of chloride, sulfate, and citrate of copper metal ion were reported as 148, 300, and 1500 mg/kg for the rat, respectively.[44,45] The WHO guideline for copper is 1 mg/L.

3.2.7 Lead (Pb)

Lead is the most significant toxin of the heavy metals, and its effects are of both toxicological and neurotoxical nature, including irreversible brain damage. Lead is a naturally occurring substance and can be found in organic and inorganic forms. Inorganic forms of lead typically affect the central nervous system (CNS), peripheral nervous system (PNS), and hematopoietic, renal, gastrointestinal, cardiovascular, and reproductive systems. Organic lead toxicities predominately tend to affect the CNS. These types of effects are observed at blood concentrations between 100 and 200 μg/L. Tumor formation in rats due to lead poisoning was reported. Other hazardous effects of lead include visual disturbances, convulsions, loss of cognitive abilities, antisocial behavior, constipation, anemia, tenderness, nausea, vomiting, severe abdominal pain, and gradual paralysis in the muscles. Various lead compounds reported to be carcinogens include PbO, $PbCrO_4$, $Pb_3(PO_4)_2$, and $PbAc_2$. The LD_{50} values of some lead compounds are given in Table 3.3.[46–51] The WHO has fixed 0.01 mg/L as the permissible limit of lead in drinking water.

Inorganic forms of lead are absorbed through ingestion or inhalation, whereas organic lead salts are absorbed through the skin. Only about 10% of an ingested dose is absorbed in adults, but the absorbed percentage may be much greater in

TABLE 3.3
LD_{50} Values of Some Lead Compounds

Lead Compounds	Animals	LD_{50} Values (mg/kg)
Pb-arsenite	Rat	100
Pb-naphthinate	Rat	5100
Pb-nitrate	Guinea Pig	1300
Pb-chloride	Guinea Pig	2000
Pb-fluoride	Guinea Pig	4000
Pb-oleate	Guinea Pig	8000

children. Lead absorption is enhanced by deficiencies of iron, calcium, and zinc. Under typical conditions, lead is absorbed and stored in several body parts. Five to ten percent is found in blood, most of which is located in erythrocytes — 80 to 90% is taken up in the bone and stored with the hydroxyapatite crystals, where it easily exchanges with blood. Some authorities list the half-life of lead in the bone as long as 30 years, while others estimate that lead half-life in bone is 105 days. Generally, the excretion of lead is slow, with an estimated biological half-life in soft tissues of 24 to 40 days. The remainder of the stored lead is found in soft tissue, notably, the kidney and the brain. The primary route of excretion is through feces (80 to 90%). To a lesser extent, lead is excreted in urine (10%). Lead passes the placental barrier and is found in breast milk. A correlation exists between lead toxicity and fetal wastage, premature rupture of membranes, and sterility.

Nunes et al.[52] determined the additive or synergistic toxic effects in adult male rats exposed to orally delivered lead carbonate (2000 mg/kg) for 9 days before trichloroethylene (TCE), 2000 mg/kg, was given concurrently for an additional 7 days. The comparisons were made with groups of vehicle-treated rats and rats given only lead or only TCE. The potential neurotoxicity was evaluated by using the Functional Observational Battery (FOB) recommended for neurotoxicity screening. The results of the FOB indicated that lead carbonate was more responsible than TCE for the changes observed. The additive or synergistic neurotoxicities were not noted. Histological examination of the kidney from lead-treated rats revealed inclusions, an increased incidence of coagulated proteins, and tubular dilation that was generally more severe in the medullary segments. Gastric necrosis and testicular necrosis were found in rats given lead carbonate both with and without TCE (15/20 and 6/20 treated, respectively). The results suggested that, even when given concurrently, the toxicities of lead carbonate and TCE were expressed only as though one toxicant was given.

3.2.8 Mercury (Hg)

Mercury occurs in the forms of inorganic and organic compounds. Although the use of organomercury compounds, particularly as fungicides and pesticides, has decreased in recent years, abiotic and biotic methylation and transmethylation

TABLE 3.4
LD_{50} Values of Some Mercury Compounds

Mercury Compounds	Animals	LD_{50} Values (mg/kg)
HgCl	Rat	210
HgI	Mouse	110
$HgNO_3$	Mouse	388
HgO	Rat	8
MeHg	—	10
EtHg	—	40

Sources: Zollinger, H.U., *Vir. Arch. Pathol. Anat. Physiol.*, 323, 694, 1953; Boyland, E., Dukes, C.E., and Grover, P.L., *Br. J. Canc.*, 16, 283, 1962; van Esch, G.J., van Genderen, H., and Vivik, H.H., *Br. J. Canc.*, 16, 289, 1962; van Esch, G.J., and Kroes, R., *Br. J. Canc.*, 23, 765, 1969; Kanisawa, M., and Schroeder, H.A., *Canc. Res.*, 29, 892, 1969; Furst, A., Schlauder, M., and Sasmore, D.P., *Canc. Res.*, 36, 1779, 1976. With permission.

processes have focused the interest of several species on the toxicity of organomercury compounds.[53–59] Inorganic mercury compounds are toxic to kidneys, leading to neurological and renal disturbances. Organic mercury compounds are very toxic to the central nervous system. Feelings of discouragement, irritability, personality changes, learning disabilities, muscle tremors, jerky gait, spasms of extremities, inflammation of mouth and gums, swelling of salivary glands, excessive flow of saliva, and loosening of teeth are other hazardous effects of mercury poisoning. The mechanisms of toxicity and, particularly, the potential carcinogenicity of inorganic mercury are still debated. The results of mutagenicity and genotoxicity testing with mercury have been inconsistent, for example, mercury induces DNA single-strand breaks at low concentrations in mammalian cells but has not proved to be mutagenic in several bacterial mutagenicity assays. The LD_{50} values of some mercury compounds in different animals are given in Table 3.4.[60,61] The WHO guideline for the permissible limit is 0.001 mg/L.

Farrell et al.[62] studied the integration of physicochemical procedures for studying mercury(II) speciation with microbiological procedures for the effects of mercury on bacterial growth with evaluation of ionic factors (e.g., pH and ligand species and concentration) that affected biotoxicity. A *Pseudomonas fluorescens* strain capable of methylating inorganic Hg(II) was isolated from sediment samples collected at Buffalo Pound Lake in Saskatchewan, Canada. The effects of pH and ligand species on the toxic response (50% inhibitory concentration [IC_{50}]) of the *P. fluorescens* isolated to mercury were determined and related to the aqueous speciation of Hg(II). It was determined that the toxicities of different mercury salts were influenced by the nature of the co-ion. At a given pH level, mercuric acetate and mercuric nitrate yielded essentially the same IC_{50};

mercuric chloride, on the other hand, always produced lower IC_{50}. For each mercury salt, toxicity was greatest at pH 6 and decreased significantly at pH 7. Increasing the pH to 8 had no effect on the toxicity of mercuric acetate or mercuric nitrate but significantly reduced the toxicity of mercuric chloride. The aqueous speciation of Hg(II) in the synthetic growth medium M-IIY (a minimal salts medium amended to contain 0.1% yeast extract and 0.1% glycerol) was calculated by using the computer program GEOCHEM-PC with a modified database. The results of the speciation calculations indicated that complexes of Hg(II) with histidine [Hg(H-His)HIS^+ and Hg(H-His)2^{2+}], chloride ($HgCl^+$, $HgCl_2$, HgClOH, and $HgCl^3$), phosphate ($HgHPO_4$), ammonia ($HgNH_3^{2+}$), glycine [$Hg(Gly)^+$], alanine [$Hg(Ala)^+$], and hydroxyl ion ($HgOH^+$) were primarily responsible for toxicity in the M-IIY medium. The toxicity of mercuric nitrate at pH 8 was unaffected by the addition of citrate, enhanced by the addition of chloride, and reduced by the addition of cysteine. In a chloride-amended system, $HgCl^+$, $HgCl_2$, and HgClOH were the species primarily responsible for observed increases in toxicity. In a cysteine-amended system, the formation of $Hg(CYS)_2^2$ was responsible for the detoxification effects observed. The formation of mercury citrate complexes was insignificant and had no effect on Hg toxicity.

Kungolos et al.[63] studied the effects of six different forms of mercury on the growth of the yeast *Saccharomyces cerevisiae* using five kinds of strains of *S. cerevisiae*. They were a wild type, a mercury-resistant type, and three mutants — mutation repair-deficient mutant, excision repair-deficient mutant, and recombination repair-deficient mutant. In terms of EC_{50} toward the wild-type strain, the toxicity order for the inorganic forms was $Hg(NO_3)_2 > HgSO_4 > HgCl_2$. Monovalent nitrate mercury [$Hg(NO_3)_2$] was more toxic than bivalent $Hg(NO_3)_2$. The toxicity of organic mercury CH_3HgCl on cell growth was two orders of magnitude higher than that of inorganic $HgCl_2$. Between the two organic forms, CH_3HgCl was more toxic than CH_3HgOH. The survival rate in the presence of a certain concentration of CH_3HgCl was about 100th of the survival in the presence of the same concentration of $HgCl_2$. On the other hand, the concentration of CH_3HgCl in the cell was about 170 times that of $HgCl_2$. The addition of chelating agents, ethylenediaminetetraacetic acid (EDTA) and methylpenicillamine, to the medium did not reduce the toxicity of mercury. Among three mutants tested, one deficient in recombination repair systems was most sensitive to mercury.

3.2.9 Selenium (Se)

Selenium is considered the essential element for growth with a daily intake of 100 to 200 μg/day. Selenium is found in the forms of Se(IV), Se(VI), dimethylselinide, dimethyldiselenide, hydrogen selenide, and selenium amino acids. It is important to mention here that the presence of selenium reduces the toxicities of arsenic, mercury, cadmium, and lead compounds. Such antagonistic effects between toxic compounds have been observed in many types of organisms. The research has been directed to understand these effects, but the overall mechanisms remain unclear. However, in high concentrations, its presence may result in nervousness,

TABLE 3.5
Toxicities of Some Selenium Compounds

Selenium Compounds	Animals	Lethality	Values
Na-selenite (IV)	Rat	MLD	3.5
Na-selenate (VI)	Rat	MLD	5.7
DL-Selenocysteine or methionine	Rat	MLD	4
Dimethylselenide	Rat	LD_{50} (mg/kg)	1600
Trimethyl-Se-Cl	Rat	LD_{50} (mg/kg)	50
Se(IV) and Se(VI)	Rabbit	LD_{50} (mg/kg)	2
Selenium disulfide	Mouse	LD_{50} (mg/kg)	48

Source: Gilon, N., and Potin-Gautier, M., *J. Chromatogr. A*, 732, 369, 2002.

depression, convulsions, vomiting, cough, dyspnea, abdominal pain, diarrhea, somnolence, decrease in blood pressure, respiratory failure and death, marked pallor, a garlic odor of breath and sweat and urine, red staining of fingers and teeth and hair, marked debility, epistaxis, gastrointestinal disturbances, dermatitis, and irritation of the nose and throat. Hydrogen selenide can cause pneumonitis and damage to liver, kidney, and spleen (arthritis, eruptions, and a yellowish tinting of the skin). Briefly, selenium is well known for its toxic and beneficial effects on living organisms. The toxicity is closely correlated with the form in which selenium is present.[64] The LD_{50} values of some selenium compounds are given in Table 3.5.[65] The WHO guideline for the permissible limit is 0.01 mg/L.

The role of selenium in biology appears from the evidence to be as a catalyst par excellence. As a unique prosthetic group of a variety of enzymes, presumably as Se(II), selenium functions with tocopherol to protect cell and organelle membranes from oxidative damage, to facilitate the union between oxygen and hydrogen at the end of the metabolic chain, to transfer ions across cell membranes, in protein synthesis in erythrocytes, and in liver organelles, in immunoglobulin synthesis, and in ubiquinone syntheses. As perhaps the most versatile and rapid nucleophile, selenium is thought to amplify and orient –SH in equilibrium –S–S– interactions involving glutathione and proteins. Its toxicity appears to be due to the overaccumulation of selenite ions, which act as oxidants to inhibit –SH interactions. Such toxicity is readily avoided or reversed in many ways. Although not yet recognized as essential for man, selenium is clearly essential for many animal species and some microorganisms. As the active selenide, selenium emerged as the target for many heavy metal toxicities and, contrarily, as a specific antidote against heavy metal toxicities. Despite this, its unusual toxicity and many preconceived notions about selenium continue to confuse attitudes toward the safe uses of selenicals. From a suspected cause of cancer, selenium metamorphosed, via evidence over many years, into something of possible anticancer value. Interrelations between selenium, vitamin E, the ubiquinones, and various chronic diseases appear as beckoning research areas. The reported veterinary

TABLE 3.6
LD_{50} Values of Some Organotin Species in Rats

Organotin Compounds	LD_{50} (mg/kg)
Bis(tributyltin) oxide	150–234
Trimethyltin hydroxide	540
Triphenyltin hydroxide	125
Trimethyltin acetate	9.1
Triethyltin acetate	4.0
Tributyltin acetate	>4000

Source: Merian, E., *Metal and Their Compounds in the Environment*, VCH, Verlagsgesellschaft, Weinheim, 1991.

values of selenium tocopherol combinations in animals, together with clinical evidence, plus human and animal evidence for safety, offer promise for intensive medical investigation. The historical confusion and misunderstandings regarding selenium must be corrected, however, before advantage can be taken of its potential for human welfare. Many misjudgments about selenium, ever since 1900 and more obviously since the 1930s, have involved other trace elements. Unrealistic regulations stemming from these misunderstandings prevail worldwide. Evidence suggests that once the nutrition biochemistry and toxicology of selenium are sufficiently understood and appreciated, major breakthroughs in agriculture, medicine, and public health can result. Much has been accomplished along these lines in New Zealand in animal agriculture, in the United States and other countries in veterinary medicine, and in Mexico in human medicine.[66]

3.2.10 Tin (Sn)

Tin has been suggested as an essential element for growth, but its role as health promoter is under debate, and it was found to be toxic at higher concentrations. Tin occurs in the environment in the form of inorganic and organic compounds. Inorganic tin has low toxicity, while organotin species are quite toxic. Organotin compounds have been used as fungicides, bactericides, and insecticides, and therefore, they are toxic in respect to human beings. The most commonly found tin compounds are R_4Sn, R_3SnX, R_2SnX_2, and $RSnX_3$, where R is the alkyl or aryl group, and X is the anionic group (halogens, hydroxyl, etc.). Most toxic organotin compounds found in aquatic organisms are tributyltin and phenyltin. The LD_{50} values of some organotin species are given in Table 3.6.[67]

3.2.11 Miscellaneous Metal Ions

In addition to the above-discussed toxic heavy metal ions, other metal ions are also toxic to humans and other organisms. These metal ions include antimony,

TABLE 3.7
LD_{50} Values of Some Antimony Compounds

Antimony Compounds	Animals	LD_{50} Values (mg/kg)
Sb(III) or Sb(V)	Rat	100
Sb_2O_3	Rat	3250
Sb_2S_3	Rat	1000
Sb_2O_5	Rat	4000
Sb_2S_5	Rat	1599

Source: Furst, A., in *Toxicology of Metals: Clinical and Experimental Research*, Ellis Horwood Ltd., Chichester, 1987, 17. With permission.

nickel, and vanadium. Antimony species found in the environment are Sb(III), Sb(V), monomethylastibonic acid, and dimethyl stilbinic acid. All these species have different degrees of toxicity.[14] The elemental antimony is more toxic for rats than the sulfides or oxides. The different LD_{50} values for antimony are given in Table 3.7.[1] The toxic gas hypothesis proposes that exposure to stilbine (antimony trihydride) generated from the microbial contamination of mattress materials is a possible cause of sudden infant death syndrome (SIDS) as a consequence of cholinesterase inhibition.[68] Hussain et al. measured the direct effects of antimony compounds including stilbine on the activity of plasma cholinesterase, red blood cell acetylcholinesterase (AChE), and mouse neuronal AChE *in vitro*. Nickel is supposed to be carcinogenic in nature, and various compounds are reported carcinogens with different degrees of toxicity. Ni_3S_2 and NiS are 100% carcinogenic, while the percentages of carcinogenicity of other nickel compounds are low, as given in Table 3.8.[1] Pane et al.[69] studied the effect of nickel on *Daphnia magna*, and they reported that the animal suffered from respiratory function failure.

Silver is an element with low toxicity; however, in high concentrations, it may result in convulsions, coma, blood dyscrasias, dermatitis, argyria, heart (especially aortitis or aneurysm), liver, or renal disease, gastroenteritis, paralysis, and nephritis. Vanadium is not an essential dietary element, and its high concentration may result in respiratory disorders. There is no report on acute toxicity due to the presence of vanadium metal ion. A committee formed to study the medical and biological effects of environmental pollutants met in Washington DC and fixed LD_{50} values of ortho-, pyro-, tetra-, and hexavanadium compounds.[70] Hexavanadate was found to be less toxic in comparison to other compounds. For guinea pigs, ortho- and pyrovanadate are found to be ten times more toxic than other compounds. For rats, all compounds were of equal toxicity with the exception of sodium vanadate. All vanadium compounds were found to show equal toxicities for mouse models; therefore, no generalization was made about the toxicities of vanadium metal ion species. Atsmon et al.[71] reported the first case of thallium poisoning to occur in Israel in almost 30 years. A 40-year-old man was apparently poisoned by a business associate when, on several

TABLE 3.8
Percent Activities of Nickel Compounds as Carcinogens

Nickel Compounds	Percent Activity
Ni_3S_2	100
NiS	100
NiO	65
NiSe	50
Nickelocene	36
$NiAc_2$	22
$NiCrO_4$	6
NiAs	0
$NiCl_2$	0
$NiTiO_3$	0

Source: Furst, A., in *Toxicology of Metals: Clinical and Experimental Research*, Ellis Horwood Ltd., Chichester, 1987, 17. With permission.

occasions, he unknowingly drank an alcoholic beverage containing the toxic substance. Delayed admission and recurrent thallium ingestion resulted in both acute and chronic symptoms being present concomitantly. Conventional treatment modalities (Prussian blue and forced diuresis) were employed. The patient survived, although neurological sequelae ensued. The problems encountered in diagnosis and treatment of this relatively uncommon entity were discussed. Hoffman[72] reviewed the toxicity of thallium and the role of Prussian blue in the therapy of thallium toxicity. The exact mechanisms of thallium toxicity are not known; however, it is believed that thallium interferes with energy production at essential steps in glycolysis, Krebs cycle, and oxidative phosphorylation. Additional effects include the inhibition of sodium-potassium-adenosine triphosphatase and binding to sulfhydryl groups. In the last 10 years, research about vanadium effects on living beings has increased substantially, due to its presence in the environment from different sources. Interest for vanadium and its compounds is shown because of its toxic effects and its uses in biomedical areas, such as antineoplastics, cholesterol and glucose blood levels, diuretics, and oxygen hemoglobin affinity. Hoffman reported that the toxic effects of vanadium are due to its property of inhibiting many enzymatic systems. Vanadate and vanadyl ions make chemical complexes exhibiting the property of inhibiting or increasing the activity of the enzymes participating in DNA and RNA synthesis. They also induce mutagenic and genotoxic effects. Biochemical assays show cytotoxic effects, increase in cellular differentiation, gene expression alterations, and other biochemical and metabolic alterations. Research has been done with *in vitro* systems but with few laboratory animals.[72] The classification of metals

TABLE 3.9
Classification of Metals According to Toxicity and Availability

Noncritical	Toxic but Very Insoluble	Very Toxic and Relatively Accessible
Na	Ti	Be
K	Hf	Co
Mg	Zr	Ni
Ca	W	Cu
Fe	Nb	Zn
Li	Ta	Sn
Rb	Re	As
Sr	Ga	Se
Al	La	Te
Si	Os	Pd
	Rh	Ag
	Ir	Cd
	Ru	Pt
	Ba	Au
		Hg
		Tl
		Pb
		Sb
		Bi

according to their toxicity and availability is presented in Table 3.9. The permissible limits of metal ions in drinking water are given in Table 3.10.

3.3 BIOTRANSFORMATION OF METAL IONS

To know the exact toxicities of different metallic species, it is very important to understand the biotransformation of metal ions. Most of the trace and toxic metal ions have empty d orbitals in their atomic structures and, therefore, have very good reactions with other compounds, especially with complexing reactions. Many metal ions are reported to form complexes with different biomolecules in biological systems.[74–76] All biological activities are controlled by certain enzymes and hormones, which are regulated by different genes. All enzymes and hormones are protein molecules, which have very good chelating tendencies with metal ions. Heavy metals bind to sulfhydryl groups in proteins, resulting in alterations of enzymatic activity. The pathophysiology of the heavy metal toxidromes remains relatively constant. Nearly all organ systems are involved in heavy metal toxicity. Some metal ions are reported to interact with DNA, resulting in several kinds of lethal diseases, including carcinogenesis.

TABLE 3.10
Permissible Limits of Toxic Metal Ions (μg/L)

Metal Ions	WHO	US EPA	EEC
Al	—	—	0.20
Ag	—	0.05	0.010
As	0.01	0.05	0.05
Cd	0.003	0.01	0.005
Cu	1.00	1.00	—
Fe	0.30	0.030	0.20
Hg	0.001	—	—
Mn	0.05	0.10	0.05
Ni	—	0.05	0.05
Pb	0.01	0.05	0.05
Sb	—	—	0.01
Se	0.01	0.01	0.01
Zn	3.0	5	5

Sources: WHO Guidelines, 1997; Environment and Science Directorate, Water Quality and Laboratories Group, UK, 1993; and European Economic Community @ Nham TT, 1995. With permission.

Williams et al.[77] studied the mechanisms of metal ion transfer into biological systems by studying heavy metal ions such as copper, cobalt, iron, manganese, nickel, and zinc. Thus, mechanisms must exist to satisfy the requirements of cellular metabolism and to protect cells from toxic effects. The mechanisms deployed in the acquisition of essential heavy metal micronutrients were not clearly defined, although a number of genes have now been identified that encode potential transporters. Three classes of membrane transporters implicated in the transport of heavy metals in a variety of organisms were studied. Heavy metal (CPx-type) ATPases, the natural resistance associated macrophage protein (Nramp) family, and members of the cation diffusion facilitator (CDF) family are important in this concern. Williams et al. attempted to describe an overview of the main features of these transporters in plants in terms of structure, function, and regulation drawing on information from studies in a wide variety of organisms. Briefly, metal ions interact with many biological constituents and modulate their activities toward negative directions, which results in various serious and lethal diseases. Krizek et al.[78] used DNA microarrays in toxicant-specific gene discovery. After combining results from two DNA microarray experiments, the authors identified genes from the model plant *Arabidopsis thaliana* that are induced in response to one but not other heavy metals. The promoters of these genes should be useful in developing metal-specific transgenic biomonitors. To test this idea, the authors fused the promoter of one of the newly identified nickel-inducible genes (AHB1) to the β-glucuronidase (GUS) reporter gene. Arabidopsis plants containing the AHBI::GUS transgene show reporter gene activity when they are grown on media containing nickel but not when grown on media containing cadmium, copper, zinc, or without added metals. Thus, this approach resulted in the creation of a transgenic strain of Arabidopsis that can report on the presence and concentration of nickel in plant growth media. Such transgenic models can serve as inexpensive and efficient biomonitors of bioavailable heavy metal contamination in soils and sediments. Metal salts can inhibit cell activity through direct toxicity to critical cellular molecules and structures. On the other hand, they can also change cell behavior by inducing specific genes (including genes encoding members of the metallothionein [MT] gene family). Therefore, transition metals may affect cell functions either by acting as a toxin or by transmitting or influencing signals controlling gene expression. To explore the latter possibility, Koropatnick et al.[79] measured the ability of low, nontoxic metal pretreatment to alter immune cell behavior. The authors found that pretreatment of human monocytes with zinc-induced metallothionein gene expression altered their capacity to undergo a bacterial lipopolysaccharide-induced respiratory burst. Furthermore, the authors showed that cadmium and mercury salts, at concentrations that exert no discernible toxicity, inhibit the activation of human monocytic leukemia (THP-1) cells. A treatment with $CdCl_2$ (1 μ*M*), $ZnCl_2$ (20 to 40 μ*M*), or $HgCl_2$ (2 μ*M*) for 20 h induced MT-2 mRNA and total MT protein accumulation. It was also reported that no affect on proliferation potential or metabolic activity was observed. However, cadmium and mercury salts significantly inhibited the ability of subsequent lipopolysaccharide treatment to induce the oxidative burst, increased adhesion to plastic, and MT-2 and interleukin-1 beta (IL-1) mRNA accumulation. The phenomenon of metal-induced suppression of monocyte activation, at metal concentrations that have no effect on cell viability, has important implications for the assessment of acceptable levels of human exposure to cadmium, zinc, and mercury.

3.4 BIODEGRADATION OF METAL IONS

Many metal ions, such as Mn(II), Fe(II), Co(II), Ni(II), Cu(II), and Zn(II), are similar in size and have ionic diameters between 138 and 160 pm. Cells usually do not differentiate between physiologically required and toxic metals during uptake. Some metal ions are transported across cell membranes through an unspecific and fast chemiosmotic-driven uptake system (e.g., the CorA membrane integral protein, which is part of the MIT [metal inorganic transport] family, or by the Mg uptake system). These represent unspecific transporters. Hence, when the extracellular concentration of a toxic metal is high, it will be accumulated to even higher concentrations; this is the reason for heavy metal toxicity. The oxyanions such as AsO_4^3 and CrO_4^2 are taken up unspecifically by the PIT (phosphate inorganic transport) and the sulfate uptake system. At lower concentrations, cells contain inducible metal transport systems that are typically expressed under metal limited conditions (these are ATP or proton-motive-force-driven) that may also cotransport heavy metals.

In bacteria, metal ions can bind to glutathione, which results in the formation of bisglutathionato complexes, and these interact with molecular oxygen to form bisglutathione (GS-SG). As the reduction of bisglutathione back to glutathione requires NADPH, and heavy metals can thus impose considerable oxidative stress, Rawlings et al.[80] reviewed the microbial action with metal ions. Biomining is the use of microorganisms to extract metals from sulfide- or iron-containing ores and mineral concentrates. The iron and sulfide are microbially oxidized to produce ferric iron and sulfuric acid, and these chemicals convert the insoluble sulfides of metals such as copper, nickel, and zinc to soluble metal sulfates that can be readily recovered from solution. Although gold is inert to microbial action, microbes can be used to recover gold from certain types of minerals, because as they oxidize the ore, they open its structure, thereby allowing gold-solubilizing chemicals such as cyanide to penetrate the mineral.

Thiobacillus species oxidizes Fe(II) and other metals and solubilizes them. The most abundant and notable exception is Fe(III) reduction by strain, although others are known as well. Other examples of metals reductively transformed are the reduction of Se to Se(0), Te to Te(0), Cr(III) to Cr(0), Hg to Hg(0), and As(III) to As(0) by *Geobacter metalloreducens*. Also, several *Desulfovibrio* species can reduce U(VI) to U(IV). Selenium can exist in several oxidation states, namely, Se(VI) in SeO_4^{2} (selenate), Se(IV) in SeO_3^{2} (selenite), elemental selenium Se(0), and Se(-II) selenide. The first three can serve as electron acceptors for microbial transformations. This provides for a natural bioremediation of agricultural drainage waters and groundwaters. This reduction is favored under conditions of oxygen depletion. Recently, it was documented that several selenate-reducing strains have both selenate reductase and nitrate reductase (e.g., *Thauera selenatis*, *Sulfospirillum barnesii*) enzymes.[81] Hg(II) has been used as a fungicide in organomercurials in paper making and agriculture, and as a disinfectant in hospitals. Unlike many other heavy metals, Hg(II) is not known to be required for any biological activity. The further hazard is associated with its ability to bioaccumulate as methylmercury in the food chain. Hg(II) is used as a substrate for methylation; therefore, reduction mitigates the formation of methylmercury. The formation of methylmercury is also catalyzed microbiologically due to the interaction of methylcobalamin (coenzyme B12) and Hg(II), which can form monomethyl and dimethylmercury. This is, therefore, a nonspecific Hg transformation, as is the formation of HgS during sulfate reduction. Hg(II) reductase is cytosolic and consumes $NADPH^{+}$ and acts on RS-Hg-SR, which is formed in the presence of excess thiols. Resistance to mercuric ions in bacteria is conferred by mercuric reductase, which reduces Hg(II) to Hg(0) in the cytoplasmic compartment. Specific mercuric ion transport systems exist to take up Hg(II) salts and deliver them to the active site of the reductase.[82]

A wide variety of microorganisms can enzymatically reduce Cr(VI) to Cr(III). The reaction is probably fortuitous by an enzyme, which carries another physiological function. So far, energy conservation linked to Cr(VI) has not been shown. CrO_4^{2} resistance was found in several bacterial strains, and it is probably due to the active export of the CrO_4^{2} which is taken up by the SO_4^{2} uptake system. Note that even cells that are CrO_4^{2} reducers must have some type of CrO_4^{2} resistance, because the reduction rates are typically lower than the accumulation rates. A lot of work has been done with *Enterobacter cloacae* HO1 [up to 10 mM of Cr(VI)], but its application to the real waste stream is dubious because of the high toxicity of heavy metals and sulfate. This suggested the use of the *Desulfovibrio* species, which could also reduce Cr(VI) using cytochrome C_3. The reduction of CrO_4^{2} by *Desulfovibrio vulgaris* resting cell assayed and its cytochrome C_3 in the presence of H_2 was reported. Similarly, Oliver et al.[83] reported the reduction of hexavalent chromium [Cr(VI)]. Native microbial communities in subsurface sediments with no prior Cr(VI) exposure were shown to be capable of Cr(VI) reduction.

Arsenic typically occurs as the oxyanions, such as AsO_4^{3-} (arsenate) and AsO_2 (arsenite). Resistance against AsO_4^{3-} by microorganisms involves a combination of AsO_4^{3-} reduction to AsO_2^-, which is thought to be more toxic. There are several arsenite efflux systems — some rely on chemiosmosis and some are ATP-binding cassettes.[84] The work of Newman et al.[85] illustrates that there may be diversity in the AsO_4^{3-} respirers. The organisms they isolated from the same Mystic Lake in Minneapolis Minnesota, USA, as MIT-13 was acquired from were sulfate-reducing G+ *Desulfotomaculum* species that were able to reduce both SO_4^{2-} and AsO_4^{3-}. All AsO_4^{3-} respirers known to date produce AsO_2^-, which is still toxic, so AsO_4^{3-} respiration is probably still contingent on having an AsO_2 efflux mechanism.

Inoue et al.[86] reported the biodegradation of triphenyltin (TPT) using *Pseudomonas chlororaphis* CNR15. Furthermore, the authors found that the fluorescent pseudomonads showed TPT degradation activity in the solid extracts of their culture supernatants. The copA gene of *Escherichia coli* encodes a copper transporter, and its promoter is normally regulated by Cu(I) ions and CueR, a MerR-like transcriptional activator. Stoyanov and Brown[87] showed that CueR could also be activated by gold salts and that Cys(112) and Cys(120) were involved in the recognition of gold, silver, and copper salts. Gold activation is unaffected by copper chelating agents but was affected by general metal chelators. Bhattacharyya-Pakrasi et al.[88] studied the bacterial regulation of manganese acquisition by both the biochemical activity of the transporters and the transcriptional regulation of gene expression. Structural analysis suggested that calcium ions might regulate the function of a manganese ATP-binding cassette (ABC)-permease in Synechocystis 6803, a cyanobacterium, as well as in a number of other bacteria. The expression of genes encoding the manganese transporter in Synechocystis 6803 is regulated by a two-component signal transduction mechanism that was not previously observed for manganese and zinc transport in bacteria.

Some metal ligand complexes are easily biodegradable by many bacteria and microbes. EDTA forms stable and water-soluble complexes with metals, hindering their adsorption to soil particle surfaces and, hence, making them freely available for biodegradation. Scientists at the U.S. Environmental Protection Agency (EPA) conducted a series of computer-modeling trials and laboratory experiments for a system consisting of EDTA–metal complexes and an EDTA-degrading microorganism (BNC1, DSM 6780). They modeled EDTA biodegradation in the presence of metal ions with CCBATCH, a biogeochemical computer model developed at Northwestern University, Illinois. CCBATCH couples equilibrium EDTA speciation reactions with kinetically controlled EDTA biodegradation reactions to predict the concentration profiles of all the biologically affected species over time. They also experimentally determined the rate and extent of EDTA biodegradation in batch reactors using cultures of BNC1. Cells were grown at 35°C in agitated flasks containing a defined mineral and vitamin medium until exponential phase growth. The cultures were then washed and resuspended in media containing EDTA and metals. Biomass was measured by correlating optical density measurements with biomass dry weight. Total EDTA concentration was measured by high-performance liquid chromatography (HPLC). Carbon dioxide produced from EDTA biodegradation was monitored by acidification of the culture medium and analysis of the carbon dioxide concentration in the reactor headspace. In filtered culture media, dissolved oxygen was measured by an oxygen microelectrode, ammonium by an ammonium ion selective electrode, and metals by spectroscopy.

3.5 ESSENTIALITY VERSUS TOXICITY OF METAL IONS

Mostly all metal ions present in our environment are essential to human beings at low concentrations. For example, iron is the main component of hemoglobin, while cobalt is the central metal ion in vitamin B_{12}. Similarly, other metal ions contribute greatly to biological systems. Metal ions also play an important role in other biological activities by catalyzing the reaction rates of certain enzymes. The most important, useful metal ions are iron, cobalt, copper, zinc, selenium, manganese, calcium, sodium, and molybdenum. Metals also serve a chemically important role as essential components of many enzymes. These metalloenzymes are involved in the synthesis, repair, and degradation of biological molecules, the release and recognition of certain biological signaling molecules, and the transfer of small molecules and electrons in crucial processes such as photosynthesis and respiration.

The problem with certain heavy metals is that they tend to form very stable and long-lasting complexes with sulfur in biological molecules, which can disrupt their biological functions. In some cases, this allows these metals to become concentrated at higher levels of the food chain. Basically, the toxicities of heavy metal ions depend on their concentrations and vary from one organism to other. Therefore, useful metal ions may be toxic if their concentrations exceed permissible limits. Therefore, permissible limits are decided by the authorities for each metal ion. The toxicity of metal ions may be decreased by the presence of other metal ions, as discussed above. The selenium metal ion is the best example, as it reduces the toxicities of many metal ions. Metals that inhibit the toxicities of

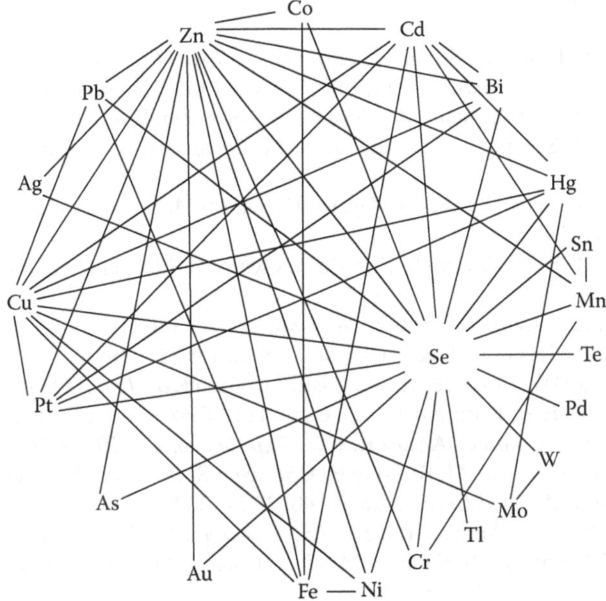

FIGURE 3.2 The reactivity of selenium with other metal ions.

other metal ions act as antagonists, and this sort of association of some metal ions with respect to selenium in biological systems is shown in Figure 3.2. Briefly, the essential metal ions may be lethal if their doses exceed the required values.

3.6 CONCLUSION

Although metal ions are essential for the proper growth of humans and other animals, they can be extremely toxic at high concentrations and can even act as carcinogens. As discussed above, heavy metal ions occur in different oxidation states and have different levels of toxicity, and despite this, many reports published in the literature deal with toxicities due to overall concentrations. Therefore, these data are not correct and need to be revised. There are only a few reports published on the toxicities of different species of metal ions. Therefore, there is an urgent need to explore the toxicities of all inorganic and organic metal ion species of different metal ions. The health regulatory authorities, especially the WHO, should describe the permissible limits of toxic metal ions in terms of different species. Briefly, much remains to be done in this area, particularly in the determination of toxicities of different species of all the toxic metal ions in relation to human beings and other animals. The effects of these different metallic species should be studied on the enzymatic reactions of some useful microbes present in our environment. Definitely, these findings will be useful in preserving the health of human beings and other animals.

REFERENCES

1. Furst A, in Brown SS, Kodama Y (Eds.), *Toxicology of Metals: Clinical and Experimental Research*, p. 17, Ellis Horwood Ltd, Chichester (1987).
2. Stoeppler M, *Hazardous Metals in the Environment*, Elsevier, Amsterdam (1992).
3. Kunito T, Saeki K, Oyaizu H, Satoshi M, *Ecotoxicol. & Environ. Safety*, 44: 174 (1999).
4. Kunito T, Senoo, K, Saeki K, Oyaizu H, Satoshi M, *Ecotoxicol. & Environ. Safety*, 44: 182 (1999).
5. Kunito T, Saeki K, Goto S, Hayashi H, Oyaizu H, Matsumoto S, *Biores. Technol.*, 79: 135 (2001).
6. Wataha JC, Hanks CT, Sun Z, *Dent. Mater.*, 10: 156 (1994).
7. Stohs SJ, Bagchi D, Free Radic. *Biol. Med.*, 18: 321 (1995).
8. England MW, Turner JE, Hingerty BE, Jacobson KB, *Health Phys.*, 1: 115 (1989).
9. de la Fuente H, Portales-Perez D, Baranda L, Diaz-Barriga F, Saavedra-Alanis V, Layseca E, Gonzalez-Amaro R, *Clin. Exp. Immunol.*, 129: 69 (2002).
10. Hartwig A, Asmuss M, Blessing H, Hoffmann S, Jahnke G, Khandelwal S, Pelzer A, Burkle A, *Food Chem. Toxicol.*, 40: 1179 (2002).
11. Wolterbeek HT, Verburg TG, *Sci. Total Environ.*, 12: 279 (2001).
12. Qu W, Kasprzak KS, Kadiiska M, Liu J, Chen H, Maciag A, Mason RP, Waalkes MP, *Toxicol. Sci.*, 63: 189 (2001).
13. Seoane AI, Dulout FN, Mut. Res., 20: 99 (2001).
14. Cornelis R, Crews H, Donard OFX, Ebdon L, Quevauviller Ph, *Fresenius' J. Anal. Chem.*, 370: 120 (2001).
15. Venturini-Soriano M, Berthon G, *J. Inorg. Biochem.*, 85: 143 (2001).
16. Illmer P, Marschall K, Schinner F, *Lett. Appl. Microbiol.*, 21: 393 (1995).
17. Illmer P, Schinner F, *FEMS Microbiol. Lett.*, 155: 121 (1997).
18. Illmer P, Schinner F, *FEMS Microbiol. Lett.*, 170: 187 (1999).
19. Illmer P, Erlebach C, *Antonie van Leeuwenhoek*, 84: 239 (2003).
20. Illmer P, Mutschlechner W, *J. Basic Microbiol.*, 44: 98 (2004).
21. Ochmanski W, Barabasz W, *Przegl Lek*, 57: 665 (2000).
22. Hotta N, *Jpn. J. Const. Med.*, 53: 49 (1989).
23. Hsueh YM, Cheng GS, Wu, MM, Yu SH, Kuo TL, Chen CJ, *Br. J. Canc.*, 71: 109 (1995).
24. Borgono JM, Greiber R, *Trace Substances in Environmental Health*, Vol. 5 (1971).
25. Goldsmith JR, Deane M, Thom J, Gentry G, *Water Res.*, 6: 1133 (1972).
26. Yamamura Y, Yamauchi H, *Industr. Health*, 18: 203 (1980).
27. Kiping MD, in *Arsenic, the Chemical Environment, Environment and Man*, Lenihan J, Fletcher WW (Eds.), Glasgow, Vol. 6: p. 93 (1977).
28. World Health Organisation, *Environmental Health Criteria, 18: Arsenic*, WHO, Geneva (1981).
29. Pershagen G, in *The Epidemiology of Human Arsenic Exposure*, Fowler BA (Ed.), Elsevier, Amsterdam, p, 199 (1983).
30. Csanady M, Straub I, *Health Damage due to Pollution in Hungary*, Proc. of the Rome Symposium, September, 1994, IAHS Publ. 233, (1995).
31. Tabor MW, in *International Conference on Arsenic in Ground Water: Cause, Effect and Remedy*, Jadavpur University, Calcutta, p. 38 (1995).
32. Andreae MO, in Craig PJ (Ed.), *Organometallic Compounds in the Environment: Principles and Reactions*, John Wiley & Sons, New York (1986).

33. Penrose WR, *CRC Crit. Rev. Environ. Control*, 4: 465 (1974).
34. Kaise T, Fukui S, *Appl. Organomet. Chem.*, 6: 155 (1992).
35. Petrick JS, Ayala-Fierro F, Cullen WR, Carter DE, Vasken Aposhian H, *Toxicol. Appl. Pharmacol.*, 163: 203 (2000).
36. Omura M, Tanaka A, Hirata M, Zhao M, Makita Y, Inoue N, Gotoh K, Ishinishi N, *Fundam. Appl. Toxicol.*, 32: 72 (1996).
37. Sandrint TR, Maier RM, *Environ. Toxicol. Chem.*, 21: 2075 (2002).
38. John De Zuane PE, *Hand Book of Drinking Water Quality: Standards and Control*,
39. Lison D, De Boeck M, Verougstraete V, Kirsch-Volders M, *Occup. Environ. Med.*, 58: 619 (2001).
40. Mulyani I, Levina A, Lay PA, *Angewend. Chem. Int.*, Ed. 43, 4504 (2004).
41. Kim SD, Park KS, Gu MB, *J. Hazard. Mater.*, 22: 155 (2002).
42. Dayan AD, Paine AJ, *Hum. Exp. Toxicol.*, 20: 439 (2001).
43. Zayed A, Lytle CM, Qian JH, Terry N, *Planta*, 206: 293 (1998).
44. Owen CA, *Am. J. Physio*l., 207: 1203 (1964).
45. Fairchild EJ, Lewis RJ, Tatkson RL, *Registry of toxic effects of chemical substances*, Vol II, US DHEW (NIOSH # 78-104-B), Washington DC, USA (1977).
46. Zollinger HU, *Vir. Arch. Pathol. Anat. Physiol.*, 323: 694 (1953).
47. Boyland E, Dukes CE, Grover PL, *Br. J. Canc.*, 16: 283 (1962).
48. van Esch GJ, van Genderen H, Vivik HH, *Br. J. Canc.*, 16: 289 (1962).
49. van Esch GJ, Kroes R, *Br. J. Canc.*, 23: 765 (1969).
50. Kanisawa M, Schroeder HA, *Canc. Res.*, 29: 892 (1969).
51. Furst A, Schlauder M, Sasmore DP, *Canc. Res.*, 36: 1779 (1976).
52. Nunes J, Ehrich M, Robertson J, *Toxicol. Pathol.*, 29: 451 (2001).
53. Kramer JR, Allen HE, *Metal Speciation: Theory, Analysis and Application*, Lewis, Chelsea (1988).
54. Krull IS, *Trace Metal Analysis and Speciation*, Elsevier, Amsterdam (1991).
55. Donard OFX, Martin FM, *Trends Anal. Chem.*, 11: 17 (1992).
56. Hill JS, Haswell SJ, *Trace Element Speciation*, Wiley & Sons, Chichester (1993).
57. Quevauviller Ph, *Appl. Organomet. Chem.*, 8: 715 (1994).
58. Quevauviller Ph, Morabito R, *Fresenius' J. Anal. Chem.*, 351: 343 (1995).
59. Schurz F, Sabater-Vilar M, Fink-Gremmels J, *Mutagenesis*, 15: 525 (2000).
60. Cassidy DR, Furr A, in Oehme FW (Ed.), *Toxicity of Heavy Metals in the Environment — Part I*, p. 803, Marcel Dekker, New York (1978).
61. EPA, Ambient water quality criteria for mercury, US EPA, Washington DC (EPA 440/5-80-058) (1980).
62. Farrell RE, Germida JJ, Huang PM, *Appl. Environ. Microbiol.*, 56: 3006 (1990).
63. Kungolos A, Aoyama I, Muramoto *S, Ecotoxicol. Environ. Safety*, 43: 149 (1999).
64. Gilon N, Potin-Gautier M, *J. Chromatogr. A*, 732: 369 (2002).
65. Fishbein L, in Goyer RA, Mahlman MA (Eds.), *Advances in Modern Toxicology*, Vol. II, John Wiley & Sons, New York (1977).
66. Frost DV, Lish PM, *Annu. Rev. Pharmacol.*, 15: 259 (1975).
67. Merian E, *Metal and Their Compounds in the Environment*, VCH, Verlagsgesellschaft, Weinheim (1991).
68. Hussain SA, Jane DE, Taberner PV, *Hum. Exp. Toxicol.*, 17: 140 (1998).
69. Pane EF, Smith C, Mcgeer JC, Wood CM, *Environ., Sci., Technol.*, 37: 4382 (2003).
70. *Medical and Biological Effects of Environmental Pollutants*, A committee on medical and biological effects of the environmental pollutants, Washington DC (1974).

71. Atsmon J, Taliansky E, Landau M, Neufeld MY, *Am. J. Med. Sci.*, 320: 327 (2000).
72. Hoffman RF, *Toxicol. Rev.*, 22: 29 (2003).
73. Altamirano-Lozano M, *Invest. Clin.*, 1: 39 (1998).
74. Lenz GR, Martell AE, *Biochem.*, 3: 745 (1964).
75. Sorenson JR, *J. Med. Chem.*, 19: 135 (1976).
76. Laurie SH, in Wilkinson G, Gillard RD, McCleverty JA (Eds.), *Comprehensive Coordination Chemistry: The Synthesis, Reactions, Properties and Applications of Coordination Compounds*, Vol. 2, Pergamon Press, Oxford (1987).
77. Williams LE, Pittman JK, Hall JL, *Biochim. Biophys. Acta*, 1465: 104 (2000).
78. Krizek BA, Prost V, Joshi RM, Stoming T, Glenn TC, *Environ. Toxicol. Chem.*, 22: 175 (2003).
79. Koropatnick J, Zalups RK, *Br. J. Pharmacol.*, 120: 797 (1997).
80. Rawlings DE, Dew D, du Plessis C, *Trends Biotechnol.*, 21: 38 (2003).
81. Oremland RS, Blum JS, Bindi AB, Dowdle PR, Herbel M, Stolz JF, *Appl. Environ. Microbiol.* 65: 4385 (1999).
82. Brown NL, Shih YC, Leang C, Glendinning KJ, Hobman JL, Wilson JF, *Biochem. Soc. Trans.* 30: 715 (2002).
83. Oliver DS, Brockman FJ, Bowman RS, Kieft TL, *J. Environ. Qual.*, 32: 317 (2003).
84. Nies DN, Silver S, *Appl. Microbiol. Biotechnol.*, 51: 730 (1999).
85. Newman DK, Kennedy EK, Coates JD, Ahmann D, Ellis DJ, Lovley DR, Morel FM, *Arch. Microbiol.*, 168: 380 (1997).
86. Inoue H, Takimura O, Kawaguchi K, Nitoda T, Fuse H, Murakami K, Yamaoka Y, *Appl. Environ. Microbiol.*, 69: 878 (2003).
87. Stoyanov JV, NL Brown, *J. Biol. Chem.*, 278: 1407 (2002).
88. Bhattacharyya-Pakrasi M, Pakrasi HB, Ogawa Aurora R, *Biochem. Soc. Trans.*, 30: 768 (2002).

4 Sample Preparation

Trace elements are present in very low concentrations in environmental and biological samples, and, therefore, their speciation is a tedious job. Sometimes, two or more species of the same metal ion may have similar properties, often unstable, again making speciation difficult. Speciation becomes more serious and critical when the analyses of natural samples are to be carried out. Thousands of compounds, as impurities, are present along with metal ions in unknown samples. The samples containing a high ionic matrix cause problems in capillary electrophoresis (CE), as high ionic strength imparts a low electric resistance resulting in very poor and broadly shaped peaks. Electroosmotic flow (EOF) in the capillary is altered by the influence of the sample matrix, which may result in poor separation. Additionally, the detector baseline is perturbed when the pH of the sample differs greatly from the pH of the background electrolyte (BGE). The samples containing ultraviolet (UV)-absorbing impurities may be problematic in the detection of metal ions. It is important to mention here that the detection limits of certain metal ions are sometimes increased by using preconcentration methods.[1] Under such circumstances, sample preparation is an essential requirement before starting speciation analysis. One of the most important trends to simplify these complications is the generation of simple, rapid, and reliable procedures for sample preparation. The method development and the setup require the use of materials of known compositions, e.g., certified reference materials. Therefore, spiking experiments have to be performed for quality control of the method. In such cases, emphasis must be placed on the spiking procedures, as they exert an influence on recovery values. Even if the recoveries of spiked metal ions equilibrated in a certain matrix are complete, there is no evidence that the incurred metal ions were extracted with the same efficiency. However, it can be concluded that although the present scientific knowledge is not perfect, the use of spiking experiments helps to minimize errors. The integration and automation of all the steps between the sample preparation and detection significantly reduce the time of analysis, increasing reproducibility and accuracy. For this reason, the development of a new single device for the complete speciation analysis of the samples is required. Some review articles appeared in the literature on this issue.[2–12] Gomez-Ariza et al.[5] reviewed the extraction, concentration, and derivatization strategies of their environmental samples before loading them onto a chromatographic column. Recently, Timerbaev[13] reviewed sample preparation methods used before metal ion speciation by capillary electrophoresis.

In 1995 (May 7 through 10), in Lund, Sweden, the seventh symposium on the handling of environmental and biological samples in chromatography was held. This symposium was a continuation of the series started by the late Dr. Roland Frei, one of the early visionaries in sample preparation technologies in analytical sciences. A survey of the papers presented at this symposium indicates that five points were considered as essential and must be highlighted during sample preparation. First, a need for a continuous search for new technologies was realized so that the high costs due to chemicals and experimental labor may be reduced. Second, there is a need for increased sensitivity with better and more selective concentration techniques. This has driven scientists to examine affinity and immunoaffinity support that can selectively remove compound classes for further investigation. Third, development of multidimensional chromatographic techniques allowing online sample cleanup, which provides several advantages including automation, better reproducibility, and closed-system capability, was advocated. The fourth point considered was the development of better sample preparation techniques enabling the more effective use of biosensors and other sensors, because exposure to raw matrices can foul many sensors. The last point, which requires considerable attention from scientists, was that a quality movement needs to extend to sample handling.[14] Therefore, extraction, purification, and preconcentration of natural samples are essential processes in the speciation analysis of metal ions. We carried out an extensive literature review on the sample preparation technologies and found many reports on this subject in metal ion speciation studies. A general protocol of the sample preparation is given in Scheme 4.1. In view of all of these points, in this chapter, we describe the art of sample preparation, including the sampling, extraction, purification, and preconcentration processes necessary before conducting speciation analysis of metal ions in the natural samples.

4.1 SAMPLING

Before starting the sampling, it is very important to design and decide the goal of the study. Sampling is the first and most important step in metal ion speciation, and its design and implementation have a decisive influence on the final analysis results. Due to the low concentrations of metal ions in the environmental and biological samples, sampling requires special attention; hence, it is very important to take special care when collecting samples. Before sampling, one should have complete knowledge and information about the surroundings in the case of environmental samples and about the biological systems in the case of biological samples. Before sampling, changes in the physicochemical properties (transfer, transformation, deposition, accumulation, etc.) of the target metal species have to be considered in more detail. By considering these points, useful samples can be composed that may significantly improve the analytical results.

4.1.1 Environmental Samples

In the case of environmental sample collection, the sampling strategy includes the selection of sampling sites, type of samples (grab, mixed, or composite),

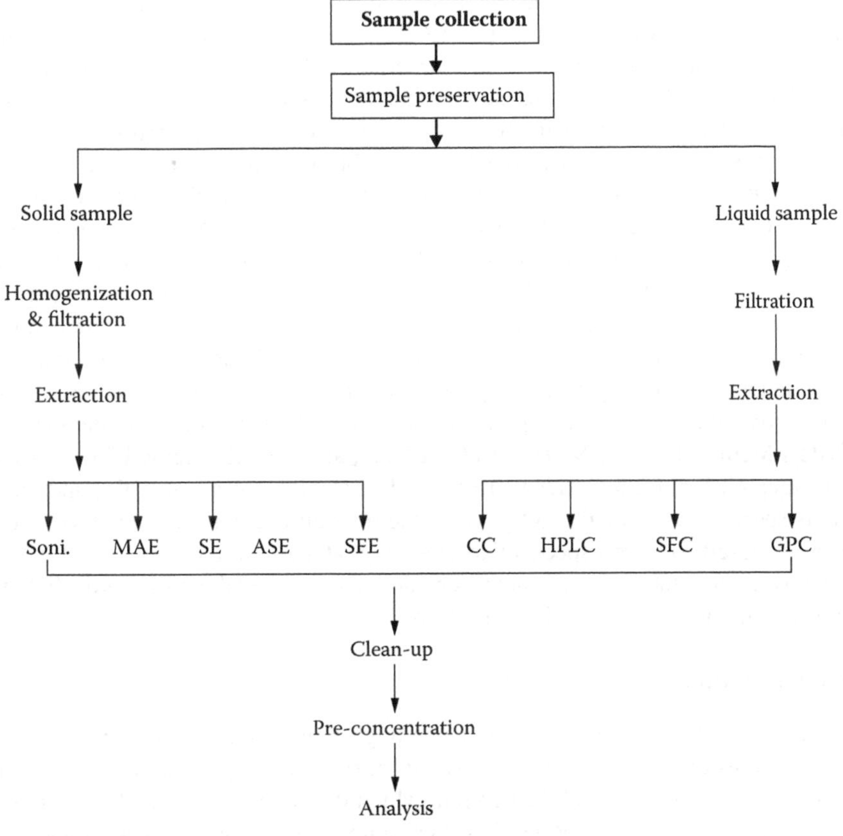

SCHEME 4.1

sample container, volume collection, sample handling, transportation, preservation, and storage. The selection of the sampling sites and the type of the samples are based on the objective and the nature of the study. The environmental sampling strategies of air, water, soil, and sediment are summarized below.

4.1.1.1 Air

Normally, metalloid compounds are analyzed in the gas phases. Hydrides of arsenic, lead, mercury, and selenium have been sampled from terrestrial sources, such as waste deposits or from aquatic ecosystems (i.e., ocean water surfaces). Mainly cryotrapping, solid adsorbent cartridges, polymer bags, and stainless steel canisters are used for obtaining metal ion samples from these deposits. Air samples are collected into balloons, cylinders, canisters, and glass traps. The vacuum filling of stainless steel containers is the official sampling method of the U.S. Environmental Protection Agency (EPA) for volatile compounds in monitoring urban air.[15] Aerosols are collected by using impactors, filters, denuders, and electrostatic separators.[16] Air particles are sampled using filters, glass fiber,

Hi-Vol sampler, Institute of Occupational Medicine sampler, and the seven-holes sampler. Personal exposure monitors (PEMs) are sampling devices worn on the body to estimate an individual's exposure to air pollution. Porous polymers such as Tebax-GC, XAD resins, and polyurethane forms have been used in gas sampling. Other sorbents used for this purpose are charcoal and carbon molecular sieves. It is very important to mention here that the inner wall should be inert (coated with polymers such as polytetrafluoroethylene [PTFE]). The gas phases can be transformed into a preevacuated sampling device.[17] Mostly air is passed through the 0.45 μm filter to remove solid particles and aerosols up to a predefined size from the sample. Generally, sample collection involves an air inlet, a particulate separator or a collecting device, an air flow meter, and a flow rate control valve air pump. The parameters important for consideration during sampling include temperature, light intensity, humidity, oxygen content, and aerosol concentration. Boiano et al.[18] compared the sampling efficiency of three methods — OSHA Method ID-215, NIOSH Method 7605, and NIOSH Method 7703 — for obtaining chromium hexavalent from air. The stability of arsenic, tin, and antimony species was studied as a function of temperature, and reasonable recoveries were achieved for these species that were kept at 20°C in the dark for 24 h.[19] It is advisable that the collected samples be delivered to the laboratory immediately, and the samples be analyzed within 24 h.

4.1.1.2 Water

In water sampling, techniques vary depending on the nature of metal ion species and the types of water — ground, waste, lake, river, sea, tap, industrial, soil, rain, and snow. The location and the depth to which the water sample collection point also influence the technique used. Water samples with volatile species are collected through an online combination of purging water with an inert gas such as helium followed by cryogenic trapping. Polycarbonate or polyethylene bottles are recommended for most of the metal species, while mercury should be collected in a dark glass bottle. Borosilicate glass containers may be used for this purpose. To avoid any adsorption on glass bottles, these should be pretreated with nitric acid. The containers should be washed properly with tap water, acids (no plastic bottles), detergents again tap water and double distilled water. Sample collection varies from the simplest of hand sampling procedures at a single point to the more sophisticated multipoint sampling techniques, such as the equal discharge increment (EDI) method or equal transit rate (ETR) method. Different types of water samplers can be used to collect water samples from rivers, seas, lakes, and other water bodies. According to Blumer,[20] DHI,[21] and a high-volume water sampler were designed to pump water from a defined depth below the water body surface outside the wake of the survey vessel are suitable as samplers.[22] Soil and sediment samples may be collected with the help of Ponar grab, Orange peel, Smith McIntyre grab, Shipek grab, Petersen grab, vanVeen grab, and Ekman grab samplers.[23] Roinestad et al.[24] improved the sampling of indoor air and dust using a Tenax TA sampling pump. When sampling a river, normally, up- and downstream samples and the points where some tributaries or waste drains join the main stream under study, are included. The rigorous cleanup of all the parts of the sampler and apparatus, solvent, adsorbent materials, and other chemicals should be carried out very carefully. Possible contamination due to the penetration of the sampler through the surface layers of the water body that may be highly enriched by the pollutants has to be excluded. The collected environmental samples should be transported to the laboratory immediately, and analysis should start as early as possible.

4.1.1.3 Sediments and Soils

Investigations made on sediment and soil mainly reported fractionation rather than speciation.[25] Representativeness is very important in the sampling of sediment and soil. The texture of soil and sediment is heterogeneous, and an extensive screening of each sampling site is necessary for accurate scientific analyses. The number and distances of the lateral and vertical sampling points are selected on the basis of geological information by applying statistical models. Related sampling grids and procedures are available in the literature.[26–28] The sampling of soil and sediment are carried out by shovel, corer, spoon, or knife, made of titanium or ceramics or plastics. According to the interest, the soil can be collected by initial separation of the mineral layers or as intact depth profiles. The latter approach is common in the sampling of lake or marine sediments, which is achieved by using different corer types. Sediment can also be collected by grab or core samplers. Sediment traps are used in dynamic flow systems such as rivers, culverts, and canals. Special procedures for the species retaining samplings of soil and sediment are still lacking in the literature. Problems also occur from changes in oxygen concentration during sampling, especially for originally anoxic sediments. The interaction between solid particles and water, biofilms, and pore water occurred, and hence, definition of the sample composition is difficult for speciation purposes. Therefore, the development of *in situ* methods and new approaches to species retaining sampling are required.

4.1.2 Biological Samples

Before starting biological sampling, it is very important to know the history of the animal, the treatment given, if any, the dose, and the surroundings in the case of plants. Biological rhythmicity is very important to consider at the time of biological sampling. During sampling, some necessary precautions should be adopted, including avoiding cloth contact (wool, etc.), closing door windows, not using acids at the sampling site, and keeping all reagents and apparatuses neat and clean. The biosphere varies to a greater degree in its physical and chemical properties relevant to species distribution and transformation than the abiotic environmental identities, such as air, water, sediment, and soil. A detailed sampling protocol must be developed and adopted for each specific study depending on the particular requirement. Metal ion speciation in biological fluids have been reported rarely, however, liquid samples can be collected as per requirements and stored in polyethylene bags or bottles. Some sampling strategies and procedures of collecting biological samples are summarized below.

4.1.2.1 Blood

Reliable methodology is critical in obtaining representative blood samples. The area of skin from where blood will be collected should be cleaned with deionized water followed by ethanol or methanol, and this should be allowed to evaporate properly. Blood samples are collected by the ETS systems, and plastic syringes have also been used. Microdialysis is a newly developed technique used for the collection of blood samples. Due to the risk of contamination, the use of syringes in sampling should be avoided. The addition of anticoagulant into a collected blood sample should be carried out very carefully, as these have good capacities to complex with metal ions, and, therefore, it is advisable to run the blank for each metal ions. Generally, blood samples are stored at –20°C (for 24 h) hemolyzed, resulting in the release of lead, manganese, zinc, and so forth.[29] For this reason, serum should be collected from the blood sample before freezing and storing.

4.1.2.2 Urine

The collection of a urine sample requires more skill than does the collection of a blood sample due to more chances of contamination. A urine sample can be collected after urination or directly from the urinary bladder. The samples can be collected with the help of an ETS system or syringe or measuring cylinder made of plastic. The collected samples can be kept in polyethylene containers that can be closed with an airtight lid.

4.1.2.3 Tissues

The liver and kidney are the most widely studied organs in metal ion speciation. There are greater chances for contamination in conducting this type of tissue sampling due to the long processes involved. The sampling of tissues is carried out by microdialysis, needle aspiration technique, and biopsies (with surgical cutting apparatus). Normally, tissues are washed with 1.0% salt solution and kept in plastic containers at low temperature.

4.1.2.4 Plants

There are few reports dealing with the issues involved in sampling plant materials. However, the sampling of phytoplanktons can be carried out using a net. The samples of other aquatic plants can be collected with the help of scissors, knives, blades, and so forth. Samples from terrestrial plants can be obtained after washing the outer cell wall of the plant with deionized water. Required samples are collected by dissecting specific plant parts. These particular samples, such as those from xylem and phloem, are collected carefully by dissecting the plant stem. Samples from the root are collected by digging, washing, and dissecting the roots properly.

4.1.2.5 Foodstuffs

In the case of foodstuff sampling, deciding on the amounts of samples and the sampling methods depends on the metal ions to be analyzed and the purpose of the study. Sample collection from foodstuffs should be carried out carefully, as there are greater chances of contamination. The samples of beverages are collected with the help of syringes or cylinders made of inert plastics. Samples from solid foodstuffs are collected with the help of spoons, knives, blades, and so forth. Liquid samples can be preserved by adding certain acids, and they can be stored at low temperature for a few months. For more details, consult the sampling protocol of foodstuffs given by the Official Control of Foodstuffs Directives.[30–32]

4.2 PRESERVATION

Sometimes, laboratories and instruments are not available immediately after sampling, especially when sampling is carried out at remote areas. Under such conditions, careful storage of the collected samples is required. Preservation of the chemical identities of the sample is the prime objective. Selection of the preservation method depends on the type of sample, storage time, size, and number of the samples. Air samples cannot be stored directly for future speciation analyses; however, the preservation of gaseous components in a cold trap is possible for 1 week or more in liquid nitrogen. The storage of gaseous samples for an extended time is not recommended. Certain metal ions are lost due to adsorption on the container or ion exchange with the glass container. Types of metal ions that can be affected are aluminum, cadmium, chromium, copper, iron, zinc, lead, manganese, and silver. To avoid adsorption and exchange, it is recommended that concentrated nitric acid (1.5 mL/L or $pH < 2$) be added to water samples prior to storage. But, 5.0 mL/L concentrated nitric acid should be used in samples of high buffer capacity. These water samples should be stored in an incubator at about 4°C to avoid evaporation. The samples containing microgram-per-liter concentrations should be analyzed immediately. Sometimes, certain preservatives may be used to store the samples for metal ions, but this practice is rare in use. The acidification of water or other liquid sample is done to stop adsorption and exchange processes, but it is not suitable if the sample contains organometals. Organometals may degrade in the presence of acidic conditions. This type of study was carried out by Feldmann et al.[33] for arsenicals in urine samples. The authors collected control and volunteer urine samples and divided them into two classes — one having no acid and the second having 0.1 *M*/L HCl — and preserved them at different temperatures. The authors also studied the effects of some additives (sodium azide [NaN_3], benzyltrimethylammonium chloride [$BzMe_3NH_4Cl$], benzoic acid [Bz acid], cetylpyridinium chloride [Cetyl], and methanol [MeOH]) on the percent recoveries of the arsenicals stored at different temperatures. The effect of HCl on the recoveries of arsenicals (stored at –20, 4, and 25°C) is shown in Figure 4.1, which indicates the degradation of all arsenicals. However, greater degradation was observed for organoarsenicals. Similarly, Figure 4.2a and Figure 4.2b dictate the effects of the above-mentioned preservatives on the percentage recoveries at 4 and –20°C, respectively. A perusal of these figures clearly indicates that the degradation of the arsenicals is greater at 4°C in comparison to –20°C. Furthermore, it may be seen from these figures that maximum degradation was observed at 4°C with sodium azide as the preservative, while maximum degradation was seen with cetylpyridinium chloride preservative at –20°C. Recently, Gallagher et al.[34] reported a preservation method for drinking water samples for arsenicals. The authors tried various treatment and storage conditions, including ethylenediaminetetraacetic acid (EDTA)/acetic acid treatment, acetic acid treatment alone, and no treatment, with the effects of temperature and aeration determined on all combinations. Among various combinations, the addition of acetic acid (10 mL of 8.7 M) and EDTA (10 mL of 50 mg/mL) to 1 L of water sample was found to be the best method for water sample preservation at 5°C. Therefore, the addition of acid to water samples prior to storage is not a universal practice. The presence of organometallic species in a water sample should be kept in mind before storing collected water samples. It is not safe to comment on the addition of acids or preservatives to collected environmental samples before knowing the fate of metal ions present in the collected sample. Therefore, there is a great need to research the storage of samples. Briefly, no standard or universal method is available for the storage of water samples, and the method varies from one type of sample to another.

Sediment and soil samples are normally stored at 0°C. Temperature conditions for biological samples vary from zero to –130°C or more depending on the type of sample collected, and samples can be stored for several years.[35] Such storage can be achieved by keeping the samples in special cryocontainers in the gas phase above liquid nitrogen. Sometimes, water is removed from biological tissues to check bacterial growth, but this method of storage is not universally effective, as some metal ion species may move with water. Sometimes, the precipitation of urine salts occurs, and hence, these samples should be stored in the presence of some preservative (0.03 *M*/L HCl or HNO_3, sulfamic acid triton-X-100). Care should be taken to adjust the pH of the urine, as some metal–protein complexes destabilize at lower pH. Briefly, the storage of environmental and biological samples at low temperatures (0 to –130°C) is the best method to preserve samples.

4.3 FILTRATION

The collected environmental and biological liquid samples cannot be analyzed directly; therefore, they sometimes require processing. Water samples may contain solid impurities, and the filtration process is required before proceeding with extraction. The filtration of these samples can be carried out by using a vacuum pump and buccaneer funnel assembly using Whatman's filter paper. Weigel et al.[36] described polytetra fluoro ethylene (PTEF) tubing fitted with a gear pump

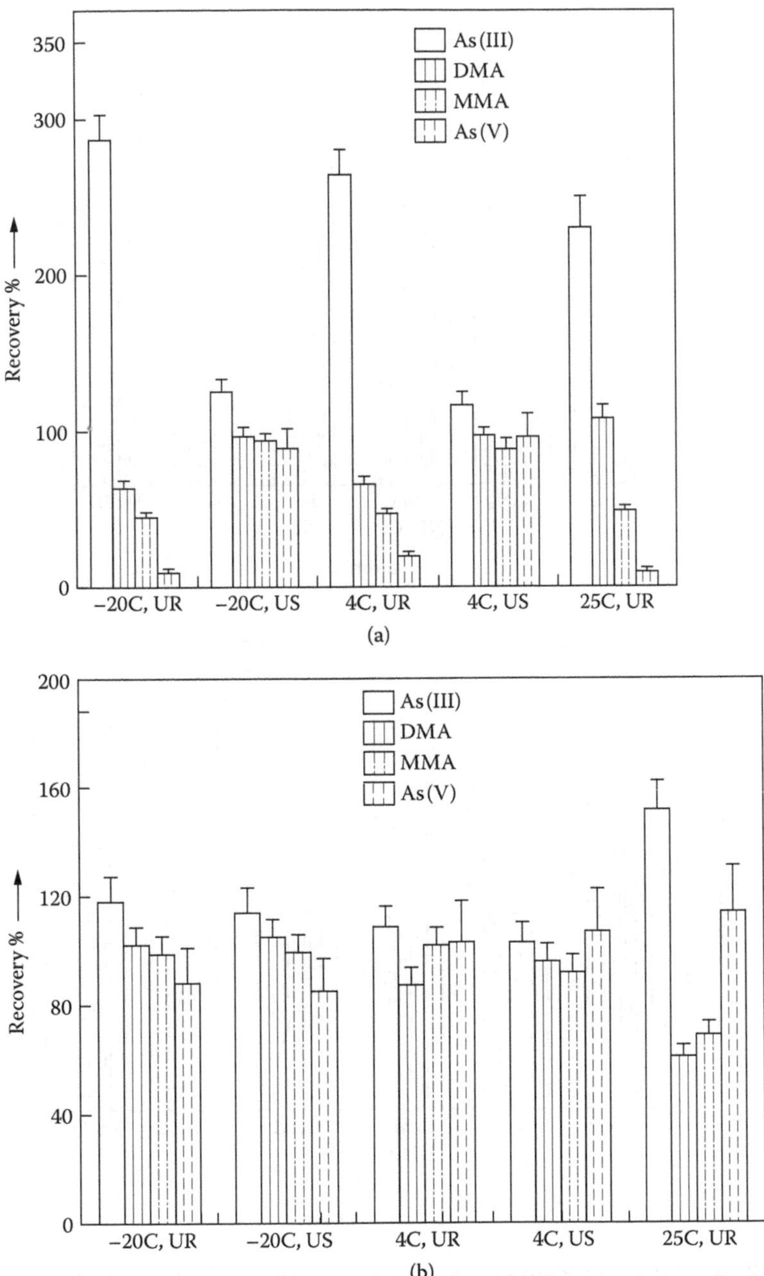

FIGURE 4.1 Recovery of four arsenic species in standard urine (US) and volunteer urine (UR) samples after storage for 2 h at different temperatures. (a) All samples contained 0.1 M HCL/L. (b) No HCl was added to the samples. (From Feldmann, J., Lai, V.W.M., Cullen, R., Ma, M., Lu, X., and Le, X.C., *Clin. Chem.*, 45: 1988 (1999). With permission.)

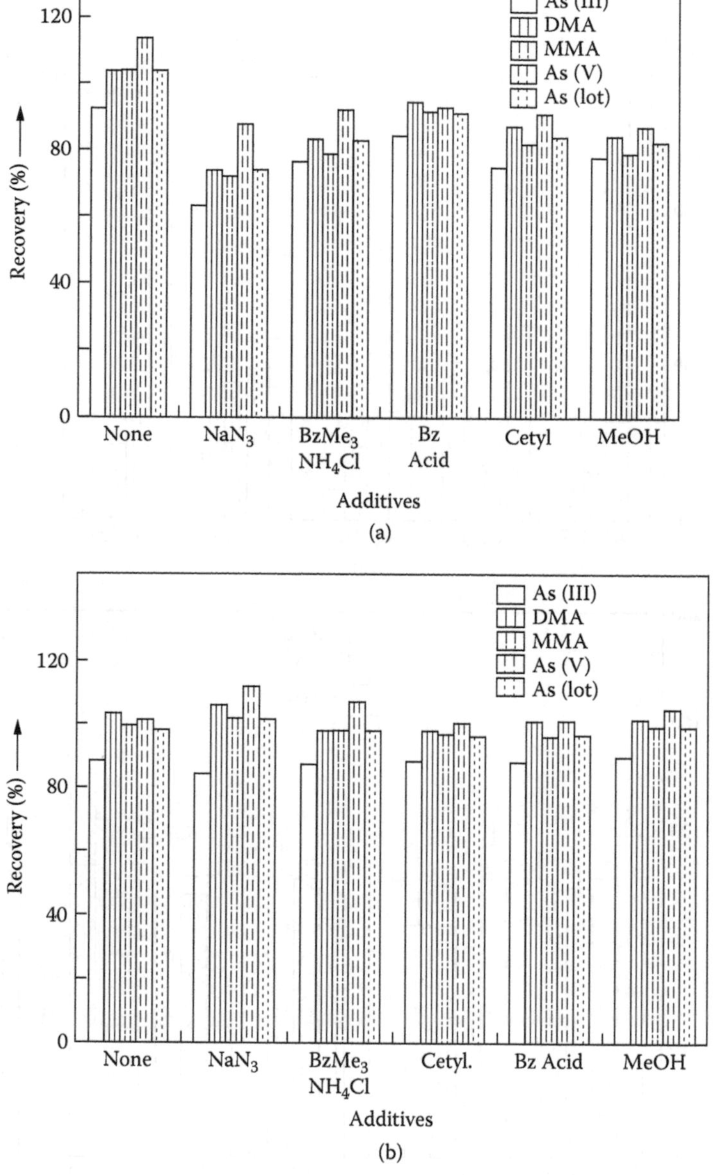

FIGURE 4.2 Effect of additives on the recovery of four arsenic species in urine samples stored at (a) 4°C and (b) 20°C for 1 month. (From Feldmann, J., Lai, V.W.M., Cullen, R., Ma, M., Lu, X., and Le, X.C., *Clin. Chem.*, 45: 1988 (1999). With permission.)

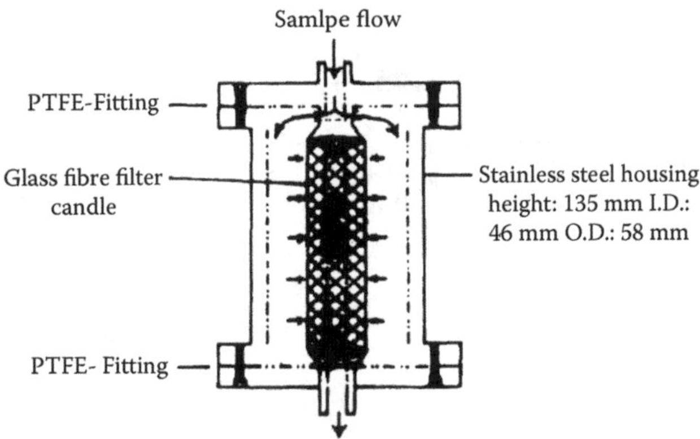

FIGURE 4.3 A filtration unit for use with filter candles. (From Weigel, S., Bester, K., and Hühnerfuss, H., *J. Chromatogr. A*, 912: 151 (2001). With permission.)

(Model MCP-Z pump head Z-120 with PTFE gears, magnet 66, Ismatec, Wertheim, Germany) for the filtration of water samples. The pump can be placed behind the extraction unit for calculating the recovery rates from purified water. The pump should be placed between the filtration and extraction units for the handling of environmental samples with a high load of particulate matter, because with increasing flow resistance, air may be drawn in at the exit of the extraction unit. This may lead to the malfunctioning of the pump. The rearrangement of the pump before the extraction unit did not cause contamination, as checked by procedural blanks. For online filtration of the sample glass fiber, filter candles (height 82 mm, outer diameter 26 mm, inner diameter 14 mm) were used in a stainless steel housing as shown in Figure 4.3.[36] As such, filtration is not required for solid samples. But, if solid samples contain other solid materials, they can be manipulated. If dust particles are present in an air sample, they should be removed before loading the collected air into the machine. The filtration of an air sample is carried out by using perch membranes under pressure. Similarly, Arce et al.[37] developed a flow injection analysis (FIA) system (Figure 4.4) for the online filtration of water samples prior to CE analysis. They also constructed a pump-driven unit for the extraction and filtration of soil samples combined with CE in an online mode (automated sample transfer between the pre-CE sample preparation step and the CE).[38] The method was precise and was four times faster than conventional off-line methods of sample preparation. Blood samples are centrifuged immediately to remove red blood cells (RBCs), and the serum is stored as discussed above. Sometimes urine samples also contain precipitates that are removed by the centrifuge technique.

4.4 DIGESTION/HOMOGENIZATION

In the past few decades, sample preparation from solid environmental and biological samples has been an active area in analytical science. The treatment of solid

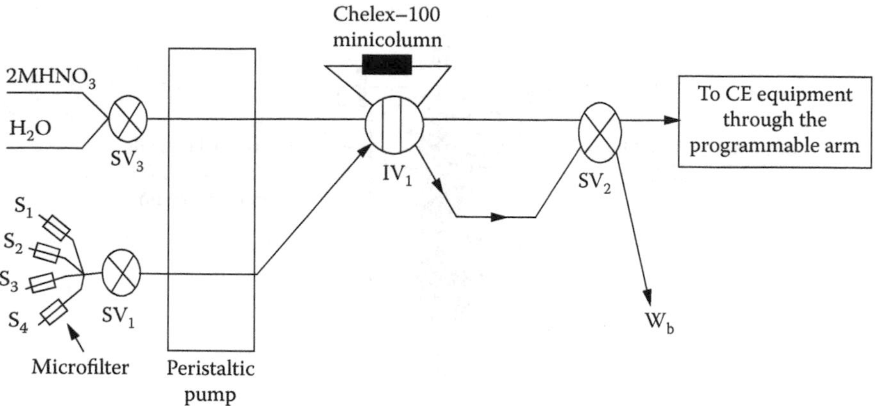

FIGURE 4.4 Single FIA system used for sample preparation/introduction in a capillary electrophoresis system. IV: injection valve, SV: selecting valve, and W: waste. (From Haddad, P.R., Doble, P., and Macka, M., *J. Chromatogr. A*, 856: 145 (1999). With permission.)

samples is more complicated in comparison to the treatment of liquid or gas samples. In biological samples, metal ions remain bonded with protein and other biomolecules, which requires application of a rigorous extraction procedure. Since 1970, acid homogenization methods have been used for the extraction of pollutants from solid environmental and biological samples.[3] Normally, concentrated nitric acid is used for the homogenization of soil and sediment samples, but sometimes, mixtures of nitric acid, sulfuric acid, hydrochloric acid, and perchloric acid, are used to digest the samples. Toluene and methanol are suitable solvents for use in the homogenization of biological samples, and dilute acids along with the above-mentioned solvents may be used under special conditions. Normally, for organometallics, environmental and biological samples are digested with an acid or a base that is then extracted by organic solvents. Generally, digestion is carried out in a flask or beaker by heating at a hot plate, but presently, an homogenizer is used for this purpose. An homogenizer contains small sets of blades and generator probes that cause vigorous mixing and turbulence that optimize sample contact and preparation. The homogenization of biological and environmental samples is tedious and time consuming. The homogenization is aided by shaking these samples at different speeds. Sometimes, mild heating also affects the extraction of metal ions during the homogenization procedure. The direct or indirect sonication of the samples greatly assists in extracting metal ions from solid samples. Sonication causes a disruption of the cell wall in biological samples (i.e., plant cells, bacteria, fungi, and yeast), and it helps in the release of the pollutants from these biological solid samples. After disruption, the proteins, DNA, RNA, and other cellular components can be separated for additional purification or processing.

Westöö[39,40] described the digestion of mercury species using hydrochloric acid. Cardellicchio et al.[41] reported the leaching of tributyltin (TBT), dibutyltin (DBT), and monobutyltin (MBT) from sediments with an HCl and methanol mixture. Dirkx et al.[42] obtained leached organotin from the matrix by the action of acetic acid and extracted it into hexane after formation of an extractable complex with diethyldithiocarbamic acid (DDTC). Jiang et al.[43] homogenized organotin species from sediment using hydrochloric acid. Similarly, Tutschku et al.[44] used hydrochloric acid for the digestion of organotin species from the sediment samples. Bettmer et al.[45] used methanolic sodium hydroxide for the release of organomercury species from soil and sediment samples. Välimäki and Perämäki[46] used potassium hydroxide for the homogenization of mercury species from dogfish liver. The acid digestion procedures are not successful, as the harsh conditions of digestion with acids may change the oxidation states of the metal ions. Acid leaching with conventional heating, steam distillation, and ultrasonic bath have been used, but these techniques require considerable time to obtain a quantitative recovery.

4.5 EXTRACTION

Extraction is the next step toward filtration in the case of water or liquid biological samples, while it comes after homogenization in the case of solid samples. Many devices have been used for extraction procedures, and the most important techniques in this field are classical methods of extraction — liquid–liquid extraction, solid-phase extraction, and liquid chromatographic techniques. The extraction of solid and liquid samples of environmental and biological origins is described herein.

4.5.1 Extraction of Solid Samples

The extraction of solid samples begins during the homogenization procedure, which involves the use of a variety of solvents as discussed above. The best solvent is one in which the solid sample is insoluble, while metal ions of interest are soluble. Besides, several new and improved technologies are used for the extraction of metal ions from solid matrices. For the best extraction of solid samples, they should be homogenized properly, creating a homogenous and finely ground form so that they can pass through a 1 mm sieve. The finely homogenized samples provide greater surface area to the solvent, which results in maximum extraction. Biological samples require complete decomposition of the tissues, as metal ions remain incorporated in the matrix. The most important methods for this purpose are Soxhlet extraction, sonication, microwave-assisted extraction, accelerated solvent extraction, and supercritical fluid extraction (SFE).

4.5.1.1 Homogenization Extraction

The extraction of metal ions from solid samples can be carried out during the homogenization procedure by using different extracting solvents. Mainly, digestion procedures have been used for the release of metal ion. Solid samples are ground to the powdered or paste forms as per the physical state of the collected solid samples. The extraction of metal ions can be achieved either by aqueous solvents or organic solvents or a mixture of organic solvents. The most commonly used solvents are water or water–methanol or water–methanol–chloroform mixtures.

4.5.1.2 Sonication Extraction

As discussed above, sonication helps in the homogenization of the solid samples; consequently, it can be used for the rapid and easy extraction of metal ions from the samples. Direct sonication uses a specially designed acoustical tool known as a horn or probe. The horns and probes are available with different cross-sectional areas and lengths depending on the volume of solution to be processed and the required intensity. The complete assembly including the horns and tips is called a sonotrode and is constructed of an inert material such as titanium alloy. The sonotrode is placed directly into the sample containing extracting solvent. Ultrasound frequencies, higher than the range of human hearing, are used in the sonication process. Ultrasonic energy generated by a piezoelectric transducer at the rate of 20,000 to 40,000 Hz creates cavitation in liquids. This cavitation is the formation and collapse of countless tiny cavities or vacuum bubbles in a liquid. The ultrasonic energy causes alternating high and low pressure waves within liquid, and these waves compress and expand liquid, which fosters the breaking of solid matrices and the extraction of the pollutants into the liquid. The effectiveness of sonication depends on the frequency and amplitude of the ultrasound and the temperature, surface tension, vapor pressure, viscosity, and density of the liquid. It is interesting to note that sonication generates heat into liquid, which augments the extraction process. EPA method SW-846-3550 recommended sonication as the effective method for the extraction of nonvolatile and semivolatile organic pollutants from soils, sludges, and other wastes.

Extraction with polar solvents, such as aqueous hydrochloric acid,[47–51] hydrochloric acid, or acetic acid in polar solvents (methanol, acetone),[50–59] or polar solvents in basic medium[60] performed under sonication has been used for organotin analysis from solid matrices.[61–63] This extraction is followed by another extraction with nonpolar solvent such as benzene, cyclohexane, or toluene [58,64–66] to recover the species for derivatization. The salting out reagents and tropolone were also used to increase the species recovery.[67–69] The extraction of lead species is similar to that for organotin compounds. Volatile nonpolar tetraalkyllead compounds can be directly extracted from the environmental samples with nonpolar solvents such as benzene and hexane. Ionic alkyllead compounds [$R_2Pb(II)$ and $R_3Pb(I)$] can be extracted and preconcentrated from water samples in the presence of sodium chloride and sodium diethyldithiocarbamate in hexane,[70] benzene,[71] and pentane.[72] Wang et al.[73] used ultrasonic extraction in combination with a strong anion-exchange solid-phase extraction technique for the extraction of Cr(VI) in the presence of Cr(III) species. Ultrasonic extraction was also used by Ashley et al.[74] for the extraction of arsenic, cadmium, copper, manganese, lead, and zinc from environmental and biological samples.

4.5.1.3 Microwave-Assisted Extraction

Microwave-assisted extraction (MAE) is a process of heating solid samples with solvent or solvent mixtures with microwave energy (2.45 GHz) and the subsequent partitioning of the compounds of interest from the sample to the solvent. It has the potential to be a direct replacement for the conventional Soxhlet extraction technique for solid samples. It is thought to be a suitable method for the environmental samples, because it involves small amounts of time and low-cost chemicals and has good efficiency and reproducibility. The EPA SW-846-3550 method describes microwave-assisted extraction for acid digestion for inorganic analytes as the best methodology. However, LeBlanc[75] described the modifications of this method, which may be used to heat solvents and samples. The heating process helps a lot in the rapid extraction of organometallic species from solid samples. Since 1986, scientists have used microwave heating to extract various metal ions from a variety of sample matrices.

Schmitt et al.[76] used the microwave technique for the release of organotin species. Similarly, Lobiniski et al.[77] used the microwave technique for the digestion of organotin species from solid sample. Schmitt et al.[78] obtained released tin and mercury species from the humic soil samples using microwave methodology. The digested species of tin were mono-, di-, and tributyltin, while the species of mercury was methylmercury. Recently, LeBlanc[75] reviewed the MAE method for solid samples with the benefits of this technology as shorter extraction time and reduced solvent consumption. In 1986, Ganzler et al.[79] reported the extraction of organic substances from soils, seeds, human-grade food, and animal feeds by using microwave energy. Many modifications were reported in MAE technology.[80,81] This technology was used in conjunction with liquid chromatography[82] and solid-phase extraction methods separately,[83] which resulted in the improved analysis of many substances. A schematic diagram of a microwave system is shown in Figure 4.5, which indicates a magnetron, isolator, wave guide, cavity, and stirrer. The most important drawback of MAE is that analytes in the polar leachate have to be later derivatized and transferred to a nonpolar solvent to produce a solution suitable for gas chromatography (GC) analysis, which increases the number of steps in the analytical procedure. This obstacle can be eliminated by integrating sample preparation for speciation analysis by combining leaching, derivatization, and liquid–liquid solvent extraction to directly produce a solution of the analytes to be analyzed by GC (microwave-assisted derivatization solvent extraction [MADSE]). Second, by integrating leaching, derivatization, and liquid–gas extraction (purge), an alternative is provided for more volatile derivatives to be obtained, which can be recovered in a cryotrapping device to avoid losses (microwave-assisted purge-and-trap [MAPT] device). Briefly, microwave methodology is the ideal technique for the homogenization of the metal species from the solid samples. This technology has good future prospects, as it can be coupled online with chromatographic techniques. Examples of MAE for some organometallic species are given in Table 4.1.

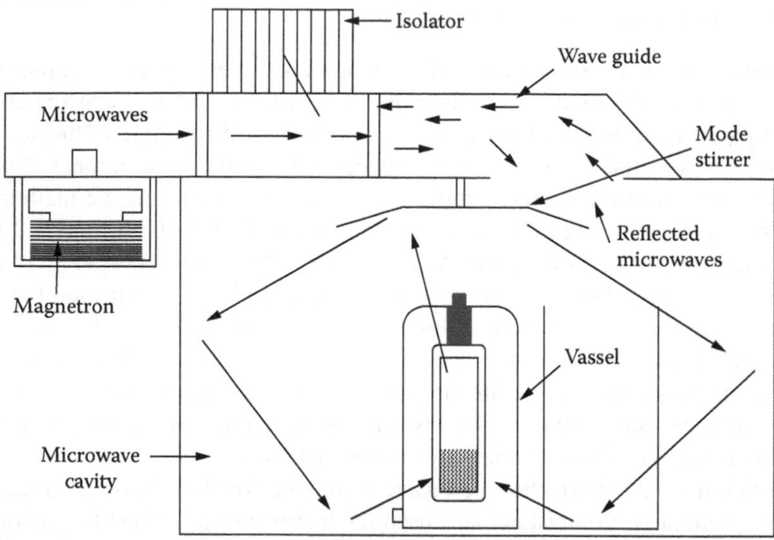

FIGURE 4.5 A microwave-assisted extraction system. (From LeBlanc, C., *LC-GC*, 17: S30 (1999). With permission.)

TABLE 4.1
Microwave-Assisted Extraction for Organometallic Species

Species	Extraction	Ref.
Organotin and MeHg(I)	MAH or KOH in MeOH, 40–60 W, 2–4 min	84, 85–87
MeHg(I)	1% $NaPhB_4$ + 17 M acetic acid + toluene, 100°C, 690 kPa, 6 min	88
MeHg(I)	2 M HNO_3, 40–60 W, 2–4 min	89
MeHg(I)	6 M HCl, toluene, 120°C, 950 W, 10 min	90
Organotin	0.5 M acetic acid in MeOH, 70 W, 3 min	91
Butyltin	Acetic acid–water (1:1), 60 W, 3 min	92
Butyltin	25% TMAH, 60 W, 3 min	92
Organoarsenicals	MeOH–water (1:1), 50 W, 5 min	94

4.5.1.4 Soxhlet Extraction

Soxhlet extraction has been used for more than 100 years as the extraction technique for various substances from different matrices. This methodology was developed by German scientist Franz Von Soxhlet in 1879. Soxhlet extraction is a good alternative for the extraction of organometallic species from solid samples, but, normally, it is not used for the extraction of such pollutants, which may decompose at higher temperatures. Therefore, Soxhlet extraction may be a good method for the extraction of organometallics with high volatility. Generally, organic solvents that have the solubility of organometallic species may be used in this methodology. Soxhlet extraction is a simple and effective but relatively expensive method, as it requires large amounts of organic solvents. Sometimes, it requires 18 to 24 h in its operation for the extraction. Automatic Soxhlet assemblies are available that work with minimum amounts of organic solvents, for example, 50 to 100 mL only, which makes their operation inexpensive.[95]

With advances in science and technology, different modifications were made in Soxhlet extraction assembly which resulted in modern and automatic Soxhlet extraction technology. Of these modifications, the procedure of Goldfisch is known as the best in which the condensed solvent is allowed to drip through the sample thimble, immediately returning to the boiling solvent. In 1996, Ali[96] presented a compact solvent extraction apparatus, which is very easy to handle. Recently, Arment[97] reviewed the automatic Soxhlet extraction procedures with an emphasis on the evolution of the technique from original design to modern automatic instrumentation. Furthermore, the author discussed several aspects of method optimization and general considerations. Many companies developed instrumentation for automatic extraction based on the Randall concept of immersing the sample in boiling solvent. Automatic Soxhlet extraction has gained worldwide acceptance. The EPA method SW-846-3541 described the extraction of analytes from soil, sediment, and sludges, and this method is also applicable for many semivolatile organic pollutants. Büchi Laboritechnik AG (Flawil, Switzerland) manufactured a model B-811 with a four-stage universal extraction system (i.e., standard Soxhlet, warm Soxhlet, hot extraction, and continuous drip). Similarly, C. Gerhardt GmbH & Co. Laboratory Instrument, Bonn, Germany manufactured the Soxtherm S-306 model with a six-stage extraction system. Foss-Tecator (Höganäs, Sweden) commercialized and licensed the Randall method in 1975 and presented an automatic Soxhlet instrument to the market (Figure 4.6). Welz[98] reported Soxhlet extraction for sample preparation in metal ion speciation. This methodology can be used for organometallic species, but it is not useful for inorganic metal ions.

4.5.1.5 Accelerated Solvent Extraction

Accelerated solvent extraction (ASE) is a technique of sample preparation, which works at elevated temperature (50 to 200°C) and pressure (500 to 3000 psi) using very little amounts of solvents for short periods of time (5 to 10 min). The restrictions and limitations of this method are similar to Soxhlet extraction technology. The method development is simple, because it involves few operational parameters. The selection of the solvents again depends on the solubility of the pollutant to be extracted. The initial reports on ASE[99–107] showed that the performance was more or less equal to the conventional method in a shorter period of time involving low amounts of solvents. Figure 4.7 shows the schematic representation of this technology, which contains solvent reservoirs, pump, extraction cell, oven, static valve, nitrogen gas cylinder, and collection vial. This technique is very popular for the extraction of organic compounds and is rarely used for the extraction of metal ions. However, Gallagher et al.[108,109] used ASE for the

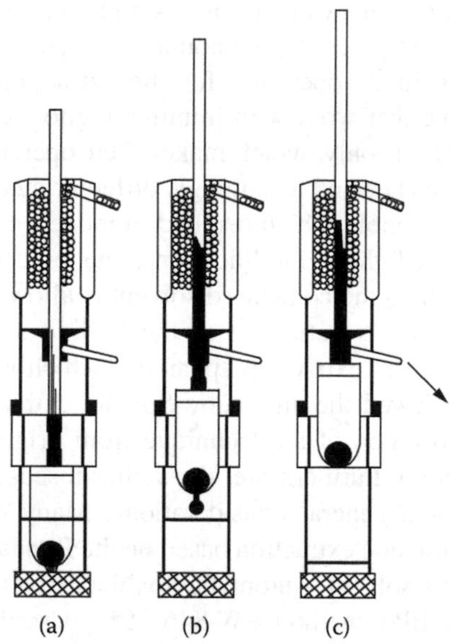

FIGURE 4.6 Three-step extraction procedure using the Foss-Tecator Soxhlet Avanti automatic extraction system: (a) boiling, (b) rinsing, and (c) evaporation steps. (From Arment, S., *LC-GC*, 17: S38 (1999). With permission.)

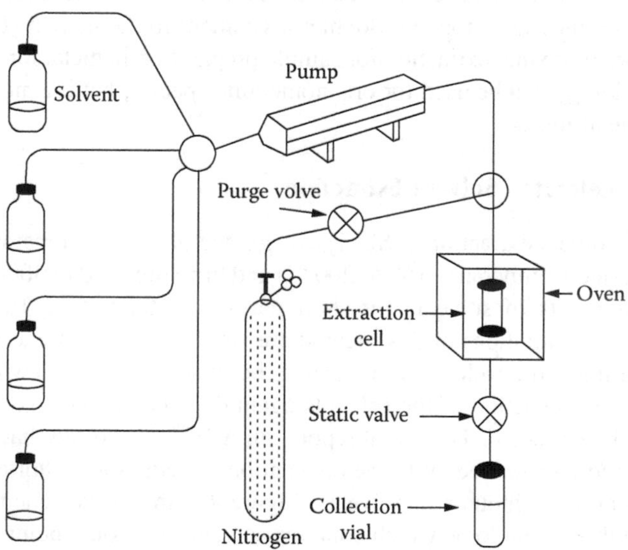

FIGURE 4.7 Accelerated solvent extraction. (From Richter, B.E., *LC-GC*, 17: S22 (1999). With permission.)

extraction of organoarsenicals from seaweeds. Recently, Richter[110] reviewed the extraction of solids by ASE methodology.

ASE method development is simple, as it requires the selection of the temperature and solvents, which governs the efficiency of the extraction. The other parameters that control the recovery of the extraction are static time, flush volume, and number of static cycles used in the extraction. It is interesting to note that the first application of this method was reported in the extraction of environmental pollutants.[99–107] Therefore, the ASE method has gained in popularity in the extraction of environmental samples.[110] Moreover, its use was expanded for the extraction of many organic and inorganic substances from food, polymers, and pharmaceutical and other types of samples. However, no observations were reported for the extraction of metal ions using this technique.

4.5.1.6 Supercritical Fluid Extraction

Supercritical fluid extraction (SFE) is a selective technique of the sample preparation, which enables the preparation of matrices by varying several physical parameters. It is supposed to be a good replacement for many extraction technologies, such as accelerated solvent, Soxhlet solvent, microwave-assisted, and so forth. It was marketed as a universal extraction tool in 1988 by Isco Inc. (Lincoln, Nebraska), Lee Scientific (Salt Lake City, Utah), and Suprex Corp. (Pittsburgh, Pennsylvania). The basic components of an SFE instrument are carbon dioxide reservoir, pump, extraction vessel, oven, restrictor (a decompression zone), and collection vial (Figure 4.8). Depending on the application, some models also contain modifier pumps, which are used to mix carbon dioxide with the organic solvents.

Practically all SFE instruments use carbon dioxide as the primary supercritical fluid. This is due to the fact that carbon dioxide at its critical temperature (31.4°C) and pressure (73 atm) can be handled easily. Moreover, carbon dioxide is readily and economically available in analytical grade form. Hexane, methylene chloride, acetone, and chloroform are suitable solvents, as they can be mixed with carbon dioxide. Several organic compounds are miscible with carbon dioxide, and Table 4.2[111,112] lists some of these modifiers and their critical parameters. The values given in this table can be used for the extraction of metallic species. Depending on the application, the use of a third modifier for additional solubility enhancement is sometimes required.[113] The efficiency of this method is increased by variation in temperature and pressure variables, which have an inverse relationship. Maximizing the solvent contact also ensures rapid and efficacious extraction. However, flow rates of the mobile phase, SFE time, and SFE mode also control extraction efficiency. Normally, analysis of the extracted samples is carried out after extraction by SFE, but in 1992, Levy et al.[114] combined the use of SFE with GC and reported good efficiency of this assembly (Figure 4.9).

4.5.2 Extraction of Liquid Samples

Normally, liquid samples of environmental and biological origins include water, blood, serum, and urine. The extraction of liquid samples is relatively easier than

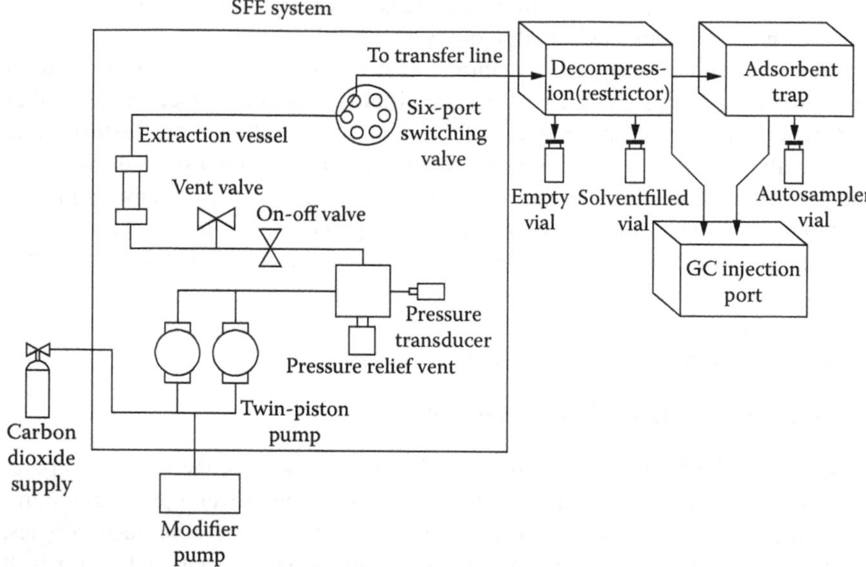

FIGURE 4.8 A supercritical fluid extraction (SFE) system with modifier pump and collection options. (From Levy, J.M., *LC-GC*, 17: S14 (1999). With permission.)

the extraction of solid samples, as the former do not require grinding or digestion procedures. The most commonly used techniques for the extraction of metal ions from liquid samples are liquid–liquid extraction and solid-phase extraction. But, with the many developments in extraction science, different kinds of chromatographic modalities have also been used as extraction methods. The chromatographic techniques used for this purpose include column chromatography (CC), high-performance liquid chromatography (HPLC), supercritical fluid chromatography (SFC), and gel permeation chromatography (GPC). If the metal ions present in solid samples are polar in nature, then the extraction is usually carried out in aqueous solvents. However, if gas chromatography is required for the analysis of such types of extracted pollutants, further extraction of the already extracted pollutant (in aqueous solvent or water) is carried out in organic solvents. This is due to the fact that aqueous solution cannot be injected into the gas chromatographic machine directly, as that may destroy the detector. Contrary to this, metal ions present in organic solvent are extracted into aqueous solvent if the analysis is carried out by capillary electrophoresis. This situation arises as the loading of the sample in organic solvent disrupts the continuity of the current in the machine.

4.5.2.1 Liquid–Liquid Extraction

Liquid–liquid extraction is the classical method used for the extraction of organometallic species only from liquid samples. This method is not applicable to the extraction of inorganic metal ions and, therefore, is used rarely for the

TABLE 4.2
Critical Parameters for Carbon-Dioxide-Based Modifiers

Modifiers	Tc(°C)	Pc (atm)	Vc (cm^3/mol)
Acetic acid	321.1	57.1	171.1
Acetone	235.2	47.2	216.5
Acetonitrile	247.7	47.7	173.0
Benzene	288.8	48.3	257.8
n-Butanol	288.2	43.6	275.1
tert-Butanol	234.8	41.7	275.0
Carbon tetrachloride	283.3	45.0	275.9
Chloroform	262.3	54.1	241.1
N,N-Dimethylformamide	376.8	54.3	—
Dimethyl sulfoxide	446.8	56.3	237.5
1,4-Dioxane	314.8	51.1	238.8
Ethanol	242.6	62.8	167.1
Ethylacetate	250.9	38.0	285.9
Ethylether	193.55	35.92	280.0
Formic acid	306.8	31.2	435.8
n-Heptanol	358.7	31.2	435.8
Hexane	234.7	29.8	368.4
Hexanol	336.8	34.3	381.0
Isobutanol	274.5	42.4	273.6
Isopropanol	243.4	28.4	219.7
Isopropyl ether	227.2	28.4	386.0
Methanol	239.8	79.1	117.9
Methyl-*tert*-butyl ether	239.7	33.3	—
Methylene chloride	235.1	60.0	180.0
2-Methoxyethanol	302.0	52.2	—
Nitromethane	314.8	62.3	173.0
Pentanol	312.7	38.4	326.0
n-Propanol	363.7	50.5	218.9
Propylene carbonate	—	—	352.0
Sulfur hexafluoride	45.5	37.1	197.0
Tetrahydrofuron	267.3	51.2	224.0
Toluene	319.9	41.5	316.6
Trichlorofluoromethane	198.0	43.2	248.0
Water	374.1	218.2	56.5

Note: Tc, Pc, and Vc are critical temperature, pressure, and volume, respectively.

Source: From Richter, B.E., LC-GC, 17, S22, 1999; Levy, J.M., LC-GC, 17, S14, 1999. With permission.

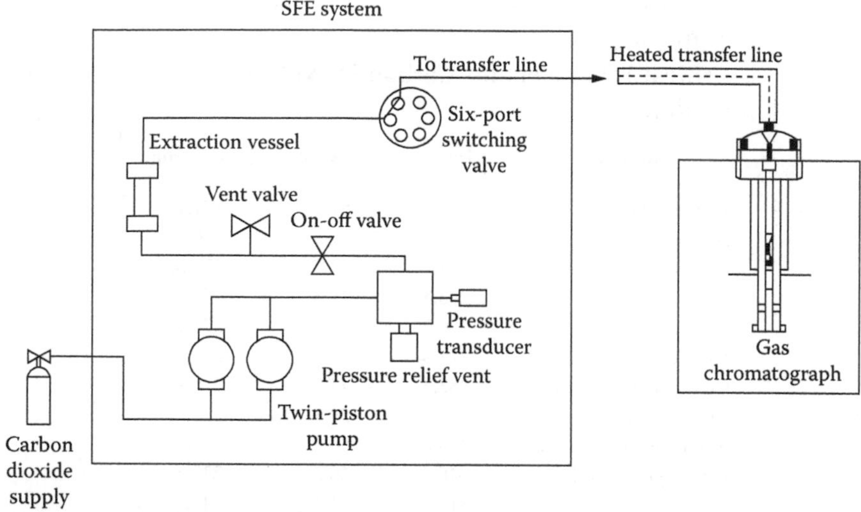

FIGURE 4.9 A supercritical fluid extraction and gas chromatography (SFE-GC) coupling. (From Levy, J.M., *LC-GC*, 17: S14 (1999). With permission.)

extraction of metal ions from unknown matrices. The extraction of organometallic species is normally carried out by simple liquid–liquid extraction (LLE) and solid–liquid extraction (SLE) methods using organic solvents. According to the distribution law, when brought in contact with an immiscible solvent pair, a solute distributes itself between these solvents on the preference basis determined by many factors but mainly by solute and solvent characteristics. In quantitative LLE of water samples, the right selection of the solvent or solvents must be carried out carefully. Although water quality characteristics such as ionic strength, turbidity, coextractives, pH, and temperature may affect the criteria for the solvent selection to some extent, the choice of solvent or mixture of solvents should be reviewed in light of the nature of the solvent, the solubility of the solvent in water, its compatibility with the detector, and its volatility. As per the general principle that like dissolves like, the polarity of the solvent to a large extent determines its extraction capability. Nonpolar solvents such as hexane, benzene, toluene, chloroform, pentane, and dichloromethane, are used for nonpolar species such as tributyltin (TBT), but highly polar mono- and dibutyltin (MBT, DBT) or mono- and dimethyltin (MMT, DMT) species need complexation or acidification of the sample. Generally, ionizable metal species may be extracted into organic solvents in the form of neutral ion pairs in the presence of an appropriate ion pairing agent, or hydrophilic compounds, which are difficult to remove from the aqueous phase. Methylene chloride is an effective solvent for the multiresidue extraction of organometallic ions from water, including those samples with considerable turbidity and organic content. It has low water solubility and is not too polar to extract excessive coextractives as compared with more polar solvents, such as diethyl ether and ethyl acetate. The other organic solvents used are dichloromethane,[115] hexane,[116] acetone,[117] acetic acid,[118] benzene,[119] toluene,[120] methanol,[121] acetonitrile,[118] petroleum ether,[121] ethyl ether,[122] isooctane,[123] and pentane.[122] These solvents are used as single, binary, or ternary mixtures for pollutants extraction.

General methods for the extraction of metal ions from water samples involve the manual or mechanical shaking, stirring with magnetic stirrers, continuous liquid–liquid extraction, and homogenization method[123] using a high-efficiency dispenser that essentially contains high-speed blades with a high-frequency ultrasonic probe coupled to it. In brief, the mechanical shaking method is efficient provided there is sufficient shaking. Different models of a shaker can provide different results, particularly with turbid waters, and thorough testing with one's own water sample is needed. The shaking method can cause severe emulsion formation for certain samples (for example, some industrial effluents). Under such conditions, the emulsion is difficult to break up, even by high-speed centrifugation, and it may be overcome by the addition of some chemicals like anhydrous sodium sulfate, sodium chloride, and alcohols. The analyst should carefully consider his or her own situation, for example, water types and choice of the solvent, before contemplating this approach.

Tutschku et al.[124] utilized hexane as an extraction solvent, after the digestion of the sediment with hydrochloric acid, for organotin species. Jiang et al.[43] used hexane and a hexane–ethyl-acetate mixture for the extraction of organotin species from water and sediment samples, respectively. Heisterkamp and Adams[125] described the use of hexane for the extraction of water and peat samples containing organolead species. Pons et al.[126] used pentane, pentane–alcohol, pentane–ethyl-acetate, pentane–salt, and hexane–salt mixtures for the extraction of alkyllead species in human urine samples. Girousi et al.[127] used hexane as the extraction solvent for the extraction of organotin species from water samples collected from Thermaïkós Kólpos (Gulf of Salonika), Greece. Lai-Feng and Narasaki[128] described the extraction methods for organotin compounds in marine sediment using a variety of organic solvents. The authors compiled different recoveries of TBT, MBT, and TPT species using acid/base digestion and organic solvents as the extraction methods. The different recoveries of these species with different extraction reagents are given in Table 4.3. It may be concluded from this table that the recoveries of the above-cited tin species varied from one reagent to another. Zufiaurre et al.[144] used hexane for the extraction of trimethyllead from rainwater. The authors studied the effect of the volume of hexane on the extent of the extraction of the organolead species in water, as shown in Figure 4.10. The information presented in Figure 4.10 indicates that the extracted amount decreases with increasing hexane volume.

Pressurized liquid extraction (PLE) constitutes an emerging and promising alternative based on the use of organic solvents at high temperatures under pressure. The extraction time by this technique is quite low, and PLE is gaining in popularity. Chiron et al.[145] presented a PLE with a methanolic mixture of 0.5 M acetic acid and 0.2% (w/v) of tropolone as the suitable extraction procedure for the quantitative extraction of butyl- and phenyltin compounds with recovery

TABLE 4.3
Comparison of Percent Recoveries of Different Organotin Species (TBT, MBT, and TPT) Using Different Extraction Solvents

Homogenization	Solvents	Recoveries (%)			Ref.
		TBT	MBT	TPT	
HCl	(1:1) Hexane-ethylacetate-0.05% tropolone	—	674[a]	—	129
HCl	CH_2Cl_2	72–99	—	—	130
HCl	Ethyl acetate	95.9 ± 2.1	73.3 ± 2.4	80.8 ± 3	131
HCl	Toluene-0.5% tropolone	90–114[a]	96–109[a]	—	132
HCl-MeOH	Isooctane	98[a]	28[a]	—	133
HCl-MeOH	Cyclohexane	79–140	—	—	134
HCl-MeOH	Toluene-0.1% tropolone	102 ± 3.4[a]	—	99 ± 8.5	135
HCl-MeOH	Toluene-isobutyl acetate tropolone	94.4 ± 4.7[a]	86.3 ± 4.2[a]	—	136
HCl-MeOH	Isooctane	95 ± 4[a]	—	—	130
	1-BuOH	97 ± 6[a]	—	—	137
HBr	Pentane-0.02% tropolone	97–100[a]	91–99[a]	98–99[a]	138
Acetic acid	MeOH-0.3% tropolone	86 ± 15[b]	60 ± 27[b]	—	139
Acetic acid	Acetic acid	—	—	—	140
Acetic acid	Acetic acid	97.9–102[a]	98–101[a]	—	141
Acetic acid	Acetic acid	96.2 ± 2[b]	111 ± 4.7[b]	127 ± 14.1[b]	142
Glacial acetic acid	Toluene-0.3% tropolone	88.1 ± 2.5[a]	84.7 ± 6.3[a]	88 ± 0.8[a]	143
Acetic acid-DDTC	Hexane	95–100[a,c]	95–100[a,c]	95–100[a,c]	136
Sodium benzoate-NaCL-KI	Hexane-0.5% tropolone	18.9[a]	—	—	77

Note: MBT: monobutyltin, TBT: tributyltin, and TPT: triphenyltin.

[a] Spiked.
[b] Reference sample.
[c] Not reported clearly.

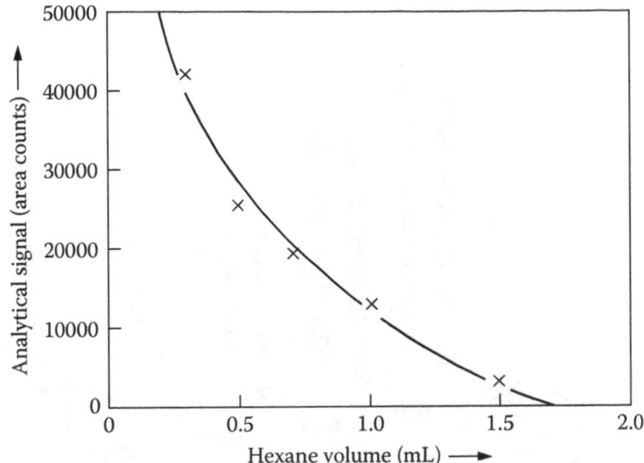

FIGURE 4.10 Effect of hexane volume on the extraction of organolead species. (From Zufiaurre, R., Pons, B., and Nerín, C., *J. Chromatogr. A*, 779: 299 (1997). With permission.)

values ranging from 72 to 102%. This analytical approach was compared to conventional solvent extraction methods making use of acids and organic solvents of medium polarity. The main advantages of PLE over conventional solvent extraction are the possibility of quantitatively extracting diphenyltin (DPhT) and methylphenyltin (MPhT) from sediments, which could not be done by a solvent extraction approach; the preservation of the structural integrity of the organotin compounds; and the reduction of the extraction time from several hours in the case of solvent extraction techniques to just 30 min. For spiked sediments, limits of the detection ranged from 0.7 to 2 ng/g of tin according to the compound. The relative standard deviations were found to be between 8 and 15%. The developed analytical procedure was validated using a reference material and was applied to various environmental samples. Schmidt et al.[146] carried out the extraction of arsenicals using PLE at different temperatures. The authors used 60 to 180°C as the working temperature and water as the extracting solvent. The recoveries of arsenicals at different temperatures are given in Figure 4.11, indicating higher recoveries at higher temperatures. Therefore, temperature is the most important experimental parameter for extraction. Maximum extraction can be achieved by varying the temperature and other experimental conditions.

Petrick et al.[147] developed a system and coupled it to a GC machine online. The extraction of the sample was performed in a liquid–liquid segment flow in a glass coil internally coated with a hydrophobic layer. The authors compared this system with the classical LLE method and reported 20% RSD for the classical LLE approach, while relative standard diviation (RSD) was found to be 10% in the online LLE system developed by them. The LLE technique is tedious, time consuming, and utilizes a large volume of costly solvents. The handling of solvents is hazardous from the health point of view. Emulsion formation in LLE is another problem of the pollutants extraction of more polar organometallic

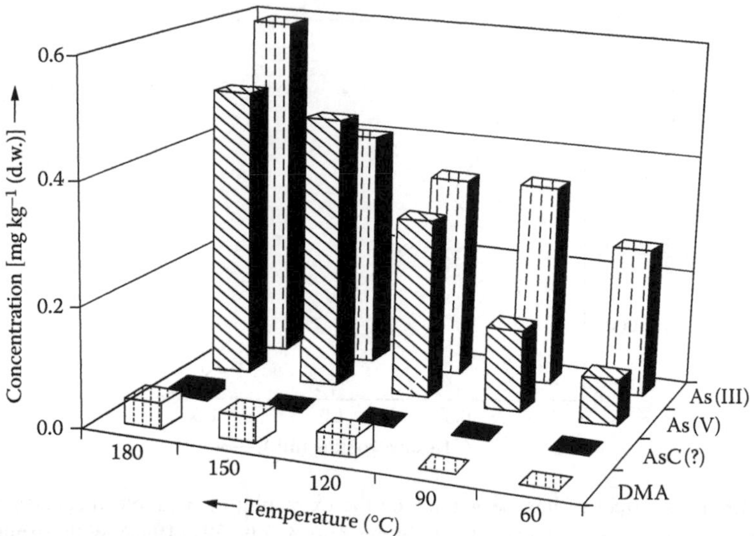

FIGURE 4.11 Extraction efficiencies for arsenicals at different temperatures in fresh and ground waters and plant materials (*Holcus lanatus)* using pressurized liquid extraction. (From Schmidt, A.C., Reiser, W., Mattusch, J., Popp, P., and Wennrich, R., *J. Chromatogr. A*, 889: 83 (2000). With permission.)

species.[148] The extraction of some metal ions by the liquid extraction methodology is summarized in Table 4.4.

4.5.2.2 Solid-Phase Extraction

On the other hand, the solid-phase extraction (SPE) technique, developed by Zhang et al.,[149,150] is free from the drawbacks of LLE and is very fast and sensitive, with the recovery of pollutants ranging from 90 to 95%. The use of SPE offers the advantages of convenience, cost savings, and minimal consumption of solvents[160]; hence, about 50% of chromatographers are using this method for sample preparation. Moreover, during the extraction of environmental samples, both biogenic and anthropogenic compounds are extracted into the organic and inorganic solvents. The complex matrix may seriously interfere with the determination of the respective analytes. Various columns, disks, and cartridges were used for extraction purposes, but cartridges are the most popular extraction devices.[160] Manufacturers developed new formats for traditional cartridges. Most SPE cartridges have medical-grade polypropylene syringe barrels with porous PTFE or metal frit that contain 40 μm d_p packings. The cartridges provide certain advantages over disks, as liquid flows faster in cartridges in comparison to disks. Most of the cartridges and disks are silica-based polymeric packings.

The important cartridges and disks are SS-401, XAD-2 (37), PL RD-S,[161] HDG-C_{18},[162] RDS-18,[163] C_{18} cartridge,[164] C_{18} silica-bonded cartridge,[165] SEP-Pak C_{18},[166] extraction fiber,[167] ODS impregnated polymers,[70] C_8 disk,[163] and C_{18} empore

TABLE 4.4
Solvent Extraction Procedures for Organotin Species

Species	Extraction	Ref.
Butyl- and phenyltin	HCl, pH 2–3, extraction with 0.25% tropolone in diethyl ether	151
Butyl- and phenyltin	NaDDC complexation, toluene-acetic acid (10:4) extraction	152
MBT and MPhT	Stirring with water-HCl (1:1), extraction with hexane	153
DBT, TBT, and TPhT	Reflux, 80°C, 30 min with HCl-MeOH (5:95)	154
Butyltin	Shaking overnight, sonication 30 min, extraction with pure acetic acid	155
Butyltin	Sonication at 30°C with 0.08% tropolone in MeOH, extraction with hexane	64
Butyltin	Extraction with 2 M HCl for 12 h	48
Butyltin	Concentrated HCl with concentrated HBr, extraction with 0.05% in pentane	156
Butyl- and phenyltin	Stiring with water–HBr mixture for 30 min, extraction with 0.04% tropolone in dichloromethane	157
Butyl- and phenyltin	Shaking with HBr, extraction with 0.07% tropolone in pentane	158
Methyl- and phenyltin	Sonication for 2 h with 6 M HCl at 50°C	159

Note: MBT: monobutyltin, DBT: dibutyltin, TBT: tributyltin, MPhT: monophenyltin, and TPhT: triphenyltin.

disk.[168,169] SPE involved the use of nonpolar C_8 and C_{18} phases in the form of cartridges and disks. Disks are preferred to cartridges, as the disks have high cross-sectional areas that provide several advantages, which are not possible in cartridges. With the disk, the decreased back pressure allows for a greater flow rate, while their wide bed decreases the chance of plugging. The embedding of the stationary phase into a disk prevents channeling and improves mass transfer.[170] Off-line or online SPE or solid-phase microextraction on various types of silica-bonded, polymeric, or carbon-type phases were used before GC application.[171] Extraction efficiency in SPE was optimized and increased by varying various parameters, such as pH, ionic strength of the sample, elution solvents, content of organic modifier in the sample, and elution gravity. Progressiveness of the technique was established using a polymer membrane containing the enmeshed sorbent particle in a web of polymer microfibrils called membrane extraction disks. A schematic diagram of a solid-phase extraction unit is shown in Figure 4.12.[172] This figure indicates a housing of a glass cartridge, and the bottom is covered with a glass fiber sheet.

Snell et al.[173] used solid-phase extraction for the extraction of mercury species. Similarly, Mothes and Wennrich[174] used a microextraction procedure for the extraction of mercury species from an aqueous sample. Furthermore, Mothes and Wennrich[175] compared the extraction concentrations of mercury on 100 μm polydimethyl siloxane (PDMS) and 75 μm carboxen/PDMS solid supports. The comparison of the extraction is shown in Figure 4.13, which indicates that the amount extracted was greater on 100 μm PDMS support in comparison to the amount

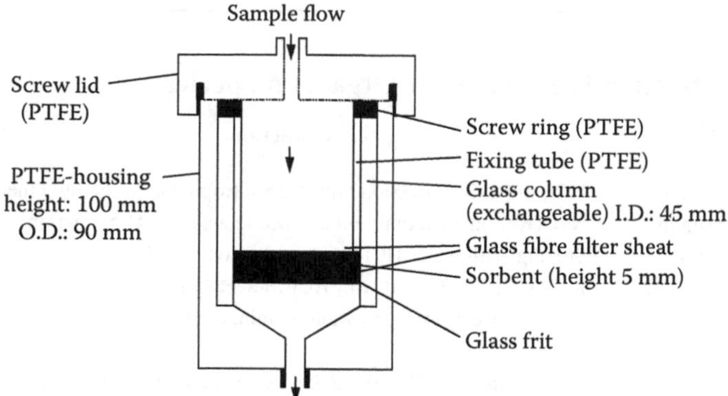

FIGURE 4.12 A solid-phase extraction unit. (From Weigel, S., Bester, K., and Hühnerfuss, H., *J. Chromatogr. A*, 912: 151 (2001). With permission.)

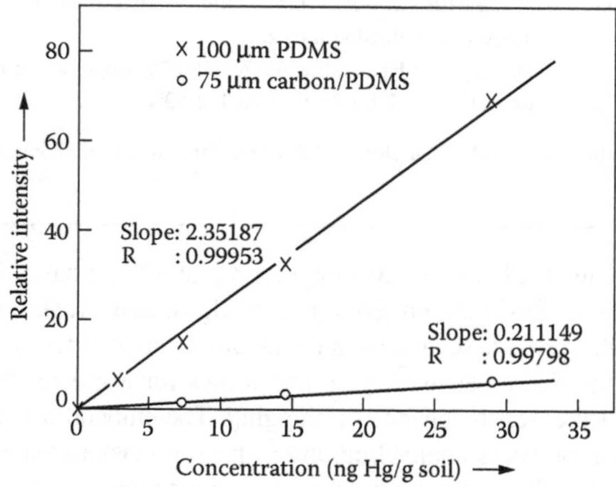

FIGURE 4.13 Effect of the solid-phase microextraction (SPME) fibers sizes on the extraction capabilities of diethyl mercury. (From Mothes, S., and Wennrich, R., *Mikrochim. Acta*, 135: 91 (2000). With permission.)

extracted on 75 μm carboxen/PDMS solid support. The authors reported that the amount extracted depends on the partition coefficient of the metal ions between the stationary and the aqueous phases. To achieve maximum extraction, a sufficient time (equilibrium time) should be applied. It is important to mention here that adsorptive-type PDMS are the most popular in speciation analysis. Mothes and Wennrich[175] equilibrated the arsenic, mercury, and lead metal ions on 100 μm PDMS solid support and reported the different equilibrium times (Figure 4.14). It is obvious from Figure 4.14 that 50 to 60 min are required for the achievement of equilibrium. It is worth mentioning here that the equilibrium time for metal ions is always greater than the equilibrium time for organic species. In addition, these authors reported the effects of temperature, pH, headspace mode

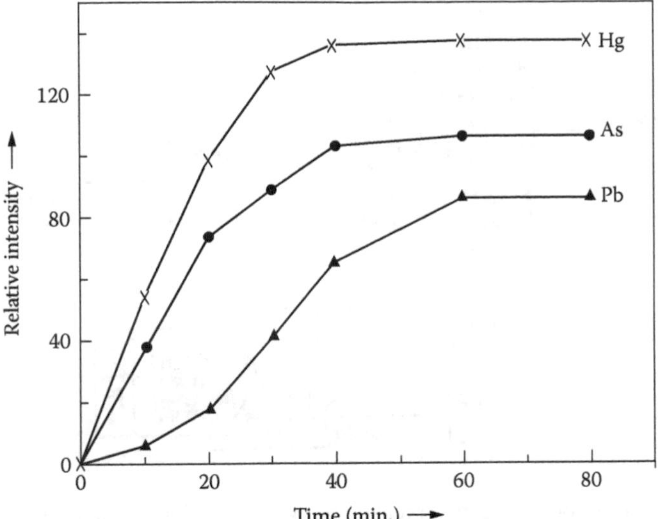

FIGURE 4.14 Effect of the exposure time on the extraction amount of mercury, arsenic, and lead metal ions using solid-phase microextraction (SPME). (From Mothes, S., and Wennrich, R., *Mikrochim. Acta*, 135: 91 (2000). With permission.)

of sampling, and storage of the loaded solid support on the extraction extent of the metal ions. Aguerre et al.[176] used solid-phase microextraction (SPME) for the speciation of monobutyl-, tributyl-, and triphenyltin species. The volatile or the semivolatile metallic species from aqueous or gaseous samples can be extracted directly by the microsolid phase extraction technique. Nerin et al.[177] used the packed columns of flurosil and silica gel, at 3% and 10% and deactivated alumina at 10% deactivated, amberlite XAD-2 and XAD-7, and Chromosorb GC for the extraction of organollead species. Not much work has been carried out on the solid-phase extraction of the metal ion species; therefore, only silica-based cartridges are popular in the literature.

Whang and Pawliszyn[178] designed an interface that enables the SPME fiber to be inserted directly into the injection end of a CE capillary. They prepared a semi-"custom-made" polyacrylate fiber to reach the SPME–CE interface. Similarly, Huen and coworkers[179] developed SPE coupling with HPLC. Brinkman[180] has been a leader in applying column switching to the real-world problem, and the author extended his work on the extraction of polar pollutants in river water using SPE coupling with liquid and gas chromatographic machines separately. van der Hoff et al.[181] described automatic sample preparation with extraction column (ASPEC) coupled with a gas chromatographic (GC) system. The assembly developed by them is shown in Figure 4.15. The authors used a loop-type interface for coupling the ASPEC with the GC machine. Yin et al.[182] reported solid-phase extraction preconcentration incorporated in a flow injection (FI) system, high-performance liquid chromatography (HPLC), and separation for speciation of MeHg, EtHg, PhHg, and Hg(II) species.

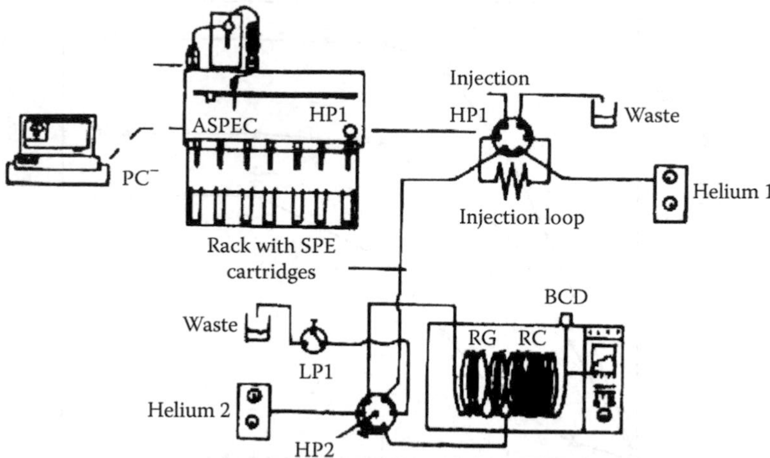

FIGURE 4.15 Automatic sample preparation with extraction column–gas chromatography–electron capture detector (ASPEC-GC-ECD) equipment: HP1 and HP2: high-pressure six-way valves, LP11: low-pressure three-way valves, RG: retention gap, RC: retention column, T1: press-fit connection between RG and RC, and T2: three-way, press-fit connection for solvent vapor exit. (From van der Hoff, G.R., Gort, S.M., Baumann, R.A., van Zoon, P., and Brinkmann, U.A.Th., *J. High Resolut. Chromatogr.*, 14: 465 (1991).

Many modifications were made in SPE, and the new version of SPE emerged as SPME, and several papers appeared on this issue in the literature.[179] In automatic SPME, a polymer-coated fiber is used instead of a GC autosampler needle, and it is then dipped into a solution of water or held in the headspace of the sample to enable the analytes of interest to diffuse into the coating. After establishing equilibrium, the system withdraws the needle and places it in the heated GC injection port, and the analytes are thermally desorbed on the GC column. Haddad and Jackson[183] proposed a SFE setup as shown in Figure 4.16. Two high-pressure switching valves and a single programmable pump with a low-pressure solvent selection valve are configured. A measured volume of the sample is loaded onto the concentrator column at a precise flow rate, with the eluent directed to waste. A wash setup can then be introduced without losing bound analytes by switching the low-pressure solvent select valve to the eluent flow path for a precisely measured time. This partially reequilibrates the concentrator column with the eluent to help minimize baseline disturbance. The valves are then switched to back-flush the analyte ions from the concentrator column and onto the analytical column. This two-valve configuration offers the advantages of unlimited and precise control over the volumes of eluent for the washing and back-flushing steps, allowing the possibility of tailoring the preconcentration procedure for different samples. Barcelo[184] reviewed the status of SPE as the extraction methodology and described it as a suitable technology for the extraction of pollutants from water samples.

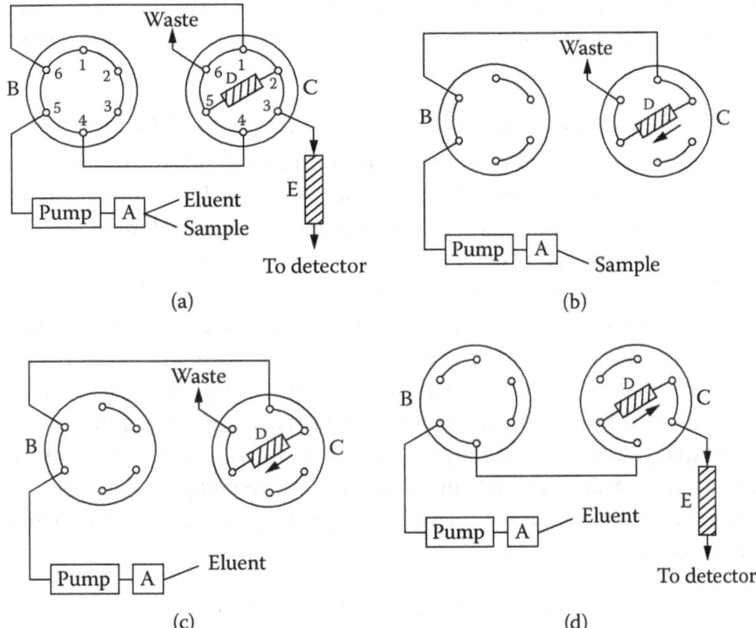

FIGURE 4.16 Solid-phase extraction (SPE) apparatus of sample preparation: (A) single pump with a low-pressure selection valve, (B and C) two high-pressure switching valves, (D) a concentration column, (E) an analytical column, (b) sample loading, (c) concentrator column washing, and (d) sample stripping. (From Haddad, P.R., Doble, P., and Macka, M., *J. Chromatogr. A*, 856: 145 (1999). With permission.)

4.5.2.3 Liquid Chromatographic Extraction

During recent years, liquid chromatography, that is, column chromatography (CC), high-performance liquid chromatography (HPLC), supercritical fluid chromatography (SFC), and gel permeation chromatography (GPC), received widespread attention for the extraction of inorganic and organic pollutants from environmental samples. The popularity of the chromatographic techniques in the extraction of the pollutants is due to many developments in this technology, which can be used for the extraction of a wide range of pollutants. Moreover, the fast speed and low consumption of costly solvents are the other assets of chromatographic methodologies. The combination of the chromatographic instrument with other analytical techniques also increased the attractiveness of this chromatographic modality.

4.5.2.3.1 Column Chromatography

Column chromatography is a classical modality of liquid chromatography and still exists due to its ease of operation and inexpensiveness. It has been used extensively for the extraction and purification of environmental and biological samples. Different types of adsorbents have been used in column chromatography, for example, silica, alumina, and Florisil®. This modality was replaced by high-pressure chromatographic technologies, and its use in sample preparation is decreasing.

4.5.2.3.2 High-Performance Liquid Chromatography (HPLC)

HPLC has been used in the extraction and purification of pollutants in environmental and biological samples. Many applications are due to the availability of various stationary phases, particularly reversed-phase columns of different dimensions. HPLC instruments used for this purpose are similar to those of the analytical machine with the use of different columns, which depends on the nature of the pollutants to be extracted and purified. Many reports are available in the literature dealing with the use of HPLC as the extraction and purification technique for metal ions in the environmental and biological samples. However, HPLC could not achieve a reputation as a universally used and applicable extraction technique due to its low range as the preparative chromatography. This is because of the insufficient advancement of the technique; therefore, further advancement is required in the technique, especially for the development of the preparative columns.

4.5.2.3.3 Supercritical Fluid Chromatography (SFC)

Supercritical fluid chromatography (SFC) is the latest development in liquid chromatographic technologies for extraction purposes (extraction close to 100%). SFC is a viable sample preparation strategy for use with liquid samples, especially for selective applications. A great diffusion coefficient, low viscosity, and the absence of surface tension of the supercritical fluid make this technology ideal for metal ions extraction. Fortunately, analysts have several new sample preparation strategies to choose from to accomplish an analytical objective. The technical and instrumental parts discussion of this methodology was presented in Section 4.5.1.6. Hawthorne[185] reviewed the status of this technology in sample preparation. Similarly, Karlsson et al.[186] reported a wide variety of applications of SFC for environmental, food, and biological samples. Barnabas et al.[187] described the main components of Carlo Erba SFC technology (Figure 4.17), which contains a carbon dioxide gas cylinder, syringe pump, oven, cartridge, and collection vial.

Emteborg et al.[188] reported the use of SFC for the extraction of methylmercury from a sediment sample. Bayona et al.[189] used the SFC method for the extraction of butyltins. Very little work was carried out on the extraction of metal ions using the SFC technique. This is due to its high cost, incapability to conduct simultaneous and parallel extraction, and good solubility of water in supercritical carbon dioxide, which make it unfit for routine applications.[190] Another problem in SFC extraction is the optimization of the experimental conditions by optimizing the temperature, pressure, amount and type of modifier, extraction time, and cell volume. Besides, SFC columns are not currently available as disposal cartridges. Moreover, SFC has a low range of application with poor reproducibility.[191] However, the coupling of SFC with SPE resulted in the fast and sensitive extraction

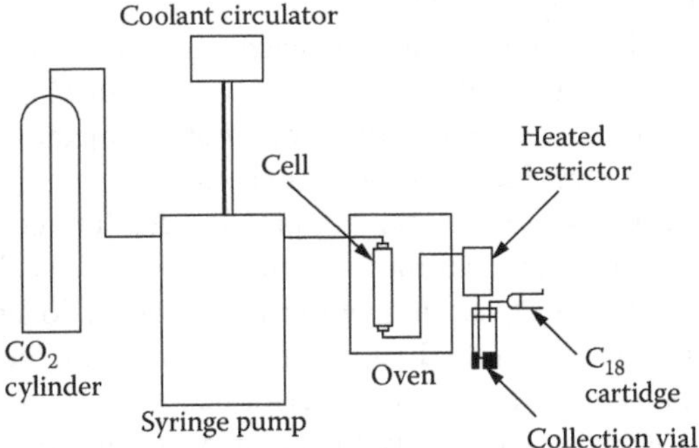

FIGURE 4.17 The main components of the Carlo Erba supercritical fluid extraction (SFE) system. (From Barnabas, I.J., Dean, J.R., Hitchen, S.M., and Owen, S.P., *Anal. Chim. Acta*, 291: 261 (1994). With permission.)

of many organic compounds with good recoveries. Recently, Kumar et al.[192] used SFE for the extraction of tributyltin and triphenyltin. The extraction conditions with carbon dioxide as the supercritical fluid (methanol modifier used) were optimized for organotins species. The total extraction time was found to be approximately 15 min. The recovery studies at optimal conditions showed a recovery of 44% for tributyltin and 23% for triphenyltin. The reproducibilities for both compounds extracted were within 2% relative standard deviation (RSD). The optimum conditions obtained were also used to extract tributyltin and triphenyltin.

4.5.2.3.4 Gel Permeation Chromatography (GPC)

The working principle of gel permeation chromatography is based on the size of the pollutants. The working column is formed of polymer beads, and the pollutants are separated by a filtration mechanism. The large-size pollutants are eluted first and then the small sizes. The smaller pollutants enter into the beads, while the larger-size pollutants pass through intraparticle spaces of the beads; hence, the latter are eluted first. The use of gel permeation chromatography (GPC) in the cleanup process is limited, as it is not capable of cleaning up metallic species due to their similar sizes.

Dabek-Zlotorzynska et al.[193] reviewed the sample pretreatment methodologies for environmental analysis. The authors also described the above-mentioned methods as the sample pretreatment methods. Some reviews on this topic appeared in the literature in the last few years.[4,194–196]

4.6 MEMBRANE METHODS IN SAMPLE PREPARATION

Membrane filtration is known as one of the most important and new developments in sample preparation for metal ion speciation. Many papers were written on this topic for the seventh symposium on the handling of environmental and biological samples in chromatography, held May 7–10, 1995, at Lund, Sweden. Electromigration injection and stacking/transient isotachophoresis (ITP) are sample handling methods that are relatively easy to combine with chromatographic and capillary electrophoretic techniques. Electromigration injection appears to be a straightforward technique but is known to suffer severe matrix dependence effects,[197] and, hence, its use for real samples is very limited. Although many workers have used this for preconcentration in organic analysis,[198] the robustness of this method for samples of varying compositions is questionable. The stacking of ions in a hydrostatically injected large sample plug is another method mostly used for sample preconcentration and is performed in CE separations. Applications in inorganic analysis are mostly in the area of preconcentration of dilute samples (typically water) having much lower conductivity than the electrolyte by exploiting the field amplification effect.[199,200] Dialysis involves the diffusion of species across a membrane, and the technique can be classified into two categories — passive dialysis (in which molecules of a certain mass are transferred across a membrane) and active (Donnan) dialysis (in which ions of specified charges are transferred across a membrane). Active dialysis is commonly employed in ion chromatography and is useful for the cleanup of samples at extreme pH. In Donnan dialysis, the sample solution is kept on one side, while the other side of the membrane contains acidic solution. Electrically driven dialysis is called electrodialysis and has been used for the off-line analysis of strongly alkaline samples containing trace amounts of common inorganic ions.[201]

Buttler et al.[202] studied various microdialysis probes used for bioprocesses monitoring. The dialysis factor of the probe was determined by several parameters, including membrane area, physicochemical properties of the membrane, flow rate of the perfusion liquid, diffusion coefficients of the analytes, and various interactions between the analytes and membrane. Kuban and Karlberg[203] reported an online dialysis/FIA sample cleanup procedure for metal ions by CE (Figure 4.18). Several reports were published on the dialysis and electrodialysis of sample cleanup prior to loading onto the CE machine.[204–207] Okamoto et al.[208] used electrodialysis to neutralize highly acidic solution prior to the determination of Mg(I) and Ca(II). The authors used an electrodialysis device consisting of a Tosoh TASN-80 anion selective membrane tube (40 cm × 1 mm, ID). Blanty et al.[209] coupled capillary isotachophoresis (CITP) with capillary electrophoresis (CE) and reported the separation of Fe(III) in water samples. The purpose of CITP was to enable a large sample volume to be used and to permit a sharp zone of Fe(III) to be formed and transferred to CE capillary for separation. A general remark on the applicability of the CITP-CE combination for cleanup is that the mobility of the analytes should differ from the matrix ions to be removed, and for such cases, some new applications in inorganic analysis can be explored. Majors[210] described the sample preparation for the environmental analysis using liquid membranes and microdialysis techniques. Recently, Chimuka[211] reviewed the sample preparation in speciation analysis in biological and environmental samples using membrane technologies. The authors advocated its suitability for

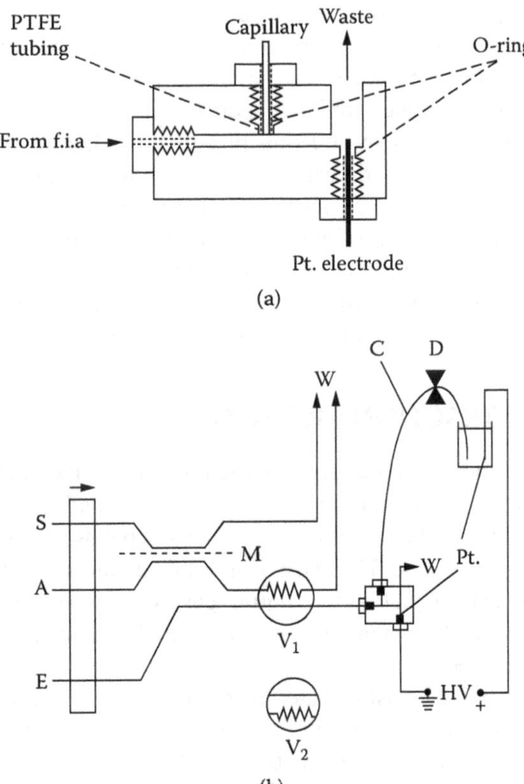

FIGURE 4.18 The setup of the online dialysis (sample purification) with capillary electrophoresis (CE): (a) cross-sectional view of the FIA-CE interface and (b) the FIA-CE system, A = acceptor stream, C = capillary, HV = high voltage, D = UV detector, E = electrolyte, M = dialysis membrane, Pt = platinum electrode, S = sample, V_1 = injection valve in filling position, V_2 = injection valve in inject position, and W = waste. (From Kuban, P., and Karlberg, B., *Anal. Chem.*, 69: 1169 (1997). With permission.)

the extraction of several metal ions such as chromium, copper, lead, cadmium, manganese, and uranium.

4.7 ENZYMATIC HYDROLYSIS

Enzymatic methods can also be used for the release of metal ions from biomolecules such as proteins, vitamin B_{12}, and hemoglobin. The main enzymes used for this purpose are protease, lipase, trypsin, pepsin, pronase, or their mixtures. Protease enzymes are used to liberate amino acids and metal ions from protein molecules. Enzymatic hydrolysis is becoming a promising tool in metal ion speciation, especially for organolead and organotin compounds.[212–218] This method has been used by Gilon et al.[219] for selenium application using protease that breaks the peptidic bonds of the proteins present in the material at pH 7.5 (phosphate citric buffer) and at incubation at 37°C for 24 h. Similar procedures were used for organoselenium[220] and organotin[212–219,221] compounds using a lipase–protease mixture under similar experimental conditions. Pannier et al.[217] used a mixture of enzymes in aqueous phosphate buffer to release organotin compounds from marine biological samples. Pronase, a protease with broad specificity, was used to isolate the levels of selenium in serum, which is digested with the enzyme at 37°C overnight.[220] Tin and selenium can be released by enzymatic digestion in several biological samples.[217,222] Several other reports are available in the literature indicating the convenience of enzymatic hydrolysis for the release of metal ions from solid and liquid biological samples.[222,223]

4.8 OTHER EXTRACTION METHODS

In addition to the above-cited extraction methods, other technologies were also used for the extraction of metal ion species from solid and liquid biological and environmental samples. The use of these methods depends on the nature of metal ions and the sample matrix. The most important methods in this category are aqueous extraction, steam distillation, liquid–gas extraction (purge and trap), field flow fractionation (FFF), sequential extraction, and derivatization. These methods are discussed briefly below.

Metal ions are extracted using water or mixtures of water and organic solvents and water and inorganic acids in an aqueous extraction method. For example, arsenic can be extracted easily using methanol (50%) and hydrochloric acid (10%) in the environmental and biological samples with ultrasound sonication.[224] Steam distillation was also used for the extraction of metal ions. It was used for the separation of methylmercury from water samples.[225–227] It was also found to give better recoveries than the other extraction methods. It is a slow technique, but the addition of ammonium pyrrolidine dithiocarbamate (APDC) catalyzed the speed of the reaction.[228]

In a liquid–gas extraction technique, inert gases such as nitrogen or helium are bubbled into aqueous solution. The nonpolar volatile species are trapped into gas and are then analyzed by an analytical technique. This technique is useful for the extraction of tetraalkyllead, tetramethyltin, and dimethylmercury. In FFF, separation is achieved by applying an external force. The force applied compels the molecules to move toward the wall of a channel. Hence, an exponential concentration distribution is built up. If the force affecting the molecules is constant, the thickness of the resulting concentration distribution depends solely on the diffusion coefficient of the molecule. In this way, slowly diffusing large molecules lie closer to the wall than smaller molecules. Similarly, the difference gradient of the molecules can be obtained by thermal FFF when applying different temperature gradients. FFF using semipermeable ceramic frit can assist in differentiating the molecules. These methods have not been used for speciation purposes, but they have good potential for future applications.

Sequential extraction is a very useful technique for speciation in which extraction is carried out by multiple extraction methods. The extractions of various types are applied successively to the sample, each successive treatment being more drastic. It is controlled by the chemical properties of leaching solutions, individual steps, nature of the matrix, heterogeneity, and so forth. The fractions of sequential extraction may be mobile exchangeable elements, elements bound to carbonates, elements bound to easily reducible fractions, elements bound to easily extractable organics, elements bound to moderately reducible oxides, elements bound to oxides and sulfides fractions, and elements bound to silicates. This technique was used successfully for the extraction of metal ions from aerosol, soil, sediment, and fly ash.[229] Recently, van Staden and Stefan[230] reviewed the instrumentation part of sequential injection analysis for metal ion speciation.

Gas chromatography is used to analyze only volatile metal species; therefore, it is essential to derivatize inorganic metallic species into volatile ones. Many methods are available for this purpose, including hydride generation (HG), cold vapors, ethylation, Grignard reaction, and other methods. Many metal ions, such as arsenic, antimony, bismuth, germanium, lead, selenium, tellurium, and tin, react with sodium borohydride and form their hydrides. Immobilized borohydrides have been used for arsenic and selenium in water.[231,232] Cold vapor is similar to hydride generation with the difference that the volatile product results from the reduction reaction. It is useful for the derivatization of cadmium and mercury. Puk and Weber[233] and Tseng et al.[87] used this technique for the derivatization of methylmercury in several biological samples. Ethylation of metal ions is carried out by using sodium tetrahydroborate, and it is useful for the derivatization of selenium, tin, mercury, and lead. Grignard reagent is very useful for the derivatization of Sn, Hg, and Pb metal ions. Other possible formations of volatile chelates include trifluoroacetylacetonates and dithiocarbamates.

4.9 CLEANUP

Natural environmental and biological samples contain thousands of other substances as impurities, and these get coextracted with metal ions of interest. Due to the similar properties of the coextractives, they usually interfere in the analysis of pollutants in any analytical technology. Therefore, generally, a cleanup procedure is required only for those samples that are not clear. Mostly blood, serum, food, plant extract, and industrial and municipal effluents samples require a cleanup procedure. If cleanup is required, the recovery of each pollutant is very important, and it should be at about 85%.[234] Cleanup of the extracted metal ions may be carried out by column chromatography, gel permeation chromatography, sweep codistillation, liquid–liquid partition, cartridges, and disks. The columns used for this purpose are nuchar carbon, silica, XAD-2, alumina, and flurosil, while the cartridges and disks used are C_8, C_{18}, and so forth. The use of membrane disks for cleanup increased significantly in

TABLE 4.5
Some Cleanup Methods for Metal Ions

Method	Coextractive	Nature
Partitioning		
Liquid–liquid partitioning	Fats, waxes, lipids, pigments, and some polar compounds	General
Partitioning adsorption column	Fats, phospholipids, waxes, and other polar substances	General
Gel permeation chromatography	Fats and lipids	General
Liquid–liquid adsorption chromatography (adsorption column)	Fats, phospholipids, pigments, and polar impurities	General
Florosil		
Alumina		
Silica gel		
Nuchar carbon		
Chemical cleanup	Fats and neutral specific molecules	—
Acidic		
Alkaline		
Acid-base		
Sweep codistillation	Comparatively nonvolatile impurities	—

the past few years. SPE membrane disks may also be used in conjunction with SFE in the preconcentration of metal ions. Online SPE may be coupled to a membrane disk for the cleanup of the metal ions species in water and other liquid biological samples. Using a SFC/SPE methodology, greater recoveries of metal ions may be achieved. The use of gel permeation chromatography (GPC) in the cleanup process is limited, as it is not capable of cleaning up the small metal ions due to their similar sizes. Some cleanup methods for metal ions are given in Table 4.5.

4.10 PRECONCENTRATION

Generally, the level of metal ions in environmental and biological samples is below the detection limits of modern detectors; therefore, preconcentration of the extracted samples is required. Normally, the volume of the extracted solvent containing metal ions is reduced to 0.5 to 1.0 mL. The classical approach of evaporation of the solvents is used to concentrate the extracted metal ions. It is recommended that nitric acid be added to the collected water samples (1 mL/L) before starting evaporation. However, some devices, such as purge-and-trap devices and cryogenically cooled capillary taps, have been used for preconcentration purposes.[235] Solvents with low boiling points are recommended for preconcentration, but the selection of the solvent depends on several factors as already discussed. The large volume of the solvent obtained after extraction is evaporated by a distillation unit. Open evaporation of extracts on a water bath may cause severe losses of the metal ions; therefore, evaporation should be carried out at a low temperature under reduced pressure. Normally, a rotary flash evaporator or a Kuderna-Danish assembly is used for this purpose. Sometimes, pure nitrogen is also allowed to pass through the evaporator assembly so that the oxidation of the metal ions may be checked. The loss of metal ions sometimes occurs during evaporation. To check this, some keeper solvent (toluene or isooctane — a solvent that has a high boiling point) is added to the extract before starting the evaporation. In addition to this conventional method, other techniques such as chromatography and SPE may also be used for preconcentration purposes. Amalgam formation, cold trap, high-temperature trap, and so forth, were also used for preconcentration purposes.

For preconcentration, a microcolumn of bonded silica with octadecyl functional groups (C_{18} reversed-phase material) was used as a sorbent for mercury complexes, formed online with ammonium pyrrolidine dithiocarbamate.[182] Retained mercury species were eluted with a methanol–acetonitrile–water mixture and subjected to separation on an octadecylsilane (ODS) column. In order to perform online measurements, the preconcentration microcolumn was mounted in a pressure-tight casing. The limits of detection for MeHg, EtHg, PhHg, and Hg(II), employing a sample volume of 58.5 mL, were 9, 6, 10, and 5 ng/L, respectively. The RSDs calculated from nine repeated measurements were found to be 3.6%, 5.5%, 10.4%, and 7.6% for MeHg, EtHg, PhHg, and Hg(II), respectively. Fritz, et al.[236] discussed the use of ion-exchange and chelating resins to preconcentrate the metal ions samples prior to CE application. Nerin and Pons[237] described the percentage recoveries of organolead species with different volumes of hexane after evaporation. The results are shown in Figure 4.19. It may be seen from this figure that a 3 to 4 mL volume of hexane gave the maximum percentage of recoveries of organolead species. Because the extraction step in metal ion speciation is very important, possible chances of error should be minimized by the spiking experiments. To minimize possible errors, the scientists in the speciation meetings[238–240] encouraged the spiking experiments. Giokas et al.[241] used cloud point extraction (CPE) for the extraction of iron, cobalt, and nickel metal ions. Recently, Giokas et al.[242] used CPE and low-temperature directed crystallization (LTDC) for preconcentration of Cd, Pb, Cr, Cu, Zn, Ni, and Fe in biological and environmental samples.

During extraction procedures, especially in the LLE methodology, some invisible moisture obviously remains in the samples along with organic solvents that may again create problems in the detection of the metal ions. Therefore, the moisture should be removed from the sample prior to its loading onto the analytical machine — especially for GC. Normally, the moisture from the organic solvent is removed by the addition of high-purity anhydrous sodium sulfate. Sometimes, the extracted and concentrated solvent is passed through a glass column containing anhydrous sodium sulfate. Preconcentrations of some metal ions are presented in Table 4.6.

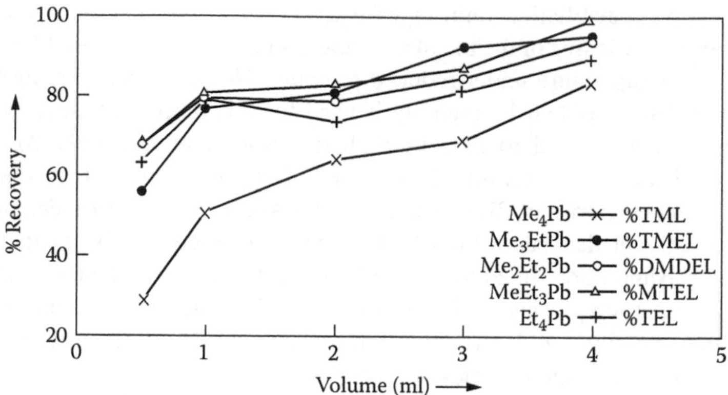

FIGURE 4.19 Effect of volume of hexane, after evaporation, on the percent recoveries of organolead species. (From Nerin, C., and Pons, B., *Appl. Organomet. Chem.*, 8: 607 (1994).)

TABLE 4.6
Preconcentrations of Metal Ions

Sample	Species	Column	Recovery (%)	Detection Limit	Ref.
River and mineral water	Transition metals	Iminodiacetate resin	80–100 resin	0.01–0.1 ppb	243
Mine wastewater and beverage	Cu(II), Al(III), and Fe(III)	Nucleosil C_{18}	—	5–40 ppb	244
Seawater	Zn, Cu, Ni, Co, and Mn	LiChrospher 100 RP-18	72–95	15–150 ppb	245
Natural waters	Organic Pb and Hg	Nucleosil C_{18}	75–90	—	246
Natural waters	Triorganotin compounds	Baker C8, C18, and phenyl Amberlite XAD-2 and XAD-4	20–100	2 ppb	247

4.11 CONCLUSION

Sample preparation in metal ion speciation is a very important issue in ensuring rapid, reliable, and reproducible results. The sampling should be carried out in such a way as to avoid changes in the chemical composition of metal ions. The quantitative values of metal ion speciation depend on the strategy adopted in sample preparation. Extraction recoveries may vary from one species to another of the same metal to another, and they should, consequently, be assessed independently for each compound as well as for the compounds together. A compromise has to be found between a good recovery and preservation of the speciation. For a detailed evaluation of extraction methods, materials with an integral analyte, that is, those bound to the matrix in the same way as the unknown, which is preferably labeled (radioactive labeling), would be necessary. As such materials are not available, the extraction method used should be validated by other independent methods. Briefly, to obtain a maximum recovery of metal ions, sample preparation should be handled very carefully. Online methods are also available for the extraction and purification of environmental and biological samples. The online coupling of the extraction and purification methods avoids possibilities for contamination. Moreover, time and costly chemicals can be saved by using online extraction technologies. However, the recovery problem still exists, and it requires more advanced modifications in these technologies.

REFERENCES

1. Li K, Li FY, *Analyst*. 120: 361 (1995).
2. Timerbaev AR, *Talanta*. 52: 573 (2000).
3. Quevauviller Ph, *J. Chromatogr. A*., 750: 25 (1996).
4. Haddad PR, Doble P, Macka M, *J. Chromatogr. A*, 856: 145 (1999).
5. Gomez-Ariza JL, Morales E, Giraldez I, Sanchez-Rodas D, Velasco A, *J. Chromatogr. A*, 938: 211 (2001).
6. Liu W, Lee HK, *J. Chromatogr. A*, 834: 45 (1999).
7. Timerbaev AR, Buchberger W, *J. Chromatogr. A*, 834: 117 (1999).
8. Gebauer P, Thormann W, Bocek P, *Electrophoresis*, 16: 2039 (1995).
9. Brumley WC, *J. Chromatogr. Sci*., 33: 670 (1995).
10. Krivankova L, Bocek P, J. *Chromatogr. B*, 689: 13 (1997).
11. Mazereeuw M, Tjaden UR, Reinhoud NJ, *J. Chromatogr. Sci*., 33: 686 (1995).
12. Valcarcel M, Rios A, Arce L, *Crit. Rev. Anal. Chem*., 28: 63 (1998).
13. Timerbaev AR, *Electrophoresis*, 23: 3884 (2002).
14. Majors RE, *LC-GC*, 13: 555 (1995).
15. Evans GF, Lumplein TA, Smith DL, Somerville MC, *J. Air Waste Manag. Assoc*., 42: 1319 (1992).
16. Broekaert JAC, Gücer S, Adams F (Eds.), *Metal Speciation in the Environment*, Springer, Berlin, p. 409 (1990).
17. Schweigkofler M, Niessner R, *Environ. Sci. & Technol*., 33: 3680 (1999).
18. Boiano JM, Wallace ME, Sieber WK, Groff JH, Wang J, Ashley K, *J. Environ. Monit*., 2: 329 (2000).
19. Haas K, Feldmann J, *Anal. Chem*., 72: 4205 (2000).
20. Clark RC, Blumer M, Raymond OS, *Deep Sea Res*., 14: 125 (1967).
21. Gaul H, Ziebarth U, *Dtsch. Hydrogr. Z*, 36: 191 (1983).
22. Sauer TC, Durrell GS, Brown JS, Redford D, Boehm PD, *Mar. Chem*., 27: 235 (1989).
23. Eaton AD, Clesceri LS, Greenberg AE (Eds.), *Standard Methods for the Examination of Water and Wastewater*, 19th ed., American Public Health Association, Washington, DC (1995).
24. Roinestad KS, Louis JB, Rosen JD, *J. AOAC Int*., 76: 1121 (1993).
25. Templeton DM, Ariese F, Cornelis R, Danielsson LG, Muntau H, van Leeuwen HP, Lobiniski R, *Pure Appl. Chem*., 72: 1453 (2000).

26. Kateman G, *Chemometrics: Sampling Strategies*, Springer, Berlin (1987).
27. Guy MP, *Sampling of Heterogeneous and Dynamic Material Systems*, Elsevier, Amsterdam (1992).
28. Einax JW, Zwanziger HW, Geiss S, *Chemometrics in Environmental Analysis*, VCH, Weinheim (1997).
29. Nygren O, Nilsson C, *J. Anal. At. Spectrom.*, 2: 805 (1987).
30. Official Control of Foodstuffs Directives, Council Directive 85/591/EEC, Off J L 372: 0050-0052 (1985).
31. Official Control of Foodstuffs Directives, Council Directive 87/524/EEC, Off J L 306: 0024-0031 (1987).
32. Official Control of Foodstuffs Directives, Council Directive 2001/22/EEC, Off J L 077: 0014-0021 (2001).
33. Feldmann J, Lai VWM, Cullen R, Ma M, Lu X, Le XC, *Clin. Chem.*, 45: 1988 (1999).
34. Gallagher PA, Schwegel CA, Parks A, Gamble BM, Wymer L, Creed JT, *Environ. Sci. & Technol.*, 38: 2919 (2004).
35. Emons H, Schladot JD, Schwuger MJ, *Chemosphere*, 34: 1875 (1997).
36. Weigel S, Bester K, Hühnerfuss H, *J. Chromatogr. A*, 912: 151 (2001).
37. Arce L, Rios A, Valcarcel M, *J. Chromatogr. A*, 791: 279 (1997).
38. Arce L, Rios A, Valcarcel M, *Fresenius' J. Anal. Chem.*, 360: 697 (1998).
39. Westöö G, *Acta Chem. Scand.*, 20: 2131 (1996).
40. Westöö G, *Acta Chem. Scand.*, 21: 1790 (1997).
41. Cardellicchio N, Giandomenico S, Decataldo A, Di Leo A, *Fresenius' J. Anal. Chem.*, 369: 510 (2001).
42. Dirkx WMR, de la Calle MB, Ceulemans M, Adams FC, *J. Chromatogr. A*, 683: 51 (1994).
43. Jiang GB, Ceulemans M, Adams FC, *J. Chromatogr. A*, 727: 119 (1996).
44. Tutschku S, Mothes S, Dittrich K, *J. Chromatogr. A*, 683: 269 (1994).
45. Bettmer J, Bradter M, Buscher W, Erber D, Rieping D, Cammann K, *Appl. Organo. Chem.*, 9: 541 (1995).
46. Välimäki I, Perämäki P, *Mikrochim. Acta*, 137: 191 (2001).
47. Francois R, Weber JH, *Mar. Chem.*, 25: 279 (1988).
48. Rapsomanikis S, Harrison RM, *Appl. Organomet. Chem.*, 2: 151 (1988).
49. Randall L, Han JS, Weber JH, *Environ. Technol. Lett.*, 7: 571 (1986).
50. Shawky S, Emons H, Dürbeck HW, *Anal. Commun.*, 33: 107 (1996).
51. Dezauziers V, Leguille F, Lavigne R, Astruc M, Pinel R, *Appl. Organomet. Chem.*, 3: 469 (1989).
52. Müller MD, *Fresenius' J. Anal. Chem.*, 317: 32 (1986).
53. Bressa G, Cima F, Fonti P, Sisti E, *Fresenius' J. Anal. Chem. Bull.*, 6: 16 (1997).
54. Pannier F, Astruc A, Astruc M, *Appl. Organomet. Chem.*, 8: 595 (1994).
55. Cai Y, Rapsomanikis S, Andreae MO, *Talanta*, 41: 589 (1994).
56. Donard OFX, Lalere B, Martin F, Lobiniski R, *Anal. Chem.*, 67: 4250 (1995).
57. Lalere B, Szpunar J, Budzinski H, Garrigues, Ph, Donard OFX, *Analyst*, 120: 2665 (1995).
58. Quevauviller Ph, Chiavarini S, Cremisini C, Morabito R, Bianchi M, Muntau H, *Mikrochim. Acta*, 120: 281 (1995).
59. Milde D, Plzak Z, Suchanek M, Collect. *Czech. Chem. Commun.*, 62: 1403 (1997).
60. Cai Y, Rapsomanikis S, Andreae MO, *J. Anal. At. Spectrom.*, 8: 119 (1993).

61. Quevauviller Ph, Donard OXF, *Fresenius' J. Anal. Chem.*, 339: 6 (1991).
62. Meinema HA, Burger-Wiersma T, Versluis-de Haan G, Gevers EC, *Environ. Sci. Technol.*, 12: 288 (1978).
63. Gomez-Ariza JL, Morales M, Ruiz-Benitez M, *Appl. Organomet. Chem.*, 6: 279 (1992).
64. Gomez-Ariza JL, Morales E, Beltran R, Giraldez I, Ruiz-Benitez M, *Analyst*, 120: 1171 (1995).
65. Sasaki K, Ishizaka T, Siziki T, Saito Y, *J. Assoc. Off. Anal. Chem.*, 71: 360 (1988).
66. Rivas C, Ebdon L, Evans EH, Hill SJ, *Appl. Organomet. Chem.*, 10: 61 (1996).
67. Chau YK, Wong PTS, Bengert GA, *Anal. Chem.*, 54: 246 (1982).
68. Chau YK, Yang F, Maguire RJ, *Anal. Chim. Acta*, 320: 165 (1996).
69. Müller MD, *Anal. Chem.*, 59: 617 (1987).
70. Heisterkamp M, DeSmaele T, Candelone JP, Noens L, Dams R, Adams FC, *J. Anal. At. Spectrom.*, 12: 1077 (1997).
71. Estes SA, Uden PC, Barnes R, *Anal. Chem.*, 53: 1336 (1981).
72. Chakrabarti D, DeJonghe WRA, VanMol WE, VanCleuvenbergen RJA, Adams FC, *Anal. Chem.*, 56: 2692 (1984).
73. Wang J, Ashley K, Marlow D, *Anal. Chem.*, 71: 1027 (1999).
74. Ashley K, Andrews RN, Cavazos L, Demange M, *J. Anal. At. Spectrom.*, 16: 1147 (2001).
75. LeBlanc C, *LC-GC*, 17: S30 (1999).
76. Schmitt VO, Szpunor J, Donard OFX, Lobiniski R, *Can. J. Anal. Sci. Spectros.*, 42: 42 (1997).
77. Lobiniski R, Dirks WMR, Ceulemans M, Adams FC, *Anal. Chem.*, 64: 159-165 (1992).
78. Schmitt VO, de Diego A, Cosnier A, Tseng CM, Moreau J, Dnard OFX, *Spectroscopy*, 13: 99 (1997).
79. Ganzler K, Salgo A, Valko K, *J. Chromatogr. A*, 371: 299 (1986).
80. Freitag W, Johan O, *Die Angew Makromol. Chem.*, 175: 181 (1990).
81. Pare J, Lapointe J, Sigouin M, U.S. Patent 5,002,784 (1991).
82. Stout S, Dacunha A, Safarpour M, *J. AOAC Int.*, 80: 426 (1997).
83. McNair M, Wang Y, Bonilla M, *J. High Resolut. Chromatogr.*, 20: 213 (1997).
84. Gebersmann C, Heisterkamp M, Adams FC, Broekaert JAC, *Anal. Chim. Acta*, 350: 273 (1997).
85. Rodriguez I, Santamaria M, Bollain MH, Mejuto MC, Cela R, *J. Chromatogr. A*, 774: 379 (1997).
86. Rodriguez I, Santamaria M, Bollain MH, Mejuto MC, Cela R, *Spectroscopy*, 13: 51 (1997).
87. Tseng CM, DeDiego A, Martin FM, Amouroux D, Donard OFX, *J. Anal. At. Spectrom.*, 12: 743 (1997).
88. Abuin M, Carro AM, Lorenzo RA, *J. Chromatogr. A*, 889: 185 (2000).
89. Tseng CM, DeDiego A, Martin FM, Donard OFX, *J. Anal. At. Spectrom.*, 12: 629 (1997).
90. Poerschmann J, Kopinke FD, Pawliszyn J, *Environ. Sci. & Technol.*, 31: 3629 (1997).
91. Donard OFX, Lalere B, Martin F, Loboniski R, *Anal. Chem.*, 76: 4250 (1995).
92. Szpunar J, Schmitt VO, Loboniski R, *J. Anal. At. Spectrom.*, 11: 193 (1996).
93. Gomez-Ariza JL, Sanchez-Rodas D, Giraldez I, Morales E, *Analyst*, 125: 401 (2000).

94. Dagnac T, Padro A, Rubio R, Rauret G, *Talanta*. 48: 763 (1999).
95. Smith RK (Ed.), *Handbook of Environmental Analysis*, Genium Pub. Corp., Schenectady, New York, 3rd ed., p. 319, 1997.
96. Ali I, *J. Chem. Edu.*, 73: 285 (1996).
97. Arment S, *LC-GC*, 17: S38 (1999).
98. Welz B, *Spectrochimica Acta Part B*, 53: 169 (1998).
99. Richter BE, Ezzel JL, Felix D, Roberts KA, Later DW, *Am. Lab.*, 27: 24 (1995).
100. Höfler F, Ezzel J, Richter BE, *Labor Praxis*, 19: 58 (1995).
101. Höfler F, Ezzel J, Richter BE, *Labor Praxis*, 19: 62 (1995).
102. Ezzel JL, Richter BE, Felix WD, Black DR, Meikle JE, *LC-GC*, 13: 390 (1995).
103. Richter BE, Ezzel JL, Felix D, Roberts KA, Later DW, *Int. Lab.*, 25: 18 (1995).
104. Höfler F, Jensen D, Ezzel J, Richter BE, *Chromatographie*, 1: 68 (1995).
105. Ezzel J, Richter BE, *Am. Environ. Lab.*, 8: 16 (1996).
106. Richter BE, Jones BA, Ezzell JL, Porter NL, Avdalovic N, Pohl C, *Anal. Chem.*, 68: 1033 (1996).
107. Jensen D, Hoefler F, Ezzell JJ, Richter BE, *Polyaromatic Compounds*, 9: 233 (1996).
108. Gallagher AP, Wei X, Shoemaker JA, Brocknoff CA, Creed JT, *J. Anal. At. Spectrom.*, 14: 1829 (1999).
109. Gallagher AP, Wei X, Shoemaker JA, Brocknoff CA, Creed JT, *Fresenius' J. Anal. Chem.*, 369: 71 (2001).
110. Richter BE, *LC-GC*, 17: S22 (1999).
111. Levy JM, *LC-GC*, 17: S14 (1999).
112. Page SH, Lee ML, *J. Microcol. Sep.*, 4: 261 (1992).
113. Levy JM, Storozynsky E, Ashraf-Khorassani M, in *Supercritical Fluid Technology*, Bright FV, McNally MEP (Eds.), American Chemical Society, Washington, DC, p. 336 (1992).
114. Levy JM, Rosselli AC, Storozynsky E, Ravey R, Ashraf-Khorassani M, *LC-GC*, 10: 386 (1992).
115. Sanchez-Brunete C, Martinez L, Tadeo JL, *J. Agric. Food Chem.*, 42: 2210 (1994).
116. Tan GH, *Analyst*, 117: 1129 (1992).
117. Tan GH, Vijayaletchumy K, *Pestic. Sci.*, 40: 121 (1994).
118. Waliszcwski MJ, Szymczynski GA, *J. Chromatogr.*, 321: 480 (1985).
119. Sericano JL, Pucci AF, *Bull. Environ. Contam. Toxicol.*, 33: 138 (1983).
120. Mangani F, Crescentini G, Bruner F, *Anal. Chem.*, 53: 1627 (1981).
121. Gomez CG, Arufe Martinez MI, Romeo Palanco JL, Gamero Lucas JJ, Vizcaya Rojas MA, *Bull. Environ. Contam. Toxicol.*, 55: 431 (1995).
122. Faeber H, Peldszus S, Schoeler HF, *Vam Wasser*, 76: 13 (1991).
123. Goosen EC, Bunschoten RG, Engelen V, de Jong D, van den Berg JHM, *J. High Resolut. Chromatogr.*, 13: 438 (1990).
124. Tutschku S, Mothes S, Dittrich K, *J. Chromatogr. A*, 683: 269 (1994).
125. Heisterkamp M, Adams FC, *Fresenius' J. Anal. Chem.*, 362: 489 (1998).
126. Pons B, Carrera A, Nerin C, *J. Chromatogr. A*, 716: 139 (1998).
127. Girousi S, Rosenberg E, Voulgaropoulos A, Grasserbauer M, *Fresenius' J. Anal. Chem.*, 358: 828 (1997).
128. Lai-Feng Y, Narasaki H, *Anal. Bioanal. Chem.*, 372: 382 (2002).
129. Jiang GB, Xu FZ, *Appl. Organometal. Chem.*, 10: 77 (1996).
130. Stephenson MD, Smith DR, *Anal. Chem.*, 60: 696–698 (1988).
131. Tsuda T, Nakanishi H, Aoki S, Takebayyashi J, *J. Chromatogr.*, 387: 361–370 (1987).

132. Narasaki H, Hou HB, Morimura Y, *Anal. Sci.*, 16: 855–858 (2000).
133. Mortensen G, Pedersen B, Pritzl G, *Appl. Organomet. Chem.*, 9: 65–73 (1995).
134. Matthias CL, Bellama JM, Olson GJ, Brinckman FE, *Int. J. Environ. Anal. Chem.*, 35: 61–68 (1989).
135. Rajendran RB, Tao H, Nakazato T, Miyazaki A, *Analyst*, 125: 1757–1763 (2000).
136. Siu KWM, Maxwell PS, Berman SS, *J. Chromatogr.*, 475: 373–379 (1989).
137. Siu KWM, Gardner GJ, Berman SS, *Anal. Chem.*, 61: 2320–2322 (1989).
138. Gomez-Ariza JL, Beltran R, Morales E, Giraldez I, Ruiz-Benitez M, *Appl. Organomet. Chem.*, 9: 51 (1995).
139. Grotti M, Rivaro P, Frache R, *J. Anal. At. Spectrom.*, 16: 270–274 (2001).
140. Desauziers V, Leguille F, Astruc M, Pinel R, *Appl. Organomet. Chem.*, 3: 469–474 (1989).
141. Szpunar J, Schmitt VO, Lobiniski R, *J. Anal. At. Spectrom.*, 11: 193–199 (1996).
142. Rodriguez I, Mounicou S, Lobinski R, Sidelnikov V, Patrushev Y, Yamanaka M, *Anal. Chem.*, 71: 4534–4543 (1999).
143. Chau YK, Yang F, Maguire RJ, *Anal. Chim. Acta*, 320: 165–169 (1996).
144. Zufiaurre R, Pons B, Nerín C, *J. Chromatogr. A*, 779: 299 (1997).
145. Chiron S, Roy S, Cottier R, Jeannot R, *J. Chromatogr. A*, 879: 137 (2000).
146. Schmidt AC, Reiser W, Mattusch J, Popp P, Wennrich R, *J. Chromatogr. A*, 889: 83 (2000).
147. Petrick G, Schulz DE, Duinker JC, *J. Chromatogr.*, 435: 241 (1988).
148. Bernal LJ, Del Nozal MJ, Jiminez JJ, *Chromatographia*, 34: 468 (1993).
149. Zhang Z, Pawliszyn J, *Anal. Chem.*, 65: 1843 (1993).
150. Zhang Z, Yang M, Pawliszyn J, *Anal. Chem.*, 66: 844A (1994).
151. Tolosa I, Bayon JM, Albaiges J, Alencastro LF, Tarradellas J, *Fresenius' J. Anal. Chem.*, 339: 646 (1991).
152. Abalos M, Bayona JM, Quevauviller Ph, *Appl. Organomet. Chem.*, 12: 541 (1998).
153. Tsuda T, Nakanishi H, Morita T, Takebayashi J, *J. Assoc. Off. Anal. Chem.*, 69: 981 (1986).
154. Hattori Y, Kobayashi A, Takemoto S, Takami K, Kuge Y, Sigimae A, Nakamoto M, *J. Chromatogr.*, 315: 341 (1984).
155. Quevauviller Ph, Donard OXF, *Appl. Organomet. Chem.*, 4: 353 (1990).
156. Martin-Landa I, Pablos F, Marr IL, *Appl. Organomet. Chem.*, 5: 339 (1991).
157. Gomez-Ariza JL, Morales E, Giraldez I, Beltran R, *Int. J. Environ. Anal. Chem.*, 66: 1 (1997).
158. Gomez-Ariza JL, Morales E, Ruiz-Benitez M, *Analyst*, 117: 641 (1992).
159. Hartik S, Tekel J, *J. Chromatogr. A*, 733: 217 (1996).
160. Hawthorne SB, Miller DJ, Krieger MS, *J. Chromatogr. Sci.*, 27: 347 (1989).
161. Noij THM, van der Kooi MME, *J. High Resolut. Chromatogr. A*, 18: 535 (1995).
162. Shi M, Zhu X, Wang G, Hu Z, *Fenxi Huaxue*, 17: 299 (1989).
163. Stan HG, *J. Chromatogr. A*, 643: 227 (1993).
164. de la Colina C, Sanchez-Raseor F, Cancella GD, Toboada ER, Pena A, *Analyst*, 120: 1723 (1995).
165. Readman JW, Kwong LLW, Grondin D, Bartocci J, Villeneuve IP, Mee LD, *Environ. Sci. Technol.*, 27: 1940 (1993).
166. Miliadis GE, *Bull. Environ. Contam. Toxicol.*, 52: 25 (1994).
167. Eisert R, Levsen K, *Fresenius' J. Anal. Chem.*, 351: 555 (1995).
168. Thompson TS, Morphy L, *J. Chromatogr. Sci.*, 33: 393 (1995).
169. Armishaw P, Millar RG, *J. AOAC Int.*, 76: 1317 (1993).

170. Triska J, *Chromatographia*, 40: 712 (1995).
171. Crespo C, Marce RM, Borrull F, *J. Chromatogr. A*, 670: 135 (1994).
172. Guenu S, Hennion MC, *J. Chromatogr. A*, 665: 243 (1994).
173. Snell JP, Frech W, Thomassen Y, *Analyst*, 121: 1055–1060 (1996).
174. Mothes S, Wennrich R, *J. High Resolut. Chromatogr.*, 22: 181–182 (1999).
175. Mothes S, Wennrich R, *Mikrochim. Acta*, 135: 91 (2000).
176. Aguerre S, Lespes G, Desauziers V, Potin-Gautier M, *J. Anal. At. Spectrom.*, 16: 263–269 (2001).
177. Nerin C, Pons B, Zufiaurre, *Analyst*, 120: 751 (1995).
178. Whang C, Pawliszyn J, *Anal. Commun.*, 35: 353 (1998).
179. Huen JM, Gillard R, Mayer AG, Baltensperger B, Kern H, presented at the *Seventh Symposium on the Handling of Environmental and Biological Samples in Chromatography*, May 7–10, Lund, Sweden (1995).
180. Brinkmann UATh, Vreuls JJ, presented at the *Seventh Symposium on the Handling of Environmental and Biological Samples in Chromatography*, May 7–10, 1995, Lund, Sweden (1995).
181. van der Hoff GR, Gort SM, Baumann RA, van Zoon P, Brinkmann UATh, *J. High Resolut. Chromatogr.*, 14: 465 (1991).
182. Yin X, Frech W, Hoffmann E, Lüdke C, Skole J, *Fresenius' J. Anal. Chem.*, 361: 761 (1998).
183. Haddad PR, Jackson PE, in *Ion Chromatography: Principles and Applications, J. Chromatogr. Library*, Vol. 46, Elsevier, Amsterdam, 1990.
184. Barcelo D, presented at the *Seventh Symposium on the Handling of Environmental and Biological Samples in Chromatography*, May 7–10, 1995, Lund, Sweden (1995).
185. Hawthorne SB, presented at the *Seventh Symposium on the Handling of Environmental and Biological Samples in Chromatography*, May 7–10, 1995, Lund, Sweden (1995).
186. Karlsson L, Björklund E, Järemo M, Jönsson JA, Mathiasson L, presented at the *Seventh Symposium on the Handling of Environmental and Biological Samples in Chromatography*, May 7–10, 1995, Lund, Sweden (1995).
187. Barnabas IJ, Dean JR, Hitchen SM, Owen SP, *Anal. Chim. Acta*, 291: 261 (1994).
188. Emteborg H, Björklund E, Ödman F, Karlsson L, Mathiasson L, Frech W, Baxter DC, *Analyst*, 121: 19–29 (1996).
189. Bayona JM, Dachs J, Alzaga R, Quevauviller Ph, *Anal. Chim. Acta*, 286: 319 (1994).
190. Pichon V, Hennion MC, *J. Chromatogr. A*, 665: 269 (1994).
191. Puig D, Barcelo D, *J. Chromatogr. A*, 673: 55 (1994).
192. Kumar UT, Vela NP, Dorsey JG, Caruso JA, *J. Chromatogr. A*, 655: 340 (1993).
193. Dabek-Zlotorzynska E, Aranda-Rodriguez R, Keppel-Jones K, *Electrophoresis*, 22: 4262 (2001).
194. Fritz JS, Macka M, *J. Chromatogr. A*, 902: 137 (2000).
195. Pedersen-Bjegaard S, Rasmussen KE, Halvorsen TG, *J. Chromatogr. A*, 896: 95 (2000).
196. Major RE, *LC-GC*, 13: 364 (1995).
197. Huang X, Gordon MJ, Zare RN, *Anal. Chem.*, 60: 375 (1988).
198. Macka M, Haddad PR, *Electrophoresis*, 18: 2482 (1997).
199. Liu Y, Avdalovic N, Pohl C, Matt R, Dhillon H, Kiser R, *Am. Lab.*, 30: 48 (1998).

200. Conradi S, Vogt C, Wittrisch H, Knobloch G, Werner G, *J. Chromatogr. A*, 791: 279 (1997).
201. Anonymous, *Supelco Appl.*, 55: 1994 (1994).
202. Buttler T, presented at the *Seventh Symposium on the Handling of Environmental and Biological Samples in Chromatography*, May 7–10, 1995, Lund, Sweden (1995).
203. Kuban P, Karlberg B, *Anal. Chem.*, 69: 1169 (1997).
204. Bao L, Dasgupta PK, *Anal. Chem.*, 64: 991 (1992).
205. Buscher BP, Hofte AP, Tjaden UR, Vander Greef J, *J. Chromatogr. A*, 777: 51 (1997).
206. Buscher BP, Tjaden UR, Vander Greef J, *J. Chromatogr. A*, 764: 135 (1997).
207. Buscher BP, Tjaden UR, Vander Greef J, *J. Chromatogr. A*, 788: 165 (1997).
208. Okamoto Y, Sakamoto N, Yamamoto N, Kumamaru T, *J. Chromatogr.*, 539: 221 (1991).
209. Blanty P, Kvasnicka E, Kenndler E, *J. Chromatogr. A*, 757: 297 (1997).
210. Majors RE, *LC-GC*, 13: 542 (1995).
211. Chimuka L, *LC-GC*, 22: 102 (2004).
212. Ceulemans M, Witte C, Loboniski R, Adams FC, *Appl. Organomet. Chem.*, 8: 451 (1994).
213. Nagase M, Hasebe K, *Anal. Sci.*, 9: 517 (1993).
214. Nagase M, Kondo H, Hasebe K, *Analyst*, 120: 1923 (1995).
215. Stäb JA, Brinkman UATh, Cofino WP, *Appl. Organomet. Chem.*, 8: 577 (1994).
216. Mackey LN, Beck TA, *J. Chromatogr.*, 240: 455 (1982).
217. Pannier F, Astruc A, Astruc M, *Anal. Chim. Acta*, 327: 287 (1996).
218. Forsyth DS, Iyengar R, *J. Organomet. Chem.*, 3: 211 (1998).
219. Gilon N, Astruc A, Astruc M, Potin-Gautier M, *Appl. Organomet. Chem.*, 9: 633 (1995).
220. Tan Y, Marshall WD, *Analyst*, 122: 13 (1997).
221. Forsyth DS, Clerous C, *Talanta*, 38: 951 (1991).
222. Moreno P, Quijano MA, Gutierrez AM, Perez-Conde MC, Camara C, *J. Anal. At. Spectrom.*, 16: 1044 (2001).
223. Potin-Gautier M, Gilon N, Astruc M, De Gregori I, Pinochet H, *Int. J. Environ. Anal. Chem.*, 67: 15 (1997).
224. Tyson JF, *J. Anal. At. Spectrom.*, 14: 169 (1999).
225. Bloom N, *Can. J. Fish Aquat. Sci.*, 46: 1131 (1989).
226. Horvat M, Bloom NS, Liang L, *Anal. Chim. Acta*, 281: 135 (1993).
227. Bowles KC, Apte SC, *Anal. Chem.*, 70: 395 (1998).
228. Bowles KC, Apte SC, *Anal. Chim. Acta*, 419: 145 (2000).
229. Cornelis R, Caruso J, Crew H, Heumann K (Eds.), *Handbook of Elemental Speciation: Techniques and Methodology*, John Wiley & Sons, Chichester (2003).
230. van Staden JF, Stefan RI, *Talanta*, 64: 1109 (2004).
231. Tesfalidet S, Irgum K, *Anal. Chem.*, 61: 2079 (1989).
232. Tesfalidet S, Irgum K, *Fresenius' J. Anal. Chem.*, 341: 532 (1991).
233. Puk R, Weber JH, *Anal. Chim. Acta*, 292: 175 (1994).
234. Pressley TA, Longbottom JE, U.S. Environ. Prot. Agency, Off. Res. Dev. [Rep] EPA-600/4-82-006: 31 (1982).
235. Faller J, Hühnerfuss H, König WA, Krebber R, Ludwig P, *Environ. Sci. Technol.*, 25: 676 (1991).

236. Fritz JS, Freeze RC, Thoronton MJ, Gjerde DT, *J. Chromatogr. A*, 739: 57 (1996).
237. Nerin C, Pons B, *Appl. Organomet. Chem.*, 8: 607 (1994).
238. Donard OFX, Quevauviller Ph (Eds.), *Mikrochim. Acta*, 109: 1–4 (1992).
239. Quevauviller Ph, Morabito R, *Appl. Organomet. Chem.*, 8: 5 (1994).
240. Quevauviller Ph, *Fresenius' J. Anal. Chem.*, 351: 345 (1995).
241. Giokas DL, Paleologos EK, Tzouwara-Karayanni SM, Karayannis MI, *J. Anal. At. Spectrom.*, 16: 521 (2001).
242. Giokas DL, Eksperiandova LP, Blank AB, Karayannis MI, *Anal. Chim. Acta*, 505: 51 (2004).
243. Motellier S, Pitsch H, *J. Chromatogr. A*, 739: 119 (1996).
244. Ryan E, Meaney M, *Analyst*, 117: 1435 (1992).
245. Sarzanini C, Sacchero G, Aceto M, Abollino O, Mentasti E, *J. Chromatogr.*, 60: 365 (1993).
246. Bettmer J, Cammann K, Robecke M, *J. Chromatogr. A*, 654: 177 (1993).
247. Pobozy E, Glod B, Kaniewska J, Trojanowicz M, *J. Chromatogr. A*, 718: 329 (1995).

5 Speciation of Metal Ions by Gas Chromatography

As discussed in Chapter 1, growth in the research of metal ion speciation increases every year. This growth is due to the fact that speciation is a very important aspect in modern analytical science. Many techniques ranging from chromatography to spectroscopy have been used in this concern. It is well known that gas chromatography (GC) may be used for the analysis of any analyte, which is volatile at the working temperature of a GC machine. Therefore, GC has been used for the speciation of many metal ions and organometallic species, which are volatile at the GC temperature. The nonvolatile metal ions are converted into the volatile ones by the derivatization process (see Chapter 4). Moreover, GC is the choice technique for speciation of organometallic species due to the development of modern detectors and their coupling with GC. Furthermore, developments in capillary GC columns increased its efficiency for metal ion speciation even at trace levels. The detection devices, such as flame ionization detector (FID), electron capture detector (ECD), flame photometric detector (FPD), mass spectrometer (MS), atomic absorption spectrometer (AAS), atomic emission spectrometer (AES), and inductively coupled plasma (ICP), make this technique a popular alternative for metal ions speciation. These detectors were used with packed and capillary columns for a wide range of metallic species, that is, organo- or halo organic, species. The transfer efficiency of GC with detectors is very good in speciation; hence, precise speciation of metal ions can be carried out at lower concentrations. In addition, speciation of metal ions may be enhanced by using temperature gradients rather than changing the composition of the mobile phase, as in liquid chromatography. In spite of the development of several techniques (liquid chromatography [LC], capillary electrophoresis [CE], etc.) for speciation analysis of metal ions, GC maintains its status in this field. This is due to the fact that GC is an effective technique, with determination of the relative peak areas of metallic species at high precision (relative standard deviations generally better than ±2%). This chapter describes metal ion speciation using GC. Various aspects of metal ion speciation, such as application, optimization, mechanisms, extraction, derivatization, and so forth, will be discussed in detail.

5.1 APPLICATIONS

The selection of speciation method depends on the nature of the metal ions in the environmental and biological samples. The physicochemical properties of the analyte (i.e., volatility, charge, polarity, molecular mass, etc.) are responsible for the speciation of metal ions. As discussed above, all metallic species are allowed to react with organo- or halo-organo compounds and are converted into volatile derivatives (derivatization). However, the volatile metal ion species are injected directly onto a GC machine. Generally, very low concentrations of metallic species are found in water; hence, it is advisable to work with a large amount of water samples, with the volume of the water sample in liters. The exact amount of the water sample should be determined by a trial-and-error approach. Similarly, it is natural to find a low amount of metal ions in the atmosphere, and again, a large volume of air sample will be required. The necessary amount of a soil and sediment sample should be optimized through trial-and-error. A general procedure for carrying out the analysis of metal ion species is presented in Scheme 5.1. The analysis of some metal ion species analyzed by GC is discussed below.

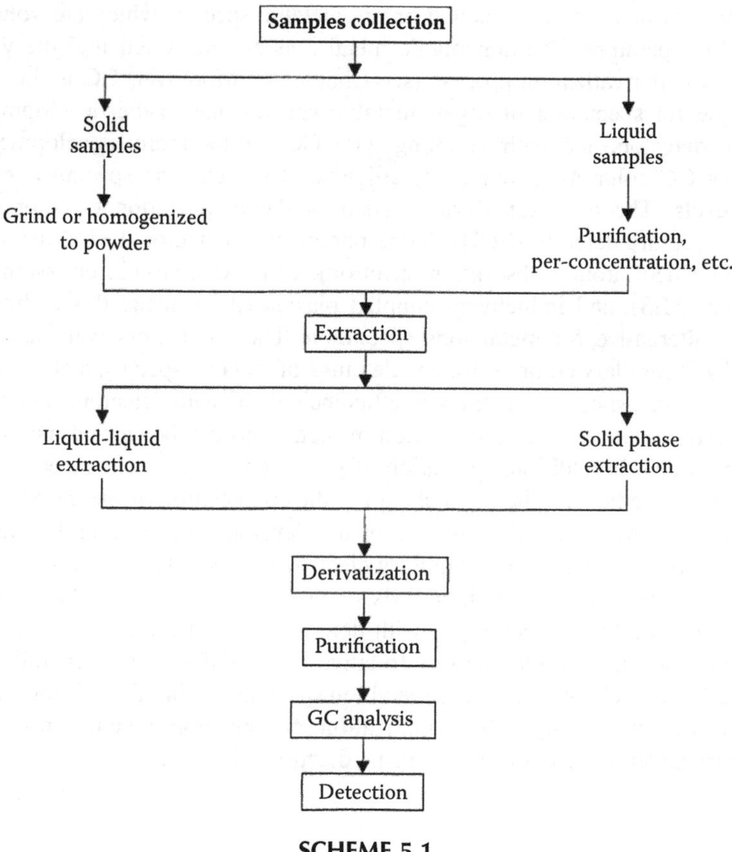

SCHEME 5.1

5.1.1 Arsenic (As)

Dix et al.[1] determined inorganic (arsenite, arsenate) and organic (monomethylarsonate, dimethylarsinate) arsenic species by capillary gas chromatography as their corresponding methylthioglycolate derivatives, which were simultaneously separated on wide-bore borosilicate glass and fused-silica columns under conditions of temperature programming. Schoene et al.[2] converted various organoarsenic halogenides, oxides, and hydroxides into the corresponding thioarsenites by reacting them with thioglycolic acid methyl ester (TGM). The yields and the chemical structures of the TGM derivatives were evaluated by GC coupled with mass spectrometry and atomic emission spectrometry. Most toxic arsenic was speciated by GC successfully by converting the metal ions into organic derivatives. Mester et al.[3] reported the speciation of monomethyl- and dimethylarsenic acid using gas chromatography–mass spectrometry (GC-MS) in human urine samples. This GC-MS method was compared with high-performance liquid chromatography–hydride generation–atomic fluorescence spectrometry. The yield of derivatization for dimethylarsinic acid (DMA) and monomethylarsonic acid (MMA) using thioglycol methylate was measured. The detection limits for DMA and MMA using this GC-MS method were 0.95 and 0.8 ng/cm^3, respectively. Furthermore, the same authors[4] analyzed the species of DMA and MMA using the solid-phase microextraction–gas chromatography–mass spectrometry (SPME-GC-MS) method with detections of 0.12 and 0.29 ng/mL, respectively. The method was linear in the range of 1 to 200 ng/mL. Prohaska et al.[5] used GC coupled to a double-focusing sector field inductively coupled plasma–mass spectrometry (ICP-MS) for speciation of arsenic of liquid and gaseous emissions from soil samples, which were equilibrated in a microcosmos experiment. A home-built and laboratory-ready transfer line from GC to ICP-MS is presented, and the quantification of arsenic in gaseous emissions was performed by external calibration via hydride generation. The microcosmos experiment revealed only low production rates of organoarsenic compounds and reflected limited capability of the biovolatilization experiment for the simulation of natural systems. Grüter et al.[6] improved the speciation technique for arsenic, enabling its identification and quantification. This study showed the results from measurements of soil samples from municipal waste deposits. The hydride generation/low-temperature gas chromatography (HG/LT-GC)/ICP-MS apparatus containing a home-built gas chromatograph enabled satisfactory separation of arsenic species with a boiling point difference of 14°C. The absolute detection limits for the elements mentioned above were below 0.7 pg. Furthermore, the same group[7] used high-generation/low-temperature gas chromatography/inductively coupled–mass spectrometer coupling for the efficient detection of arsenic species. The speciated species were AsH_3, $(CH_3)AsH_2$, $(CH_3)_2AsH$, and $(CH_3)_3As$. Killelea and Aldstadt[8] developed a novel method for arsenic speciation based on capillary gas–liquid chromatography with simultaneous quadrupole ion-trap mass spectrometric (MS) detection and pulsed flame photometric detection (PFPD). The method coupled the sensitive arsenic selectivity of PFPD with the structure elucidation capability of molecular MS detection for the determination of trace levels of unknown organoarsenicals in complex matrices. The conditions that affected the PFPD response in the presence of interfering species were optimized using the sequential simplex algorithm for three key factors — gate delay (18.3 ms), gate width (9.1 ms), and combustion gas composition (16.6 mL/min H2). The complete discrimination in the PFPD of the arsenic signal from interfering S-, C-, and OH-emitting species that were problematic in existing methods was achieved. Mothes and Wennrich[9] analyzed the species of arsenic on HP-1 (25 m × 320 μm, 0.17 μm id) as the column using nitrogen as the mobile phase.

5.1.2 Lead (Pb)

Lobinski et al.[10] developed a speciation method for the analysis of ionic organolead compounds in wine. The analytes were extracted as diethyldithiocarbamate complexes into hexane and were propylated with a Grignard reagent. The derivatized extract was analyzed by capillary gas chromatography/microwave-induced plasma atomic emission spectrometry. The method of standard additions was used for calibration to correct for variable recoveries and signal enhancements. Red, rose, and white wines from southern France were analyzed. The wines made from grapes grown close to industrial zones showed elevated concentrations of ethyllead species. The concentrations of methyllead and ethyllead found in wines were compared with the concentrations of organolead found in rainwater and plant sap collected in the viticultural regions. The ratio of methyllead to ethyllead in wines greatly exceeded the same ratio found in atmospheric deposits. Nerin et al.[11] speciated organolead compounds using GC with different detectors. These authors used the gas chromatography–mass spectrometry–selected ion monitoring mode (GC-MS-SIM) technique for the analysis of Et_4Pb, Et_3MePb, Et_2Me_2Pb, $EtMe_3Pb$, and Me_4Pb species in air samples. In another study, the authors[12] speciated BuMgCl, Me_3BuPb, and Et_3BuPb species into artificial rainwater using the same technique. Furthermore, the same group[13] analyzed BuMgCl, Me_3BuPb, Et_3BuPb, and Bu_4Pb compounds in urban and road dust particles. The authors used DB-5 (60 m × 0.25 mm, id) and SPB-1 capillary columns with helium as the carrier gas. Zufiaurre et al.[14] described an analytical procedure to determine Me_3Pb^+ in aqueous solution. The method consisted of ethylation of Me_3Pb^+ with $NaBEt_4$ in the presence of a small volume of hexane, and keeping an excess of Pb^{2+}. The organic extract was analyzed by GC-MS in the selected ion monitoring mode, using Me_4Pb as the internal standard. The method was applied for the analysis of an artificial rainwater solution as part of an international interlaboratory exercise, and the results obtained were discussed and compared with those given by the butylation method. Heisterkamp and Adams[15] described a simplified derivatization method for speciation of organolead compounds into the environmental samples followed by gas chromatography–microwave-induced plasma atomic emission detection. The detection limits at the sub-ngL^{-1} range, comparable to those achieved by Grignard alkylation, were obtained for the different species. Furthermore, the same authors[16] described the application of inductively coupled plasma–time-of-flight–mass spectrometry (ICP-TOF-MS) and GC for speciation of organolead compounds in environmental waters. The detection limits for the different organolead species were in the range of 10 to 15 fg (as Pb), corresponding to procedural detection limits between 50 and 75 ngL^{-1}. Pons et al.[17] described the speciation of $MePb^+$, Et_3Pb^+, and Pb^+ species using the GC-MS-SIM method. The authors used SGL-1 capillary column (25 m × 0.25 mm, id) with helium as the carrier gas. Hirner et al.[18] collected 13 soil samples for the speciation of organolead compounds. The analyzed species were Et_4Pb, Et_2CH_2Pb, $Et_2(CH_2)_2Pb$, $Et_2(CH_2)_3Pb$, and Me_4Pb. Grüter et al.[6] used low-temperature GC for the speciation of organolead compounds (i.e., Pb and Et_4Pb) in soil samples. The authors used hydride generation/low-temperature–gas chromatography/inductively coupled plasma–mass spectrometry apparatus, containing a home-built gas chromatograph, for the satisfactory separation of mercury species. The absolute detection limits for the elements mentioned above were below 0.7 pg.

5.1.3 Mercury (Hg)

Carro-Diaz et al.[19] used a commercial microwave-induced plasma atomic emission detector with capillary gas chromatography for mercury speciation in environmental samples. The chromatographic conditions were optimized in order to obtain an adequate resolution of the methylmercury peak versus interfering carbon signals. Under the proposed operational conditions, the detection limit (signal-to-noise ratio = 3) was 1.2 pg with a linear range of 1 to 40 ng mL^{-1} (as methylmercury in samples). The certified reference material (DORM-1) was used to evaluate accuracy. The results of the proposed procedure were compared with those obtained by means of the usual GC method with electron-capture detection. Liang and coworkers[20] developed a method for the simultaneous determination of monomethyl mercury (MMHg), inorganic mercury [Hg(II)], and total mercury (THg). The eluted organo-Hg compounds from the GC column were decomposed into Hg, and detection was completed by cold vapor atomic fluorescence spectrometry (CVAFS). The absolute detection limits were 0.6 pg and 1.3 pg of MMHg and Hg(II), respectively, corresponding to 0.3 ng and 0.6 ng/g (wet) of sample. Bettmer et al.[21] used the Ultra-1 (25 m × 0.32 mm id) column for speciation of organomercury species using helium gas. The studied species of mercury were Me_2Hg, $MeHg^+$, and Et_2Hg. Carro et al.[22] speciated organomercury species using BP-1, HP-1, BP-5, and AT-5 columns with ECD and AED detectors. The gases used for this study were helium, oxygen, and hydrogen. Rodriguez et al.[23] devised a purge-and-trap multicapillary gas chromatography for sensitive species selective analysis of methylmercury and Hg^{2+} by atomic spectrometry. The operating mode included *in situ* conversion of the analyte species to MeEtHg and $HgEt_2$ and cryotrapping of the derivatives formed in a 0.53 mm id capillary, followed by their flash (<30 sec) isothermal low-temperature separation on a minimulticapillary (22 cm) column. Mothes and Wennrich[9] analyzed the species of mercury using an HP-1 (25 m × 320 μm, 0.17 μm id) column with nitrogen as the mobile phase. Garcia Fernandez et al.[24]

developed a novel interface design for the coupling of gas chromatography and inductively coupled plasma mass spectrometry (GC-ICP-MS) for the speciation of mercury. Detection limits obtained by GC-ICP-MS ranged between 100 and 200 fg (as absolute mass) for methylmercury and between 500 and 600 fg for inorganic mercury using a 1 μL injection. The quantification of methylmercury and inorganic mercury was carried out by resorting to aqueous calibration, using ethylmercury as the internal standard for both propylation and butylation derivatization techniques. Välimäki and Perämäki[25] developed a procedure for the determination of methylmercury and inorganic mercury(II) using a capillary column DB-1. Grüter et al.[6] used low-temperature GC for the speciation of Hg and $(CH_3)_2Hg$ species in soil samples. The authors used a hydride generation/low-temperature gas chromatography/inductively coupled plasma–mass spectrometry apparatus containing a home-built gas chromatograph for the satisfactory separation of mercury species. The absolute detection limits for the element mentioned above were below 0.7 pg. Furthermore, the same group[7] used high-generation/low-temperature gas chromatography/inductively coupled–mass spectrometry for the efficient detection of mercury species. The separated species were Hg and $(CH_3)HgH$.

5.1.4 Tin (Sn)

Dowling et al.[26] developed a method for the speciation of alkyltin pollutants in seawater, which combines solid-phase extraction of organotin species from aqueous samples and an online hydride generation technique with gas chromatography, followed by element-specific detection for tin. The detection was performed using a microwave-induced plasma atomic emission detection system that provides a 0.5 pg detection limit for tin and a $3{\cdot}0 \times 10^4$ tin-to-carbon selectivity ratio. Tutschku et al.[27] developed a method for the speciation of BuSn, Bu_3SnPe, Bu_2SnPe_2, $BusnPe_3$, and Et_4Sn species using GC-MIP-AES methodology in sediment samples. The authors used a HP-1 (25 m × 320 μm, id) column with helium as the carrier gas. Jiang et al.[28] used a flame photometric detector (FPD) and quartz surface-induced luminescence with on-column capillary chromatographic determination for speciation of organotin and organogermanium compounds. The minimum detectable amounts, defined as the signal equal to three times the standard deviation (3σ) of the baseline noise, varied with the species detected and ranged from 0.7 to 2.3 pg for ethylated butyl- and phenyltin species and from 50 to 100 pg for pentylated dimethyl-, trimethyl-, and tetrabutylgermanium. The determination of butyltin compounds in water and sediments was evaluated in comparison with a GC-AAS system. Girousi et al.[29] described a rapid, sensitive, and interference-free analytical procedure for the speciation of organotin compounds in seawater. The method was based on the ethylation of ionic organotin compounds in the aqueous phase with $NaBEt_4$, followed by extraction of the derivatized species into hexane. The separation and determination of organotin species were performed by capillary gas chromatography–microwave-induced plasma–atomic emission detection (GC-MIP-AED).

Rodriguez Pereiro et al.[30] leached organotin compounds from sediment samples and followed with the simultaneous extraction and derivatization of the extracted species that were then determined by GC. Gallina and coworkers[31] described a simple, effective analytical procedure based on a gas chromatographic mass spectrometric technique for the speciation analysis of organotin compounds (OTC). The direct alkylation reaction of the organotin chlorides, in the aqueous digestion solution by $NaBEt_4$, allowed a short analysis time and a good recovery. Gui-bin et al.[32] speciated organotin compounds by gas chromatography–flame photometric detector and inductively coupled plasma–mass spectrometry. Cardellicchio et al.[33] developed a method for the determination of organotin compounds (i.e., monobutyl [MBT], dibutyl [DBT], and tributyltin [TBT]) in marine sediments by headspace solid-phase microextraction (SPME) and GC. Haas and coworkers[34] used an axial ICP-TOF-MS as a detector for fast transient chromatographic signals resulting from the coupling with capillary gas chromatography (CGC). A cryotrapping GC-ICP-TOF-MS method for the determination of volatile metal(loid) compounds (VOMs) in gases was used, and the suitability of the TOF mass analyzer for multielemental speciation analysis and multi-isotope ratio determinations was studied in terms of accuracy and precision. Dirkx et al.[35] presented a method for the determination of tri- and dibutyltin in sediments by gas chromatography–quartz furnace atomic absorption spectrometry. Tutschku and coworkers[27] determined organotin species in the environmental samples after derivatization. The results were measured with gas chromatography coupled with element-selective detection by helium atmospheric pressure microwave-induced plasma–atomic emission spectrometry (GC-MIP-AES). The sensitivity of the most intensive emission lines was in the range of 180 to 350 nm. The best results were obtained by using a 326.23 nm line. Lai-Feng and Narasaki[36] used capillary GC for the analysis of organotin compounds in marine sediment samples. The authors used a DB-1 capillary column (30 m × 0.53 mm, id) with hydrogen, helium, and oxygen as the carrier gases. Grüter et al.[6] used low-temperature GC for the speciation of SnH_4, $(CH_3)SnH_3$, and $(CH_3)_2SnH_2$ species in soil samples using a hydride generation/low-temperature–gas chromatography/inductively coupled plasma–mass spectrometry (HG/LT-GC/ICP-MS) apparatus with a home-built GC. The absolute detection limits for the element mentioned above were below 0.7 pg. Furthermore, the same group[7] used high-generation/low-temperature gas chromatography/inductively coupled–mass spectrometry (HG/LT-GC/ICP-MS) for the efficient detection of SnH_4, $(CH_3)SnH_3$, $(CH_3)_2SnH_2$, $(CH_3)_3SnH$, $(CH_3)_4Sn$, $(C_4H_9)_2SnH_2$, and $(C_4H_9)_3SnH$ species.

5.1.5 Miscellaneous Metal Ions

Hirner et al.[18] used HG/LT-GC/ICP-MS for speciation of antimony compounds. The authors speciated $SbMeH_2$, $SbMeH_2$, and $SbMe_3$ compounds in soil samples with detection limits ranging from 0.1 to 0.6 pg. Similarly, Grüter et al.[6] used HG/LT-GC/ICP-MS for the speciation of antimony species. The authors achieved the analysis of SbH_3, $(CH_3)SbH_2$, $(CH_3)_2SbH$, and $(CH_3)_3Sb$ species. Furthermore, the same group[7] used the same HG/LT-GC/ICP-MS setup for the efficient detection of SbH_3, $(CH_3)SbH_2$, $(CH_3)SbH$, and $(CH_3)_3Sb$ species. Jiang et al.[28] used flame photometric detector (FPD) and quartz surface-induced luminescence with capillary gas chromatography for the determination of organogermanium species in river water samples. The authors analyzed Bu_4Ge and Me_3PeGe as the germanium species, using an HP-1 capillary column. The minimum detectable amounts, defined as the signal equal to three times the standard deviation (3σ) of the baseline noise, varied with the species detected and ranged from 0.7 to 2.3 pg for butyl- and phenylgermanium. Grüter et al.[6] used an HG/LT-GC/ICP-MS assembly for the speciation of germanium species. The authors achieved the analysis of GeH_4, $(CH_3)GeH_3$, $(CH_3)_2GeH_2$, and $(CH_3)_3GeH$ species. Furthermore, the same group[7] used an high-generation/low-temperature–gas chromatography/inductively coupled–mass spectrometer coupling for the analysis of GeH_4 and $(CH_3)GeH_3$ species. In addition, Grüter et al.[6] again used HG/LT-GC/ICP-MS for the speciation of bismuth species, that is, BiH_3 and $(CH_3)_3Bi$. Similarly, the same authors[7] used a similar assembly for the analysis of selenium compounds, namely, $(CH_3)_2Se$ and $(CH_3)_2Se_2$. As usual, the detection and identification of the separated metal ion species are represented by the retention time (t), capacity (k), separation (α), and resolution factor (R). The speciation of different metal ion species by GC is summarized in Table 5.1. For easier understanding and to show the nature of the chromatograms of the separated metal ion species by GC, a typical example of the chromatograms of arsenic speciation is shown in Figure 5.1.

5.1.6 Isotope Ratio

The analysis of isotope dilution is an important area in speciation science that can easily be carried out by using the GC-ICP-MS assembly. Moreover, isotopic dilution methodologies may be useful for elemental speciation by using an isotope of the same element. The precision of speciation data can by calculated by isotopic dilution methods. The parameters required to be optimized in GC-ICP-MS for isotope ratios are detector dead time, mass bias, spectral interference, and chemical blanks. The most important applications of the isotope ratio are in sample preparation techniques and in environmental, biological, and geological processes. Chong and Houk[37] reported that the first report of an isotope ratio measurement was made in 1987 using the GC-ICP-MS combination. These authors measured isotope ratios of B, C, N, Si, S, Cl, and Br in different organic compounds. In 1993, Peters and Beauchim[38] determined isotope ratios in organic tin compounds. Ruiz Encinar et al.[39] speciated and differentiated trimethyl lead, ethyl, diethyl, dimethyl lead, methyl, triethyl lead, and tetraethyl lead compounds with respect to triethyl lead cation or trimethyl lead cation by GC-ICP-MS in gasoline samples. Analyses of isotope ratios by GC-ICP-MS were reported in sample preparation (digestion, extraction, cleanup, preconcentration, derivatization, separation, determination, etc.).[40–42] Basically, isotope dilution can be applied to elemental speciation using two methods: a species-unspecified spike and different

TABLE 5.1
Speciation of Metal Ions by Gas Chromatography

Metal Ions	Sample Matrix	Columns	Detection	Ref.
		Arsenic (As)		
Arsenite, arsenate and organic arsenic, i.e., monomethylarsonate and dimethylarsinate	—	Fused silica	FID	1
Thioarsenites	—	—	MS-AES	2
Monomethyl- (DMA) and dimethylarsenic acid (DMA)	Human urine	—	MS	3
MMA and DMA	—	—	MS	4
As species	Soil	—	ICP-MS	5–7
As species	—	HP-1	MS-PFPD	8, 9
		Lead (Pb)		
Diethyldithiocarbamate	Wine	—	ICP	10
Et_4Pb, Et_3MePb, Et_2Me_2Pb, $EtMe_3Pb$, and Me_4Pb			MS-SIM	11
BuMgCl, Me_3BuPb, Et_3BuPb	Artificial rainwater	—	—	12
BuMgCl, Me_3BuPb, Et_3BuPb, and Bu_4Pb	Dust	—	—	13
Me_3Pb^+ with $NaBEt_4$	Rainwater	—	MS	14
Organolead	—	—	—	15
Et_3Pb^+ and Pb^+	Environmental samples	—	MS	16
$MePb^+$, Et_3Pb^+, and Pb^+	—	—	MS-SIM	17
Et_4Pb, Et_2CH_2Pb, $Et_2(CH_2)_2Pb$, $Et_2(CH_2)_3Pb$, and Me_4Pb	Soil	—	ICP-MS	18
Pb and Et_4Pb	Soil	—	ICP-MS	6
Pb species	Fuel	—	—	47
Organolead	Plasma	—	AED	48, 49
Organolead	River water and others	—	AED	50, 51
Organolead	—	—	ICP-MS	53
Organolead	—	—	AAS	54
		Mercury (Hg)		
Organomercury	Sediment and biological samples	—	ECD, AED	19

TABLE 5.1 (Continued)
Speciation of Metal Ions by Gas Chromatography

Metal Ions	Sample Matrix	Columns	Detection	Ref.
Hg, monomethyl mercury (MMHg), and inorganic mercury [Hg(II)]	—	—	CVAFS	20
Me_2Hg, $MeHg^+$, and Et_2Hg	—	Ultra-1	—	21
Organomercury species	—	BP-1, HP-1, BP-5, and AT-5	ECD, AED	22
MeEtHg and $HgEt_2$	—	—	—	23
Hg species	—	HP-1	ICP-MS	9, 24
Hg(II) and methylmercury	—	DB-1	ICP-MS	25
Hg and $(CH_3)_2Hg$	Soil	—	ICP-MS	6
Hg species	—	—	ICP-MS	7
Organomercury	—	—	AAS	55
Organomercury	Water soil and plasma	—	AED	48, 56
Organomercury	—	—	ICP-MS	52
MeHg and EtHg	Aqueous solutions and biological samples	—	AED/ECD	52, 57, 58
Hg(II) and MeHg	Sediment and soil	—	AED	59
Hg(II), MeHg, and EtHg	Biological and water samples		AED	55, 60–65
Hg(II), MeHg, and Me_2Hg	Sediment, soil, and water	—	AED	65
Me_2Hg, Et_2Hg, and Ph_2Hg	Sediment, soil, and water	—	AED	66
Hg(II), MeHg, EtHg, and PhHg	Biological samples	—	AED	67
Hg(II), MeHg, and Me_2Hg	Natural gas	—	AED	61, 68
MeHg and Me_2Hg	Water	—	MIP	69
Hg(II) and MeHg	Biological samples	—	—	70
Hg(II) and MeHg	Water	—	AED	71
MeHg and EtHg	Sediments	—	AED	72
Hg(II) and MeHg	Water	—	AED	62
		Tin (Sn)		
BuSn, Bu_3SnPe, Bu_2SnPe_2, $BusnPe_3$, and Et_4Sn	Seawater	—	ICP-AES	26
BuSn, Bu_3SnPe, Bu_2SnPe_2, $BusnPe_3$, and Et_4Sn	Sediment	—	MIP-AES	27

TABLE 5.1 (Continued)
Speciation of Metal Ions by Gas Chromatography

Metal Ions	Sample Matrix	Columns	Detection	Ref.
Ethyl-, butyl-, and phenyltin	Water and sediment	—	FPD	28
Organotin species	—	—	MIP-AED	29
Organotin species	—	—	—	30, 31
Monobutyl, dibutyl, and tributyltin	Marine sediment	—	ICP-FPD	32, 33
SnH_4, $MeSnH_3$, Me_2SnH_2, Me_3SnH, $BuSnH_3$, and SbH_3	—	—	ICP-TOF-MS	34
Tri- and dibutyltin	Sediments	—	AAS	35
Organotin species	Environmental samples	—	MIP-AES	27
Organotin species	—	DB-1	—	36
SnH_4, $(CH_3)SnH_3$ and $(CH_3)_2SnH_2$	Soil	—	ICP-MS	6
SnH_4, $(CH_3)SnH_3$, $(CH_3)_2SnH_2$, $(CH_3)_3SnH$, $(CH_3)_4Sn$, $(C_4H_9)_2SnH_2$, and $(C_4H_9)_3SnH$	—	—	ICP-MS	7
Sn species	Sediment	—	ELAN 5000 ICP	72–74
Sn species	Sea water	—	ELAN 5000 ICP	75
Sn species	Water	—	—	76
Organotin species	Sediment	—	MS	77
Organotin species	Mussels	—	AED	78
Organotin species	Plasma	—	AED/FPD	51, 52, 79
Organotin species	Fish tissue	—	FPD, AED	80
Organotin species	Sediment and water	—	AAS, AES	81, 82
Organotin species	Water and sediment	—	FPD	83–85
Organotin species	Sediments	—	AAS	48, 86–88
Organotin	—	—	AED/ICP-MS	49
Organotin species	Water	Silica capillary DB-5	MS	89
Organotin species	Water and sediment	Trap column coated with Tenax and silica	ICP-AED	90
		Antimony (Sb)		
$SbMeH_2$, $SbMeH_2$, and $SbMe_3$	Soil	—	ICP-MS	18

TABLE 5.1 (Continued)
Speciation of Metal Ions by Gas Chromatography

Metal Ions	Sample Matrix	Columns	Detection	Ref.
SbH_3, $(CH_3)SbH_2$, $(CH_3)_2SbH$, and $(CH_3)_3Sb$	Soil	—	ICP-MS	6
SbH_3, $(CH_3)SbH_2$, $(CH_3)SbH$, and $(CH_3)_3Sb$	—	—	ICP-MS	7
		Germanium (Ge)		
Bu_4Ge and Me_3PeGe	—	HP-1	FPD	28
GeH_4, $(CH_3)GeH_3$, $(CH_3)_2GeH_2$, and $(CH_3)_3GeH$	Soil	—	ICP-MS	6
GeH_4 and $(CH_3)GeH_3$	—	—	ICP-MS	7
		Bismuth (Bi)		
BiH_3 and $(CH_3)_3Bi$	Soil	—	ICP-MS	6
		Selenium (Se)		
$(CH_3)_2Se$ and $(CH_3)_2Se_2$	—	—	ICP-MS	7

species-specific spikes. The first approach is used when the structure or identity of the species is known and no standard is available for quantification. This approach is useful in HPLC and CE applications. The second approach is used when the structure or identity of the species is known. One can synthesize these species isotopically enriched. These enriched spikes are added at the beginning of the analytical method, which provides full use of isotope dilution advantages. There are several examples of isotope dilution analyses using GC-ICP-MS, but its application is limited to mercury,[43] selenium,[44] and tin[45,46] only.

5.2 OPTIMIZATION OF METAL ION SPECIATION

The optimization of metal ion speciation is the key to obtaining reproducible, effective, and efficient separations. The speciation of metal ions by GC is controlled by a number of parameters. Some of the most important factors are the nature of the stationary phase (column), mobile phase, temperature of the column, injector and detector, derivatization, amount of loading, and sensitivity of the detector. Speciation can be optimized by the extraction procedures, for example, by using appropriate derivatizing reagents.

5.2.1 Stationary Phases

The nature and the type of stationary phase are the most important issues controlling the speciation of metal ions. Basically, many bondings of the stationary

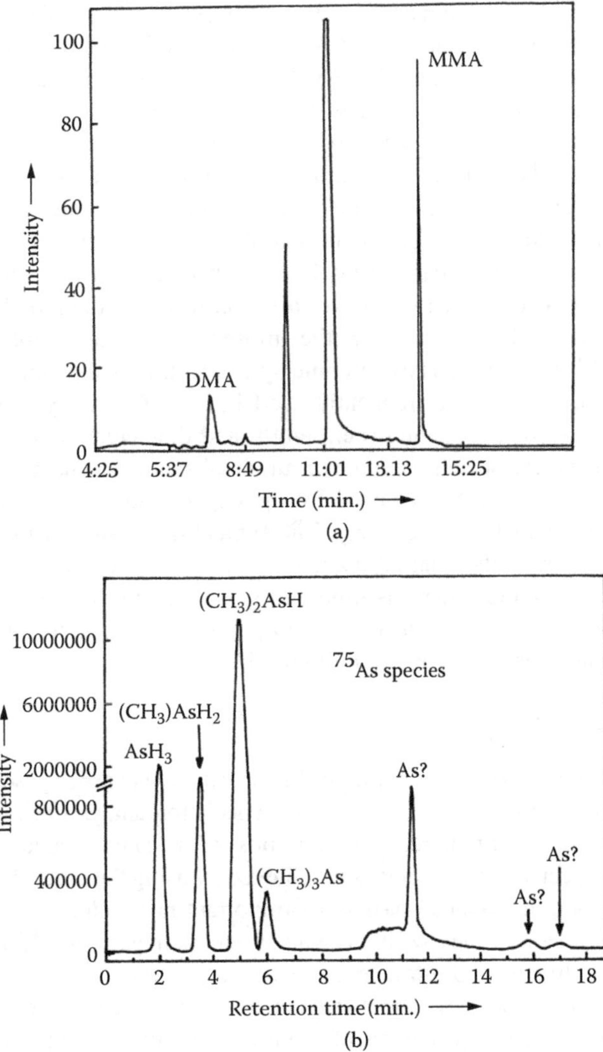

FIGURE 5.1 Chromatograms of arsenic species: (a) chromatograms of monomethyl arsenic acid (MMA) and dimethylarsenic acid (DMA) and (b) AsH_3, $(CH_3)AsH_2$, $(CH3)_2AsH$, $(CH3)_3As$, and other unknown arsenic species. (Part a from Mester, Z., Vitányi, G., Morabito, R., and Fodor, P., *J. Chromatogr. A*, 832: 183 (1999); Part b from Kresimon, J., Grüter, U.M., and Hirner, A.V., *Fresenius' J. Anal. Chem.*, 371: 586 (2001). With permission.)

phase are responsible for the speciation of metal ions. The types and the extent of the bondings vary from one stationary phase to another. Previously, packed columns were generally used in GC, but these older types of columns have been replaced by capillary columns. Generally, for packed columns, the packing material used was 5 to 10% Carbowax or OV-101 supported on Chromosorb. Capillary columns are made of silica gel. Silica-gel-based GC columns are used frequently for the analysis of metal ion species. This is due to the fact that these columns are flexible in nature and are easy to use. Their inertness and their ability to retain their coatings make them ideal columns for use in GC, even at high working temperatures. The efficiency of the capillary columns is greater than that of the packed columns due to the high theoretical plate values of the capillary columns. Moreover, the gas flow in capillary columns ranges from 1 to 5 mL/min, which increased the sensitivity due to a lower dilution factor in the mobile phase. The only limitation of these columns is the limited amount of sample that can be loaded, and this problem can be overcome by concentrating the sample. Therefore, the capillary columns were frequently used in most GC analyses of metal ion species. The size of the column (i.e., length and diameter) is also important in optimizing metal ion speciation in environmental samples. The maximum speciation of metal ions can be achieved by increasing and decreasing the column length and the diameter, respectively. The typical dimension of the GC column is 25 to 100 m, with the internal diameter in micrometers. An efficient column is one that has a shorter analysis time and good sensitivity and reproducibility. The GC columns for speciation analysis supplied by different manufacturers with their trade names are summarized in Table 5.2.

5.2.2 Mobile Phases

Optimum to slightly less than optimum linear velocities carrier gases are used to control the speciation of metal ions in GC. Speciation analysis of metal ions is a sensitive operation and requires that the utmost of precaution be taken; therefore, the optimum speciation of metal ions is achieved by using the purest form of carrier gases. Therefore, the mobile phase is an important parameter for optimizing the analysis of metal ion species by GC. The most important gases used for this purpose are helium, hydrogen, and nitrogen. As usual, the advantages of nitrogen are its low cost and high efficiency. The disadvantages of nitrogen include the optimum linear velocity of 12 cm/sec and that efficiency is rapidly lost with increasing linear velocity (the slope of the line for height equivalent theoretical plate [HETP] versus linear velocity is steep). On the other hand, helium is more expensive in comparison to other gases, but the optimum linear velocity for this is 20 cm/sec, and the slope of the line for HETP versus linear velocity is flatter. Hydrogen has an optimum linear velocity of 40 cm/sec and has the flattest slope for HETP versus linear velocity and the lowest viscosity. Generally, hydrogen forms explosive mixtures with air, and it must be used with care. For highly volatile metal ion species, low temperatures and low linear velocities are required to achieve sufficient chromatographic separations. The flow rate or the pressure of the mobile phase is adjusted

TABLE 5.2
The Most Commonly Used Gas Chromatography Columns for the Speciation of Metal Ions

GC Columns	Companies
BP-1 (bonded phase, dimethylpolysiloxane)	SGE Pty. Ltd., Ringwood, Victoria, Australia
BP-5 (bonded phase, 5% phenyl- and 95% methylsiloxane)	SGE Pty. Ltd., Ringwood, Victoria, Australia
HP-1 (bonded phase, polydimethylsiloxane)	Hewlett Packard Co., Newport, Delaware, USA
HP-5 (bonded phase, 5% diphenyl- and 95% dimethylsiloxane)	Hewlett Packard Co., Newport, Delaware, USA
DB-1 (dimethylpolysiloxane)	J & W Scientific, Folsom, California, USA
DB-5	J & W Scientific, Folsom, California, USA
AT-5 (bonded phase, 5% phenyl- and 95% methylsiloxane)	Alltech Associate, Inc., Deerfield, Illinois, USA
SGL-1 (fused silica)	
SPB-1	Bellefonte, Pennsylvania, USA
Ultra-1	J & W Scientific, Folsom, California, USA

Note: Other GC columns available with different manufacturers may be used for metal ion speciation.

and optimized by trial and error. Normally, the low flow rate favors the maximum separation of metal ion species, while the high pressure results in poor speciation, but this does not mean that high flow rate is not good for the best speciation. Basically, the flow rate depends on the types of metal ions and GC conditions, such as temperature. The head pressure of carrier gas for a particular metal ion speciation on a particular column may be calculated. To calculate the head pressure of carrier gas needed for a certain linear velocity on a 0.25 mm inner diameter column or the linear velocity at a certain column head pressure see Table 5.3. To measure average linear velocity, the following equation is used:

$$\mu_{ave} = L/t_0 \tag{5.1}$$

where L is the length of the column in centimeters, and t_0 is the time necessary for an unretained component to pass from the injector to the detector. A good unretained analyte is methane or hydrocarbon gas from a disposable lighter. It is also worthy to mention here that, sometimes, a makeup gas flow is also required in some element-specific detectors, namely, AAS and ICP.

Some workers carried out certain studies on metal ions speciation by the variation of the flow/pressure of the carrier gases. To clarify this topic, the work of some scientists is discussed herein. Girousi et al.[29] tried to optimize metal ion speciation by varying the flow of the carrier gases. The authors obtained the different peak intensities of tin metal ion species by using the different flow rates of the oxygen, hydrogen, and helium gases in the mixtures. The results of these

TABLE 5.3
Optimum and Maximum Linear Velocities and Relative Viscosities for Each Carrier Gas and Columns of Different Lengths

Carrier Gas	μ_{opt} (cm/sec)	μ_{max} (cm/sec)	Relative Viscosity (η)
Nitrogen	12	30	1.00
Helium	20	60	1.22
Hydrogen	40	120	0.497

Column Length (m)	cm/sec/psi* 50°C	100°C	150°C	200°C
20	2.78	2.51	2.29	2.12
30	1.82	1.64	1.50	1.39
50	1.29	1.16	1.06	0.98

* psi = pound per square inch. This represents relative viscosity.

findings are shown in Figure 5.2a through Figure 5.2c. It may be concluded from this figure that the maximum emission intensity was obtained at a 235 mL/min flow rate of helium (with oxygen and hydrogen makeup pressures of 160 and 400 kPa, respectively, Figure 5.2a), 350 kPa of hydrogen (with helium makeup flow rate of 240 mL/min and oxygen pressure 160 kPa, Figure 5.2b), and 165 kPa of oxygen (with helium makeup flow rate of 240 mL/min and hydrogen pressure 400 kPa, Figure 5.2c), respectively. Similarly, Heisterkamp and Adams[16] reported on the speciation of organolead species using hydrogen at 200 kPa with different flows of the makeup gas. The authors studied the effect of the makeup gas flow rate on the peak area intensity of trimethyllead. The effect of the flow rate of the makeup gas on the peak area intensity is given in Figure 5.3, which clearly shows the maximum peak area intensity at 1.18 mL/min flow rate. Välimäki and Perämäki[25] reported the effect of helium flow rate on the detector response of organomercury species. The studied mercury species were dimethylmercury, methylethylmercury, and diethylmercury. The effect of the flow of helium on these organomercury species is shown in Figure 5.4, from which it may be seen that the peak height increases with an increase in flow rate. Jiang et al.[28] carried out an interesting study on the detector response for $BuEt_3Sn$ and Bu_4Ge metal ions using different flow rates of hydrogen and air. The authors reported the different detector responses at different flow rates for both metal ion species. Similarly, Bettmer et al.[21] reported the maximum detection of mercury at a 105 mL/min flow rate of helium. In addition to these reports, other reports are available on the speciation of metal ion species with variation of the mobile

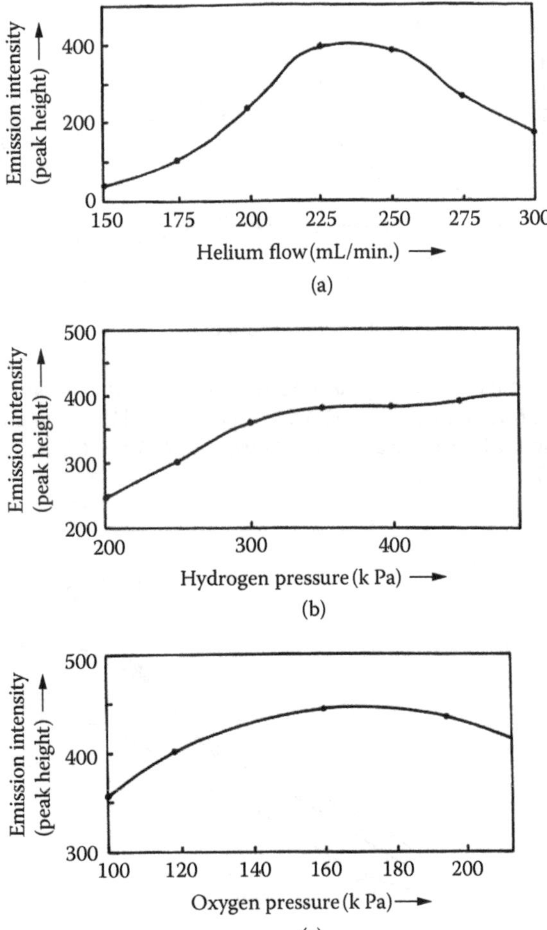

FIGURE 5.2 Effect of carrier and makeup gases flows on the separation of organotin species: (a) flow rate of helium (with oxygen and hydrogen makeup pressures of 160 and 400 kPa, respectively); (b) 350 kPa of hydrogen (with helium makeup flow rate of 240 mL/min and oxygen pressure of 160 kPa); and (c) 165 kPa of oxygen (with helium makeup flow rate 240 mL/min and hydrogen pressure 400 kPa, respectively). (From Girousi, S., Rosenberg, E., Voulgaropoulos, A., and Grasserbauer, M., *Fresenius' J. Anal. Chem.*, 358: 828 (1997). With permission.)

phase. Briefly, variation of the mobile-phase flow rate is the most important tool for optimizing metal ions speciation.

5.2.3 Temperature

As usual, temperature of the column is an important controlling factor in metal ions speciation by GC. A very slight variation in the temperature may result in

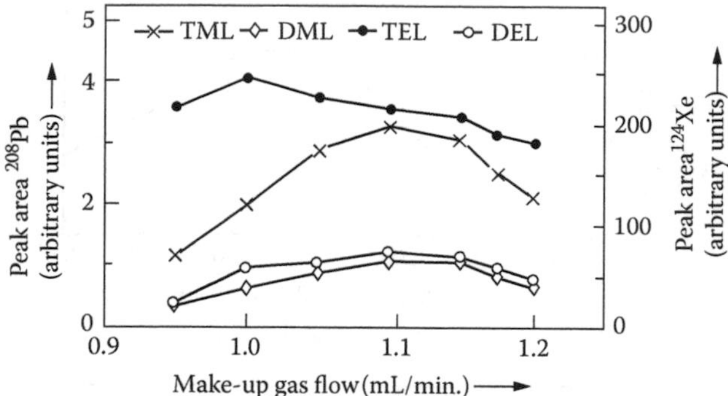

FIGURE 5.3 The influence of the makeup gas flow rate on the chromatographic signals of trimethyllead (TML), dimethyllead (DML), triethyllead (TEL), and diethyllead (DEL). (From Heisterkamp, M., and Adams, F.C., *Fresenius' J. Anal. Chem.*, 370: 597 (2001). With permission.)

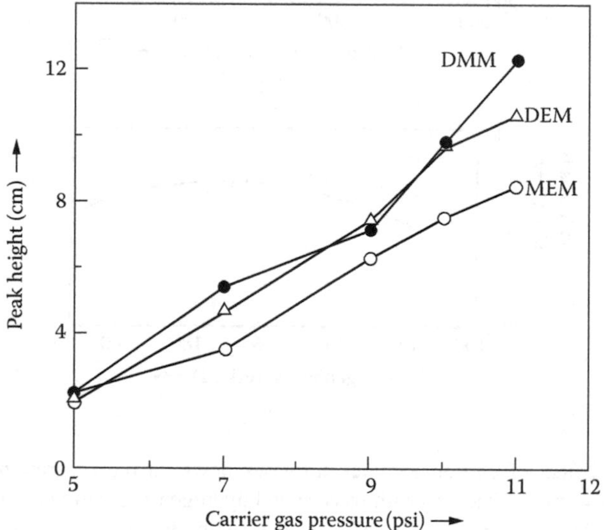

FIGURE 5.4 The influence of the carrier gas pressure on the detector responses of dimethylmercury (DMM), methylethylmercury (MEM), and diethylmercury (DEM). (From Välimäki, I., and Perämäki, P., *Mikrochim Acta*, 137: 191 (2001). With permission.)

a considerable change in retention times, which in turn affects the separation of metal ion species. Temperature variation of the injector and the detector may change the selectivity, sensitivity, and reproducibility. Maximum selectivity and separation are achieved by maximizing the energy difference between metal ion species and the stationary phase. Therefore, generally, lower temperatures (at which metal ion species are volatile) with relatively high linear velocities of carrier gas are the best choice for the good separation of metal ion species.

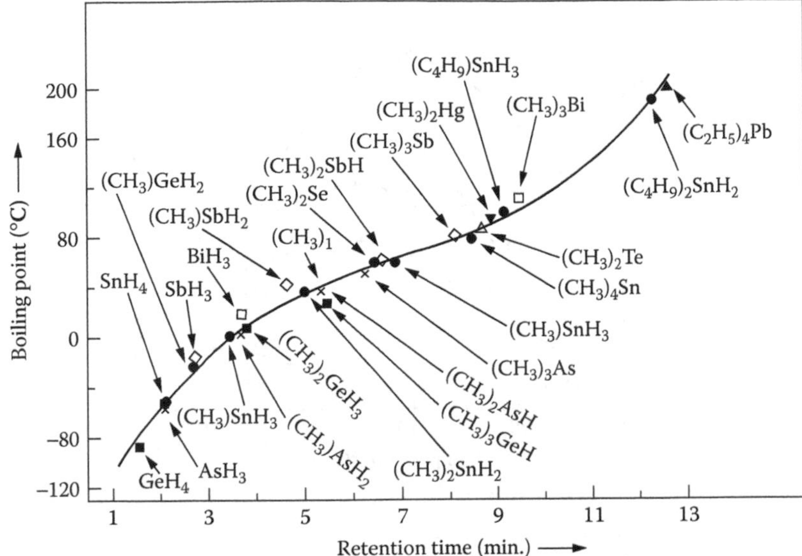

FIGURE 5.5 Boiling points and retention times correlations of some organometallic species. (From Grüter, U.M., Kresimon, J., and Hirner, A.V., *Fresenius' J. Anal. Chem.*, 368: 67 (2000). With permission.)

Generally, metal ion speciation was carried out at constant temperature of the column, but some workers also used gradient programming for the speciation of metal ions. The temperature of the GC machine for a particular metal ion species is fixed by trial and error. However, the experience may be useful to fix the temperature of GC. Grüter et al.[6] presented a graph (Figure 5.5) that indicates the boiling points and the corresponding retention times of some organometallic species. This figure may be used for temperature selection for the GC column.

Of course, temperature is a crucial parameter in the GC analysis of metal ions speciation, but only a few reports are available on the effect of temperature on the analysis of metal ion species. Jiang et al.[28] studied the effect of temperature of the detector on the peak heights of organotin and organogermanium species. The results are given in Figure 5.6a and Figure 5.6b. It may be concluded from these figures that the peak heights of organotin species first increase and then decrease with an increase in temperature. The maximum peak heights for organotin species were found at 120°C. In the case of organogermanium metal ion species, the maximum peak heights were reported at 210°C. A further increase in temperature was not useful, and the peak heights were almost constant. The authors explained this behavior in terms of the boiling point and the surface effect mechanisms. According to the authors, the increase of the peak heights (Figure 5.6a) was due to the increase in the volatility of the organotin species. Actually, the maximum response can be obtained if the temperature of the detector is higher than the boiling points of the metal ion species. The decrease

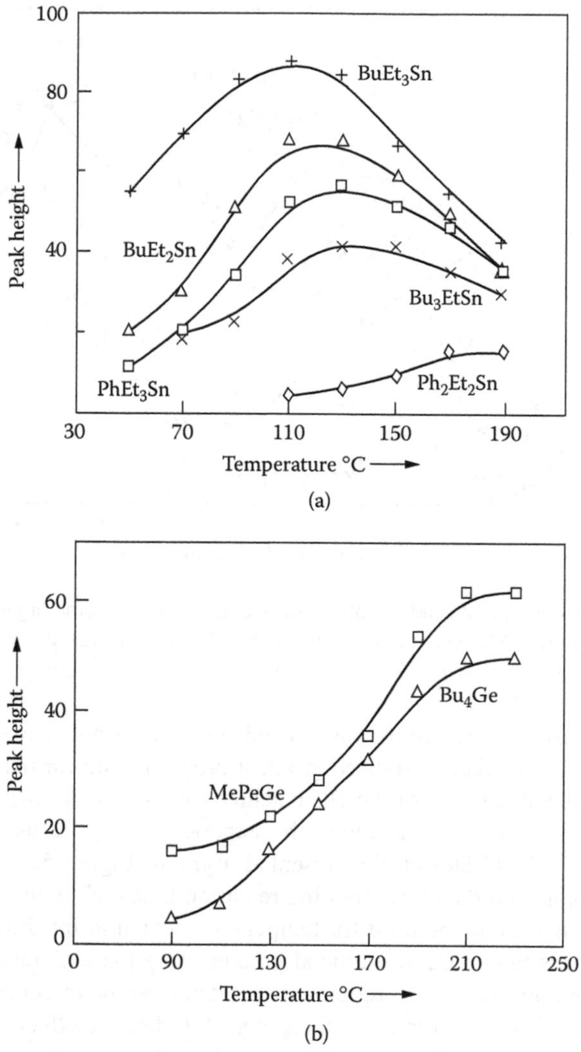

FIGURE 5.6 Effect of detector temperature on the peak heights of (a) organotin and (b) organogermanium species. (From Jiang, G.B., Ceulemans, M., and Adams, F.C., *J. Chromatogr. A*, 727: 119 (1996). With permission.)

in the peak heights at higher temperature, that is, greater than 120°C, was due to the surface effect mechanism of the blue tin emission. The behavior of the peaks in Figure 5.6b was explained by the authors as per the mechanism of germanium emission, which was different from tin compounds. It may be concluded from these figures that the experimental conditions for organotin compounds give poor detection than other species. Accordingly, this may be used as a possibility for detecting organotin species selectively in the presence of organogermanium species. Heisterkamp and Adams[16] coupled GC with an ICP-TOF-MS assembly via a transfer line unit, which was also heated. The authors studied the effect of the temperature of the transfer line unit on the peak width of the chromatograms of the organolead species, and the results are given in Figure 5.7a. It may be observed from this figure that the temperature barely affected the peak width of the signals of methylated species within the temperature range of 100 to 300°C. But the peaks of the ethylated species were noticeably broadened at temperatures below 200°C, with the peak width maximum above 250°C. Furthermore, the authors suggested 270°C as the best and maximum temperature of the transfer line unit in order to avoid decomposition of the coating of the transfer capillary and condensation of metal ion species on the column wall. Similarly, Välimäki and Perämäki[25] studied the effect of temperature of the quartz tube (interface) on the detector response for speciation of organomercury species. The results are shown in Figure 5.7b, which clearly indicates that the temperature range between 400 and 600°C was the optimum for the maximum detection of the organomercury species. Furthermore, it may be concluded from Figure 5.7b that the peak height increases with the increase of temperature (up to 600°C) due to the volatility of the species. The authors described the decrease of peak intensities by increasing the temperature (above 600°C) due to the ionization of organometallic species at higher temperatures.

5.2.4 Nature of Metal Ion Species

It may be observed from the literature that particular gas chromatographic conditions cannot be used for the speciation of certain metal ion species, due to the fact that different properties of metal ions are responsible for their separations. Therefore, the best speciation of a particular metal ion may be achieved by considering the structures of metal ions. In this regard, the chemical structures of metal ions should be determined, and then the maximum bonding site should be calculated, and a suitable GC column can then be used. Basic knowledge of chemistry may be useful to determine the maximum number of possible binding sites between metal ion species and the stationary phase (column). Moreover, molecular dynamics simulation calculations can be used to determine the suitable GC column. Derivatization should be carried out in such a way that the maximum number of bonding sites can be applied to the stationary phase. Moreover, it is worthy to mention here that by using these approaches, the costly chemicals and time spent on the procedure can be saved. To the best of our knowledge, no work was reported on this issue, but this aspect is also very important for the speciation of metal ions, and it cannot be overlooked.

5.2.5 Detection

The choice of the detector to be used in GC analysis of metal ion species is essential. Basically, the area of the chromatographic peak depends on the detector response. Metal ion species can be analyzed at trace levels if the detector

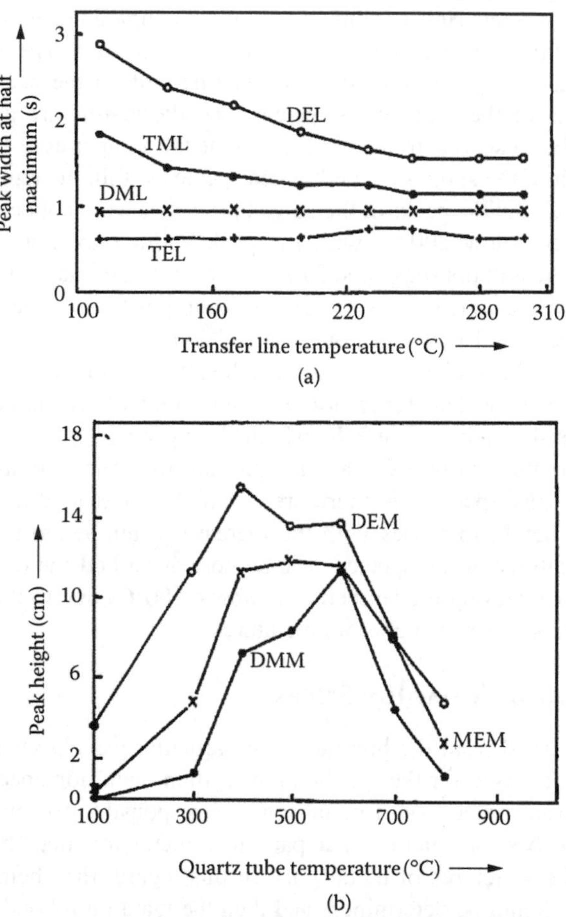

FIGURE 5.7 Effect of temperature of (a) transfer line (interface) on the peak widths (at half height) of organolead species and of (b) quartz tube on peak height of organomercury species. (Part a from Heisterkamp, M., and Adams, F.C., *Fresenius' J. Anal. Chem.*, 370: 597 (2001); Part b from Välimäki, I., and Perämäki, P., *Mikrochim. Acta*, 137: 191 (2001). With permission.)

response is very sensitive with a low limit of detection. In metal ion speciation by GC, two types of detectors are used. Those in the first category are nonelement specific and have a wide range of applications but are subject to interference due to the presence of impurities. The second class of detectors is element specific and can be used for a particular element only. Applications of these detectors can be broadened by changing the energy sources and the working wavelengths. The chances of interference in these detectors due to the presence of impurities are very low. Basically, in speciation analysis, the choice of the detector depends on the chemical forms to be determined and on the mode of the separation used. The different nonelement-specific detectors used for speciation analysis in GC are electron capture detector (ECD), flame ionization detector (FID), flame photometric detector (FPD), fluorescence detector (FD), and mass spectrometry (MS). The element-specific detectors are atomic absorption spectrometry (AAS), atomic emission detector (AED), atomic fluorescence spectrometry (AFS), inductively coupled plasma (ICP), microwave induced plasma–atomic emission detector (MIP-AED), mass spectrometry–selected ion-monitoring mode (MS-SIM), inductively coupled plasma–flame photometric detection (ICP-FPD), microwave-induced plasma–atomic emission detector (MIP-AED), cold vapor atomic absorption spectrometry (CVAFS), quartz tube–atomic absorption spectrometry (QT-AAS), ICP-TOF-MS, plasma emission detector (PED), and so forth (Table 5.1).

The nonelement-specific detectors are supposed to be good due to their universal use. Moreover, the range of the nonelement-specific detectors is very wide, and hence, they can be used for the detection of a variety of metal ion species. In addition, low cost, wide availability, competitive performance, and good detection sensitivity of the nonelement-specific detectors make them ideal detection devices for use in GC speciation of the elements. To increase the performance of nonelement-specific detectors, certain improvements were carried out. For example, Jiang et al.[28] used a special type of FPD detector in place of commercially available FPD detectors. The authors coupled the FPD detector with GC through an interface (Figure 5.8) and reported the minimization of the adsorption of metal species on the interface wall. Moreover, as per the authors, the chromatographic resolution was increased, and the peak shape was improved by directly inserting the capillary column into the innermost tube of the FPD burner at about 10 mm below the burner tip. This ensured the direct transfer of the chromatographic effluent to the burner tip, avoiding premixing of the GC effluent with the flame gas. Furthermore, the authors[28] reported the speciation of organotin and organogermanium species with good precision and sensitivity. Therefore, these detectors have been used for the detection of various elements. Despite the good performance of these detectors, they have been rapidly replaced by the element-specific detectors due to the fact that the nonelement-specific detectors experience interference due to the presence of the impurities in the environmental and biological samples. The nonelement-specific detectors also have poor sensitivity to some elements. For example, ECD has poor selectivity for organomercury species and often requires a large amount of concentration.

In the past decade, element-specific detectors have been used frequently for the speciation of metal ions in GC. Although, these detectors are very costly yet they give reliable analytical results in the analysis of element speciation. Because the working principles of these detectors are based on the type of element, the chances of interference due to impurities are very low, which outweighs the downside of the tedious extraction procedure setup. Moreover, the detection limits are very low in these types of detectors. Briefly, the element-specific detectors are supposed to be the best choice in the GC analysis of metal ion species. Of course, these detectors are the best detection tools in metal ion speciation, but the high cost is a hurdle to their use in some laboratories, especially in the laboratories

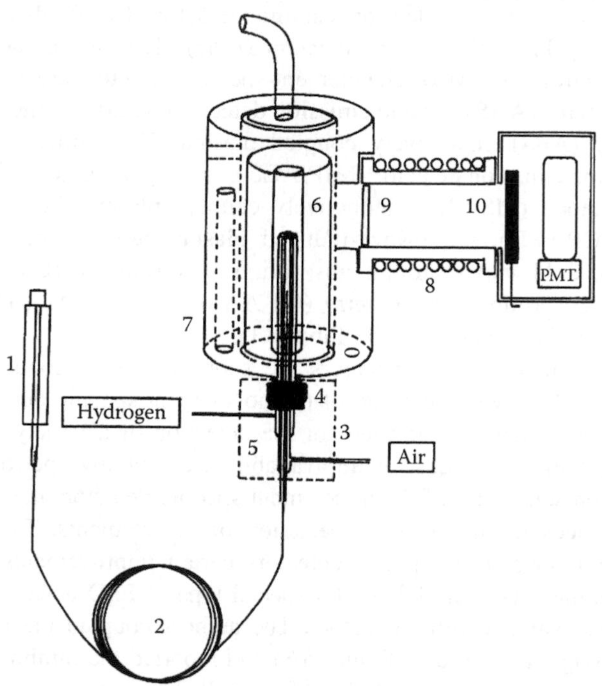

FIGURE 5.8 The gas chromatography– flame photometric detector (GC-FPD) assembly: (1) on-column injection port, (2) capillary column with retention gap, (3) temperature protecting block, (4) adjustable screw, (5) stainless steel burner, (6) quartz enclosure, (7) heating cartridge, (8) cooling water, (9) quartz window, and (10) filter. (From Jiang, G.B., Ceulemans, M., and Adams, F.C., *J. Chromatogr. A*, 727: 119 (1996). With permission.)

of developing and underdeveloped countries. Another complication in the use of these detectors is the development of the interface between GC and the detectors.

The coupling of microwave-induced, plasma–atomic emission spectrometry (MIP-AES) to GC was first commercialized by Hewlett Packard (now Agilent Technologies), and this unit has been used for many of the element species.[91] Another development in these types of detectors is the coupling of more than one detector device. The coupling of ICP and MS provides the best results in terms of detection limits and, therefore, has been used widely in metal ion speciation in environmental samples. Chong and Houk[37] in 1987 first reported the coupling of ICP and MS. The development of the capillary columns has provided a preferred status to such types of coupling in terms of sensitivity and resolution.[47] This developed unit was used for the speciation of many elements, for example, mercury, tin, and lead.[73] Moreover, the most important application of this unit was speciation of metalloporphyrins of geological interest, which exhibit high boiling points.[92] Smaele et al.[93–95] developed an important interface design for ICP and MS coupling and then connection to GC. The authors used the developed unit (GC- ICP-MS) for the successful detection of mercury, lead,

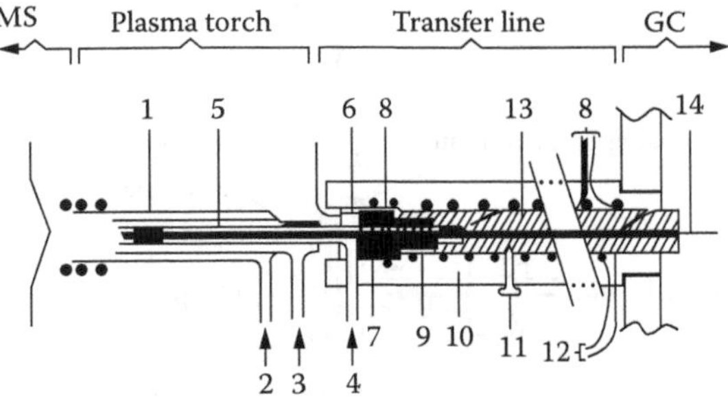

FIGURE 5.9 The gas chromatography–inductively coupled plasma–mass spectrometry (GC-ICP-MS) system. (1) demountable ICP torch, (2) cool gas, (3) auxiliary gas, (4) injector gas, (5) stainless steel tube, (6) aluminum collar, (7) graphite tape, (8) thermocouples, (9) stainless steel reducing union, (10) industrial pipe lagging, (11) grounding point, (12) heater leads to variable voltage supply, (13) aluminum bar, and (14) capillary GC column. (From Cooke, W.S., *Today's Chem. Work*, 5: 16 (1996). With permission.)

and tin metal ions in environmental samples. The authors claimed the detection limit at the parts per trillion (ppt) level by this assembly. The coupling of GC with ICP-MS involves the modification of the GC machine. Normally, a tube line of less than 1 m in length is required.[37,47,73,74,92,96–99] The transfer line is run to the base of the ICP torch, and the line is generally heated to prevent condensation of metal ion species. Similarly, Prohaska et al.[5] developed a homemade interface between GC and ICP-MS for the speciation of arsenic species in soil samples. The use of the glass[37] or quartz[25,74] tube is another development in the interface line. These materials minimize the possible adsorption of metal ions on their walls. In addition to this, sometimes the carrier gas from GC (carrying metal ions) requires a makeup gas to achieve a sufficient flow of metal ion species to the plasma. Therefore, a device to supply a makeup gas is also required in an ICP unit. A general diagram of this type of unit showing all components is shown in Figure 5.9. Recently, Montes-Bayon et al.[100] developed a nonheated flexible interface design, as shown in Figure 5.10. The different parts of this unit may be understood from this figure. This interface was coupled with GC, and the developed assembly was used for the detection of mercury, lead, selenium, and tin metal ions.[101–105] Recently, Agilent Technologies commercialized an interface design with the advantage of flexibility in terms of heating. The developed unit was found to be very fruitful in the detection of metal ions, with the detection limit in the range of femtogram.[88] Recently, Heisterkamp and Adams[16] reported ICP-TOF-MS for organolead speciation. In this detection device, ions are sampled into ion packets in the flight tube and are separated on the basis of their mass-to-charge (m/z) dependent velocities. Therefore, in this way, all ions remain available for mass analysis, showing no spectral skew effect.

Holland et al.[106] described spectral skew effect as a small change in the time that may result in a significant change in the measured intensity of the peaks during a chromatographic run. This may be categorized into two types: one

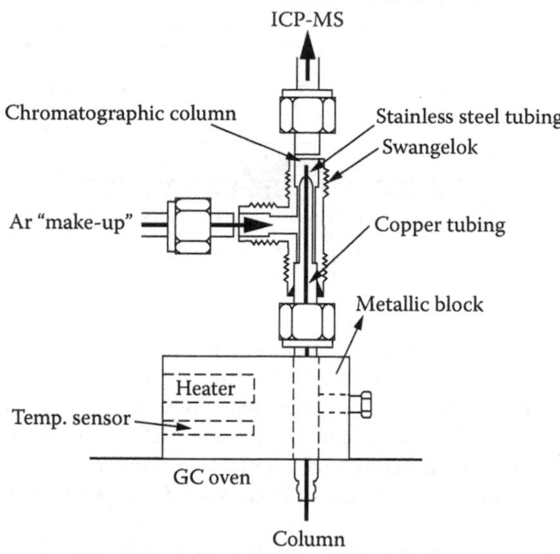

FIGURE 5.10 The flexible interface design. (From Chong, N.S., and Houk, R.S., *Appl. Spectrosc.*, 41: 66 (1987). With permission.)

caused by the transient signal and the other caused by the detection based on the scanning detectors. This effect is inherent in the profiles of transient signals and cannot be corrected by any detection system. It was reported that in the ICP-MS detection system, a quantification error occurred in the measurement of the isotopes due to spectral skew effect.[16] On this issue, the efforts of Allen et al.[107,108] are significant, as they overcame this problem by using modified MS based on scanning mass analyzers that enabled the determination of more than one isotope simultaneously, for example, twin quadrupole instruments enabled the simultaneous determination of two isotopes. Furthermore, Walder et al.[109,110] described the measurement of several isotopes simultaneously using a multicollector ICP-MS-based double-focusing magnetic selector mass analyzer with multiple Faraday cup. Besides, unlike ICP-MS, the number of isotopes determined is not limited by peak definition in ICP-TOF-MS, as the number of data points per peak is independent of the number of measured isotopes. Therefore, ICP-TOF-MS is the best detection system for the analysis of isotopes due to the high rate of data acquisition. The ICP-TOF-MS detection system developed by Heisterkamp and Adams[16] is shown in Figure 5.11, from which the different parts of this unit can be understood. Similarly, Dmuth and Hermann[105] used the isotope dilution technique in the coupling of GC with ICP-MS. The authors reported an improvement in accuracy and precision in the determination of mercury species using the species-specific spikes phenomenon. The element-specific detectors are being used frequently for the speciation of metal ions, but they suffer from certain drawbacks. The most important drawback of these detectors is that they cannot be used to speciate all the elements simultaneously. However, their use

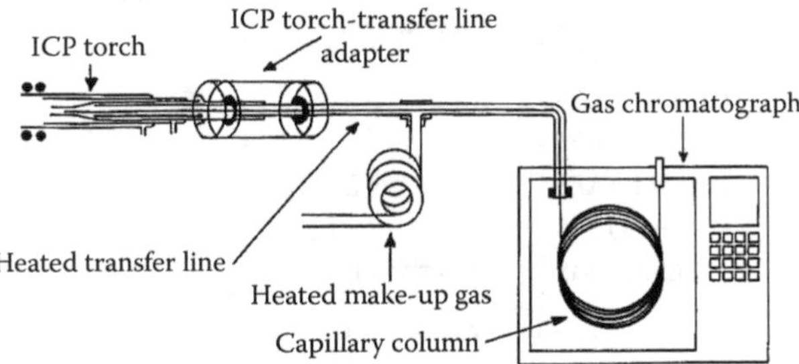

FIGURE 5.11 The gas chromatography–inductively coupled plasma–time-of-flight–mass spectrometry (GC-ICP-TOF-MS) system. (From Heisterkamp, M., and Adams, F.C., *Fresenius' J. Anal. Chem.*, 370: 597 (2001). With permission.)

for a wide range of metal ions may be possible by using different energy sources (in AAS) and wavelengths (in AES).

The detection limit is the most important measurement of the quality of a detector. A detector is supposed to be the best if it provides the lowest limit of detection. Generally, metal ion species in the environment occur at very low concentrations, and, therefore, the detector with the lowest limit of detection is required in the GC analysis of metallic species. In view of this, several authors described the limits of detection in the GC analysis of metal ion species. Liu and Lee[111] reported the limits of the detection of the nonelement-specific detectors in their reviews as ranging from nano- to picogram levels for the analysis of different metal ion species. To achieve low values of detection limits, few modifications were carried out in the nonelement-specific detectors. Jiang et al.[28] developed a special type of FPD detector (Figure 5.8) and coupled it with GC. The authors advocated an improvement in the detection limits of this detector in comparison with the commercial FPD detectors. The authors reported an improvement in detection limits due to the minimization of the adsorption of metal species on the interface wall.

In contrast to the nonelement-specific detectors, several modifications were made in the element-specific detectors to achieve the minimum limits of detection for the analysis of metallic species in GC. Jiang et al.[28] claimed that the detector response depended on gas flow rates, detector temperature, and operation mode. The linear range was up to four orders of magnitude for organotin and up to three orders of magnitude for organogermanium compounds. The proposed method is simple, highly sensitive, and reproducible. Carro-Díaz et al.[19] optimized GC conditions in order to obtain the lower detection limits for organomercury species. Under the proposed operational conditions, the detection limit (signal-to-noise ratio = 3) was 1.2 pg with a linear range of 1 to 40 ng mL^{-1} (as methylmercury in samples). The authors used certified reference material (DORM-1) to evaluate the accuracy. Rodriguez Pereiro et al.[23] reported very low detection limits (0.01 pg mL^{-1} of Hg for methylmercury). Furthermore, the authors advocated these values of limits of detection due to the narrow injection band and reduced peak broadening in a bundle of 0.038 mm capillaries at high flow rates (>60 mL min^{-1})

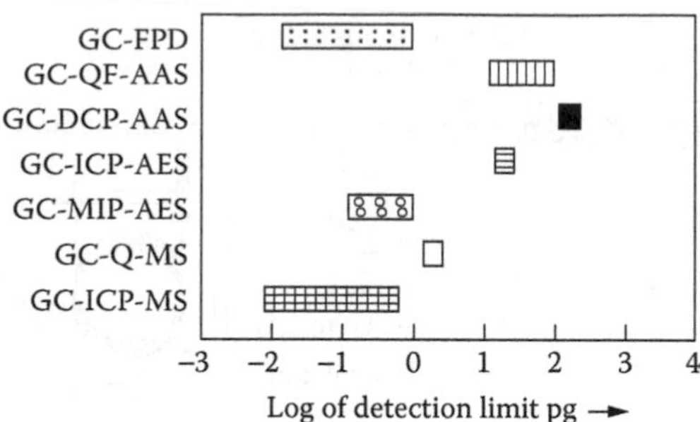

FIGURE 5.12 The detection limits (as absolute mass in pg) obtained by gas chromatography (GC) coupled with different detectors. (From Chong, N.S., and Houk, R.S., *Appl. Spectrosc.*, 41: 66 (1987). With permission.)

compatible with an MIP-AES detector (no dilution with a makeup gas is required). Välimäki and Perämäki[25] developed a purge-and-trap preconcentration technique to improve the detection limits for mercury metal ion species using Tenax-ta adsorbent. With this system, the measuring range was between 0 and 5 ng (as Hg) for the different species (dimethylmercury, methylmercury, and inorganic mercury(II)) in a 100 μL sample. The detection limits obtained for a 250 mg sample with the standard addition method were 72 μg/kg, 136 μg/kg, and 224 μg/kg for dimethylmercury, monomethylmercury, and inorganic mercury(II), respectively. Girousi et al.[29] optimized the chromatographic conditions to achieve low values of the detection limits. An amount of 100 mL of sample were used, for the analysis and the detection limits for butyltin species ranged from 17.7 to 33.4 ng/L (as Sn), and for phenyltin species from 17.8 to 22.3 ng/L (as Sn). The method was applied to the determination of organotin species in samples collected from the Thermaïkós Kólpos (Gulf of Salonika, Thessaloniki, Greece) at four sampling points and three sampling campaigns. The detection limits in GC achieved, using different detectors, are shown in Figure 5.12. It may be observed from this figure that the combination of ICP and MS resulted in an improved separation of metal ion species. The fact that element-specific detectors are superior to nonelement-specific detectors may be proven by the work of Jiang et al.,[28] who compared the detection limits of organotin species using AAS and FPD detectors in water and sediment samples. The results of the limits of the detection of organotin species by AAS and FPD detectors are given in Table 5.4. It is very interesting to note that the values of the organotin species concentrations determined by AAS are higher than the values determined by FPD detector, which clearly indicates the superiority of AAS over FPD detector. Briefly, the element-specific detectors, with or without some modifications, have become powerful detection devices in the GC analysis of metal ion species. Although early GC-MIP-AED systems had many drawbacks, the commercial MIP-AED assembly has many advantages over these earlier designs, particularly in regard to the venting of solvents to protect the plasma stability, background correction, and ease of data handling. The use of a high-performance monochromator with a photodiode array spectrometer also

TABLE 5.4
Comparison of the Determination Precision of Organotin Species in Water and Sediment Samples by Atomic Absorption Spectroscopy (AAS) and Flame Photometric Detector (FPD)

Samples	Species	AAS (μg Sn dm^{-3} or g^{-1})	FPD (μg Sn dm^{-3} or g^{-1})
River Water 1	MBT	0.207 ± 0.005	0.201 ± 0.016
	DBT	0.385 ± 0.008	0.350 ± 0.024
	TBT	0.523 ± 0.009	0.512 ± 0.028
River Water 2	MBT	0.209 ± 0.007	0.189 ± 0.004
	DBT	0.343 ± 0.009	0.298 ± 0.008
	TBT	0.373 ± 0.008	0.333 ± 0.022
Sediment	MBT	0.081 ± 0.008	0.098 ± 0.002
	DBT	0.232 ± 0.024	0.219 ± 0.007
	TBT	1.112 ± 0.012	1.255 ± 0.033

Note: MBT: monobutyltin, DBT: dibutyltin, and TBT: tributyltin.

Source: From Jiang, G.B., Ceulemans, M., and Adams, F.C., *J. Chromatogr. A*, 727, 119, 1996. With permission.

improves the sensitivity. Pereiro and Diaz[112] reviewed the speciation of mercury, tin, and lead metal ions by GC. The authors compiled the absolute detection limits of different detectors for organotin compounds. The values of these detection limits are summarized in Table 5.5. It may be observed from this table that the best detector was GC-ICP-MS in terms of low values of detection limits.

Several variables have been varied to achieve lower limit of detection by element-specific detectors. Most important are the selection of suitable wavelength, microwave power, energy angle of the light, oscillation angle, and so forth. Tutschku et al.[27] tried to optimize the limit of detection of tin metal ion using different wavelengths of MIP-AES. The authors used different wavelengths and reported lower detection limit (0.8 pg) using 326.23 nm as the wavelength (Table 5.6). Bettmer et al.[21] varied the energy angles of the light and the oscillation of the PED to obtain the best response of the detector, and these findings are given in Figure 5.13. The authors reported 7° and 9.9° as the best angles of the light and the oscillation of PED, respectively. Furthermore, the same authors reported the most sensitive detection of mercury at a microwave power of 85 W (Figure 5.14). Low values of the detection limits may also be achieved by optimizing the other working parameters of element-specific detectors. These parameters include the purity of gases, maximization of the energy of the energy sources, and proper mixing of the gases and the fuel gases.

5.2.6 Other Parameters

In addition to the above-discussed GC parameters, the optimization of metal ion speciation may be achieved by controlling the amount injected, coupling two

TABLE 5.5
Comparison of the Detection Limits of Organotin Compounds Using Different Detectors

Systems	Detection Limits (pg)
GC-AAS	40–95
GC-FPD	0.2–18
GC-QF-AAS	10–100
GC-MS	1–10
GC-ICP-MS	0.0125–0.17
GC-MIP-AED	0.01–0.03

Note: AAS: atomic absorption spectrometer, GC: gas chromatography, ICP-MS: inductively coupled plasma–mass spectrometer, MIP-AED: microwave-induced plasma–atomic emission detector, MS: mass spectrometer.

Source: From Pereiro, I.R., and Diaz, A.C., *Anal. Bioanal. Chem.*, 372, 74, 2002. With permission.

TABLE 5.6
Detection Limits of Tin Metal Ion at Different Wavelengths by Microwave-Induced Plasma–Atomic Emission Spectrometry

Wavelengths	Detection Limits (pg)
189.98	1.4
270.65	2.3
284.00	2.1
303.41	2.5
317.50	8.4
326.23	0.8

Source: From Campillo, N., Aguinaga, N., Vinas, P., Lopez-Garcia, I., and Hernandez-Cordoba, M., *Anal. Chim. Acta*, 525, 273, 2004. With permission.

columns in a series, using multicapillary columns, and utilizing proper extraction and derivatization procedures. Rodriguez et al.[23] studied the effect of volume injection on the speciation of organomercury species. Prohaska et al.[5] reported 15 and 224 pg/g concentrations of As(III) and As(V) without prereduction in hydride generation. However, the concentrations of As(III) and As(V) were

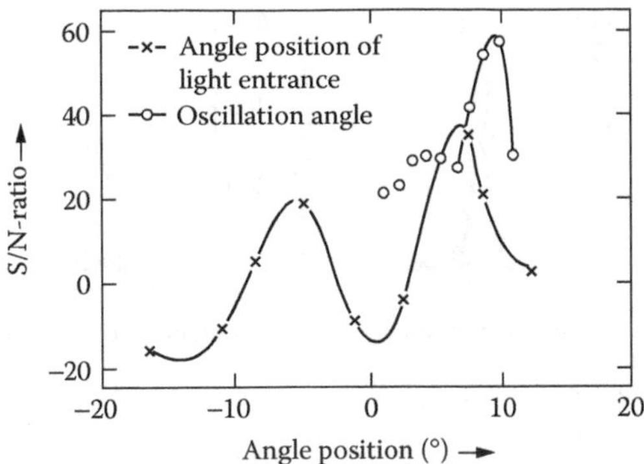

FIGURE 5.13 Effect of the entry angle of light and the oscillation angle on the detection of mercury. (From Bettmer, J., Bradter, M., Buscher, W., Erber, D., Rieping, D., and Cammann, K., *Appl. Organo. Chem.*, 9: 541 (1995). With permission.)

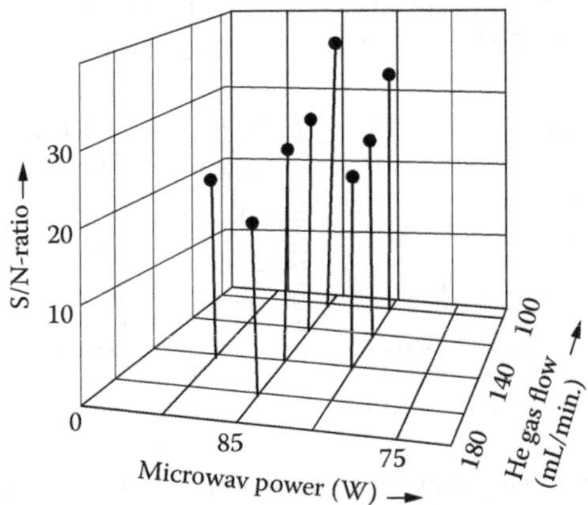

FIGURE 5.14 Effect of microwave power and gas flow on the detection of mercury. (From Bettmer, J., Bradter, M., Buscher, W., Erber, D., Rieping, D., and Cammann, K., *Appl. Organo. Chem.*, 9: 541 (1995). With permission.)

reported as 14 and 35 pg/g with prereduction before hydride generation. This finding clearly indicates that the reduction before hydride generation is affecting the concentrations of the arsenic species. The ICP-TOF-MS is able to collect more than 20,000 complete mass spectra per second, but still no computer is available to handle such a large volume of data spontaneously. Therefore, individual mass spectra are summed up to give integrated mass spectra within a particular period of time. This integration time is optimized, as it affects both peak shape and signal-to-noise ratio. In this field, Heisterkamp and Adams[16] studied the effects of the integration time on the profile of the trimethyllead peaks, which is shown in Figure 5.15. As per the authors and Figure 5.15, an integration time of 102 sec is sufficient for obtaining good-shaped peaks. Recently, Cooke[113] reported a

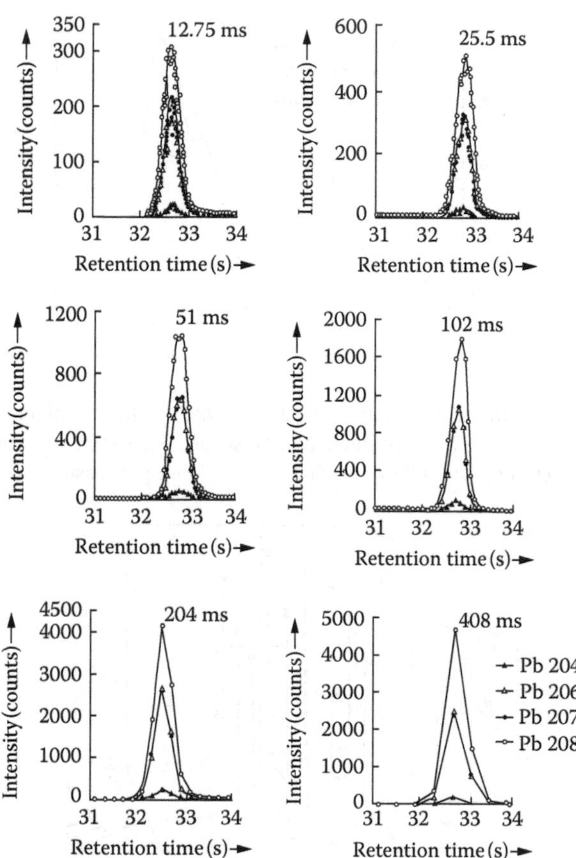

FIGURE 5.15 Effect of integration time on the profile of the trimethyllead (TML) peak. (From Heisterkamp, M., and Adams, F.C., *Fresenius' J. Anal. Chem.*, 370: 597 (2001). With permission.)

reduced analysis time without a reduction in loading amount in GC by using a bundle of 1000 capillaries of 40 μm diameter. The small length of these columns and the use of isothermal separations allows us to construct a miniaturized speciation instrument coupled to a MIS-AES detector.[114] Briefly, metal ion speciation in the environment by GC may be achieved by considering all the factors discussed above. Based on our experience and the literature available, a scheme was developed to optimize the speciation of metal ions by GC in the environmental and biological samples. The developed protocol is given in Scheme 5.2.

5.3 DERIVATIZATION OF METAL IONS

As obvious, GC can be used only for the analysis of volatile analytes; therefore, metal ions are converted into volatile species by derivatization procedures. Metal

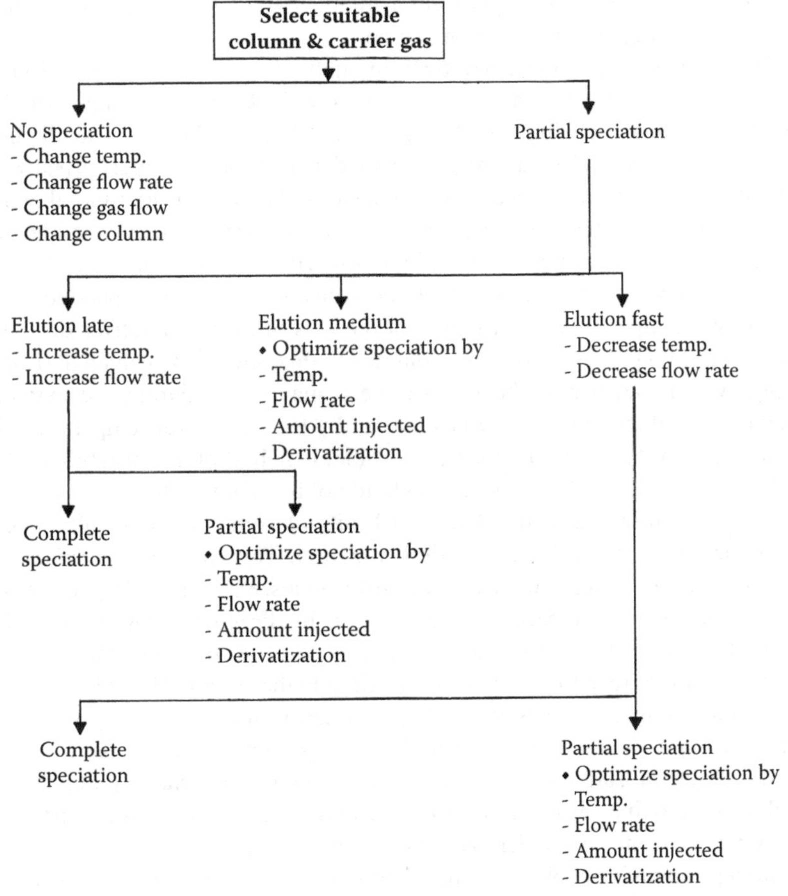

SCHEME 5.2

ions are generally nonvolatile at the temperature of GC. Therefore, the conversion of metal ions into volatile species is essential by using suitable derivatizing reagents. In addition, the derivatization changes the retention properties of metallic species, which may result in the improved resolution of metal ions by changing their volatility and polarity. The derivatization may also lower the analysis temperature and maximize energy differences in the metallic species, which results in faster analysis time, more throughput, and better separations. A derivatizing reagent should be pure and capable of forming the derivatives with a wide range of metallic species. The derivatizing reaction should not involve complex reactions and instrumentations. The derivatizing reagents are selected depending on the types of metal ions. Generally, acylation or alkylation of metal ions is carried out to make the organometallic species. It is worth mentioning here that the derivatization reagents should be selected by which no by-products may produce. In addition, complexation of metallic species with a suitable ligand can be carried out.[111] In view of this, many reports are available on the derivatization of metallic species before loading onto a GC machine.

Due to these facts, the derivatization methods of hydride generation and alkylation with Grignard reagents have been well documented since 1970 for several metal ions, including As, Bi, Hg, Pb, Sb, Se, and Sn.[111,113] The alkylation of metallic species by Grignard reagent provides the best use of the derivatization methodology in the GC speciation of metal ions. The use of different alkylating groups, for example, methyl-, ethyl-, propyl-, butyl-, pentyl-, hexyl-, and phenyl-, depends on the type of metal ion. Alkylation with longer alkyl chains, such as pentyl- or hexyl-, provides less volatile derivatives that can be separated by GC using drastic temperature conditions. But the loss of these derivatives during the preconcentration procedure is poor. Contrary to this, lower alkyl chain derivatives are highly volatile; hence, the loss of these may occur during the extraction procedure, but they may be separated easily by GC at a lower temperature. The use of tetraethylborate overcomes the problem of hydrolytic instability of the Grignard reagents, which makes the alkylation of metal ions into aqueous medium possible.[48,111] Additionally, the detection limits of some organometallic species may be better, as it is a foam-free derivatization. This method was applied for the derivatization of lead, mercury, and tin species.[114,115] Hydride generation is easier to handle in comparison to alkylation by Grignard reagent or sodium tetraethylborate.[48] This method is supposed to be better than aqueous ethylation in terms of shorter reaction time and purge time in the reactor. Hydride generation was used as the method of choice for one oxidation state of As, Ge, Sb, and Sn, while it cannot be used for lead and mercury metals due to their unstable hydrides.[116] De la Calle-Guntinas et al.[117] presented esterification and acrylation of selenomethionine in wheat sample. Lu and Whang[84] used tropolone as the derivatizing reagent for the derivatization of tin metal ions.

Garcia-Fernandez et al.[24] used three derivatization approaches for the derivatization of tin metallic species: anhydrous butylation by using a Grignard reagent, aqueous ethylation by means of $NaBEt_4$, and aqueous propylation with $NaBPr_4$. For ethylation procedures, a methylpropylmercury solution was used as the internal standard. The absence of transmethylation during sample preparation was checked using a 97% enriched Hg inorganic standard. The accuracy of three derivatization approaches was evaluated by the analyses of the certified reference material DOLT-2 (dogfish liver). Gallina et al.[31] used the direct alkylation reaction of tin metal ion in the aqueous digestion solution by $NaBEt_4$, which allowed for a short analysis time and a good recovery. The evaluation of the yield of each step constituting the analytical procedure indicated that the alkylation step is the most critical. Tutschku et al.[27] studied the influence of the measurements of sensitivity on mixtures of organotin compounds with different alkylation, which showed a dependence of the grade of pentylation. Välimäki and Perämäki[25] used sodium tetraethylborate for the ethylation of mercury species. It is interesting to mention here that the alkyl arsenic species are not volatile, and their determination by GC is difficult. For GC, these species are converted into volatile species by reacting with thioglycol methylate (TGM).[3] Mester et al.[3] described the deriva-

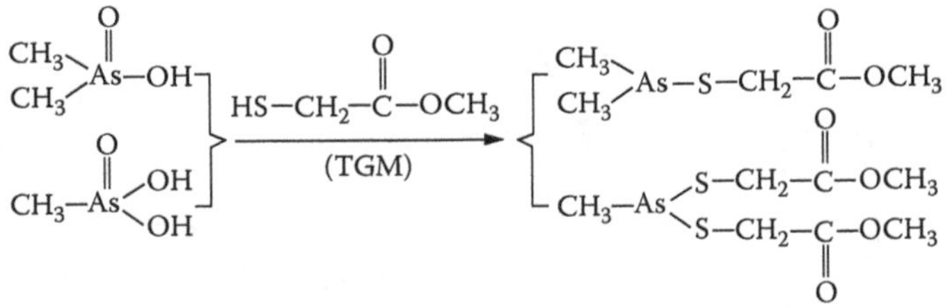

$$\left.\begin{array}{l} (CH_3)_2As(=O)-OH \\ CH_3-As(=O)(OH)_2 \end{array}\right\} \xrightarrow[\text{(TGM)}]{HS-CH_2-C(=O)-OCH_3} \left\{\begin{array}{l} (CH_3)_2As-S-CH_2-C(=O)-OCH_3 \\ CH_3-As(S-CH_2-C(=O)-OCH_3)_2 \end{array}\right.$$

FIGURE 5.16 The derivatization reaction of monomethyl arsenic acid (MMA) and dimethyl arsenic acid (DMA) with thioglycol methylate (TGM). (From Mester, Z., Vitányi, G., Morabito, R., and Fodor, P., *J. Chromatogr. A*, 832: 183 (1999). With permission.)

tization of mono- and dimethyl arsenic species by thioglycol methylate, and the reaction is shown in Figure 5.16. It may be seen from this figure that mono- and dimethyl arsenics were substituted by two and one units of thioglycol side chains, separately, and respectively. As a result, the difference between the polarities of the two derived species was noted, and speciation of these became an easy task. The rate and the extent of the derivatization reactions depend on many parameters. The most important controlling factors are pH of the medium, metal ion concentration, and temperature. Zufiaurre et al.[14] studied the effect of pH on the alkylation of lead metal ion. The authors studied a pH range from 2 to 12 and reported the major derivatization yield at pH 4 (Figure 5.17). Similarly, Mothes and Wennrich[9] described the derivatization of lead metal ions as pH dependent. The results are shown in Figure 5.18, and it may be concluded that the derivatization of lead metal ion with $NaBEt_4$ is pH dependent. The authors used different concentrations of lead metal ions, and the derivatization was carried out at dif-

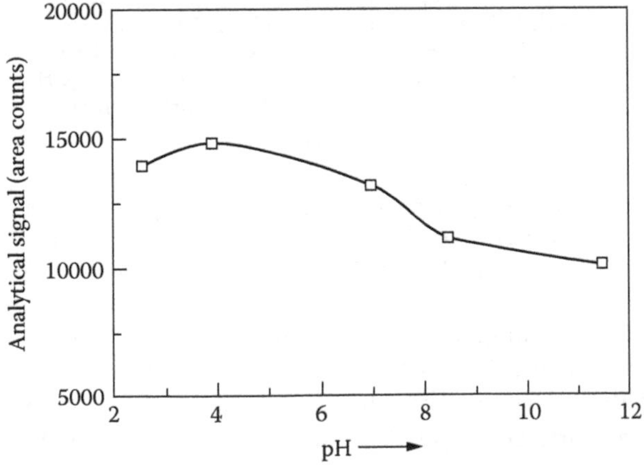

FIGURE 5.17 Effect of pH of the reaction medium on the alkylation of lead metal ion. (From Zufiaurre, R., Pons, B., and Nerín, C., *J. Chromatogr. A*, 779: 299 (1997). With permission.)

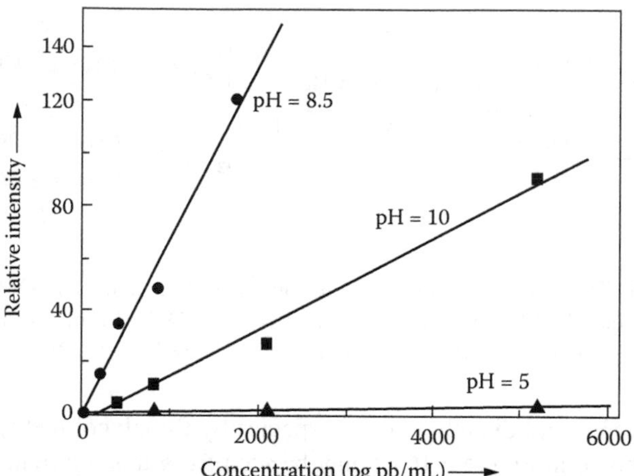

FIGURE 5.18 Effect of pH of the medium and the concentration of lead metal ion on the derivatization. (From Mothes, S., and Wennrich, R., *Mikrochim. Acta*, 135: 91 (2000). With permission.)

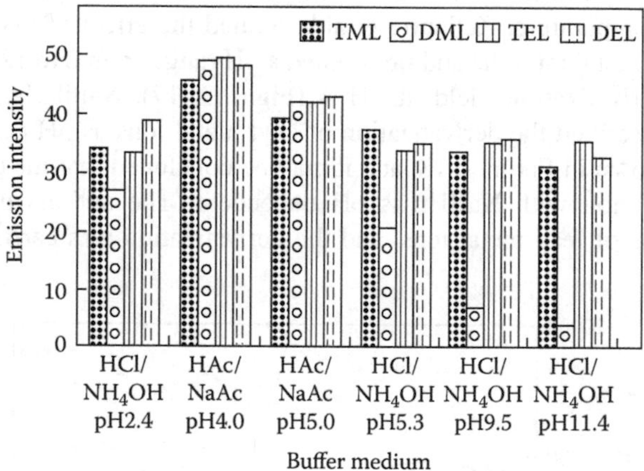

FIGURE 5.19 Effect of the reaction medium on the recovery of different organolead species: HCl: 0.1 *M* citric acid, HAc: acetic acid, and NaAc: sodium acetate. (From Heisterkamp, M., and Adams, F.C., *Fresenius' J. Anal. Chem.*, 362: 489 (1998). With permission.)

ferent pH levels. According to the authors, the most suitable pH for derivatization was 8. Heisterkamp and Adams[15] described the derivatization of lead metal ion with $[Bu_4N]^+[Bbu_4]^-$. The authors studied the effect of reaction medium on the recovery of the different organolead species (Figure 5.19). Using an acetic acid/sodium acetate buffer at pH 4, a maximum recovery was achieved. Generally, the derivatization is carried out after the extraction, purification, and concentration steps. However, the possibility of direct derivatization (with untreated samples) was reported.[112]

Despite the critical need for derivatization in GC, the risk of producing the wrong derivatives increases with the number of steps and with their complexity in a determination. The derivatization yields, matrix dependent, are very difficult to determine due to the lack of appropriate high-purity calibrants. Moreover, some derivatives are unstable, leading to a number of unnecessary by-products. These issues create the problems encountered in the GC analysis of metallic species. The exact mechanisms of the derivatization are not known; hence, the chances of the formation of by-products cannot be eliminated. Therefore, if derivatization can be avoided, it is worthwhile to consider such possibilities. In view of all of these facts, a roundtable discussion was carried out at a speciation workshop, and recommendations for improving the derivatization procedure were given.

5.4 MECHANISMS OF METAL IONS SPECIATION

The basic principle of GC is based on the distribution of analytes between the stationary and the mobile phases. It was described in many general and research books on GC. However, the exact mechanisms of the separation of metal ion species on the GC column are not known. However, a brief discussion on the separation of different metallic species of the same metal ion is presented herein. Various species of the same metal ions differ in physical and chemical properties. Moreover, the derivatization of metallic species resulted in the differentiation of their physicochemical properties. Therefore, these species are distributed to different extents between the mobile and the stationary phases. The basic processes responsible for their different distributions are adsorption and the partition that depend on the nature of the stationary phase and metal ion species. The different species are adsorbed and partitioned to different extents in the stationary phase. At the same time, in the mobile phase, these metallic species are carried forward, while on the stationary phase, the metallic species are retained, and as a result, different metallic species are eluted at different elution times and separation occurs. There are many forces and bondings responsible for the adsorption and the partition of metallic species. The most important interactions are hydrogen bonding, dipole-induced dipole interactions, and π–π forces. The steric effect also plays a crucial role in the separation of different metallic species. Some other forces, such as Van der Waals forces, and ionic interactions, are also responsible for the separation phenomenon. As discussed previously, most metal ion speciation was carried out on silica gel or silica-gel-based GC columns. Therefore, different types of interactions, as mentioned above, occurred between the hydroxyl groups of silica gel and metallic species.

5.5 ERRORS AND PROBLEMS IN METAL IONS SPECIATION

Of course, GC has been used frequently for metal ion speciation, but the chances for errors and problems still exist. Significant possible sources of errors may be due to poor extraction and purification procedures; decomposition of metal ions at the high temperatures required for GC; coelution of impurities, spuriously increasing peak areas; interference due to the presence of other metal ions; peak distortion caused by inadequate instrumentation; adsorption of metal ions on the column wall, interface, and other parts of the detectors; and nonlinear and poor detector response. The first type of errors may be reduced by applying advanced extraction and purification techniques. Solid-phase extraction and chromatographic techniques are the techniques of choice for the extraction and purification of environmental and biological samples. Errors due to purification and extraction methods may be calculated by applying blank and standard runs. Errors due to the decomposition of metal ion species at high temperatures may be handled using a low temperature for the GC machine with high flow rates. Moreover, metallic species may be converted into such derivatives, which are volatile at low temperatures. Furthermore, the analysis time may be reduced by decreasing the column length at ambient temperature. The problem of coelution of the impurities may be removed by changing some GC parameters, such as temperature and mobile phase, and by using pre-columns with different polarities in multidimensional GC. The problem due to the presence of other metal ions may be overcome by using masking reagents. Moreover, the element-selective detectors detect only the desired metal ion species. The fifth type of error may be reduced by using high-quality instrumentation with the most pure gases and the most sensitive detectors. The possible adsorption of metal ion species at the column may be reduced by changing the stationary phase material and also by optimizing GC parameters. The adsorption at the interface and the detectors may be removed by heating these parts externally and separately. Heisterkamp and Adams[16] studied the effects of temperature on the peak width of organolead species (Figure 5.3). Similarly, many authors carried out metal ion speciation by heating the interface and the other parts of the GC and detector assembly. The problem due to the nonlinearity of the detector response can be removed by using a good-quality detector and running the GC machine overnight or for a long period of time. Of course, these types of errors are possible in nonelement-specific detectors, and, therefore, these detectors were replaced by the advanced element-specific detectors, as discussed in the detection section of this chapter.

5.6 CONCLUSION

Gas chromatography is a major analytical tool for the speciation of metal ions. The ability to combine GC with advanced detectors, such as AAS, ICP, MS, and so forth, has resulted in their wide range of application in metal ion speciation in any of the matrices. The development of element-specific detectors eliminated the chances of error during detection due to the presence of impurities and other metal ions. The coupling of GC with ICP-MS is an ideal combination, as it provides metallic species in the form of gas to ICP-MS, which can be detected easily by the detector. The high separation power of GC columns in combination with the excellent detection limits of the modern detectors provide an excellent tool for metal ions speciation. A variety of derivatizing reagents are available by which a wide range of metal ions can be converted into the volatile derivatives that can be easily separated by GC. But, GC still suffers from certain drawbacks that have shifted the attention of scientists from GC to other analytical techniques. The derivatization procedure, especially in highly contaminated samples, is another problem in metal ion speciation by GC. Moreover, derivatization is a tedious job that consumes costly chemicals. In addition to these points, extraction of the samples requires careful attention or a problem or error may occur in the derivatization procedure. Every day, many unknown species are being discovered. To identify these species, we need to discover new detection techniques that can provide molecular information, such as electron impact MS, electrospray MS, and so forth. Much research remains to be considered with regard to the systematic studies for the evaluation of extraction recoveries, yield of derivatization procedure, optimal chromatographic conditions, and appropriate coupling of GC with suitable detectors.

REFERENCES

1. Dix K, Cappon CJ, Toribara TY, *J. Chromatogr. Sci.*, 25: 164 (1987).
2. Schoene K, Steinhanses J, Bruckert HJ, König A, *J. Chromatogr. A*, 605: 257 (1992).
3. Mester Z, Vitányi G, Morabito R, Fodor P, *J. Chromatogr. A*, 832: 183 (1999).
4. Mester Z, Pawliszyn J. *J. Chromatogr. A*, 873: 129 (2000).
5. Prohaska T, Pfeffer M, Tulipan M, Stingeder G, Mentler A, Wenzel WW, *Fresenius' J. Anal. Chem.*, 364: 467 (1999).
6. Grüter UM, Kresimon J, Hirner AV, *Fresenius' J. Anal. Chem.*, 368: 67 (2000).
7. Kresimon J, Grüter UM, Hirner AV, *Fresenius' J. Anal. Chem.*, 371: 586 (2001).
8. Killelea DR, Aldstadt JH III, *J. Chromatogr. A*, 918: 169 (2001).
9. Mothes S, Wennrich R, *Mikrochim. Acta*, 135: 91 (2000).
10. Lobinski R, Szpunar-Lobinska J, Adams FC, Teissedre PL, Cabanis JC, *J. AOAC Int.*, 76: 1262 (1993).
11. Nerin C, Pons B, *Appl. Organomet. Chem.*, 8: 607 (1994).
12. Nerin C, Pons B, *Quimica Analytica*, 13: 209 (1994).
13. Nerin C, Pons B, Zufiaurre R, *Analyst*, 120: 751 (1995).
14. Zufiaurre R, Pons B, Nerín C, *J. Chromatogr. A*, 779: 299 (1997).
15. Heisterkamp M, Adams FC, *Fresenius' J. Anal. Chem.*, 362: 489 (1998).
16. Heisterkamp M, Adams FC, *Fresenius' J. Anal. Chem.*, 370: 597 (2001).
17. Pons B, Carrera A, Nerin C, *J. Chromatogr. A*, 716: 139 (1998).
18. Hirner AV, Grüter UM, Kresimon J, *Fresenius' J. Anal. Chem.*, 368: 263 (2000).
19. Carro-Díaz AM, Lorenzo-Ferreira RA, Cela-Torrijos R, *J. Chromatogr. A*, 683: 245 (2002).
20. Liang L, Bloom NS, Horvat M, *Clin. Chem.*, 40: 602 (1994).
21. Bettmer J, Bradter M, Buscher W, Erber D, Rieping D, Cammann K, *Appl. Organo. Chem.*, 9: 541 (1995).
22. Carro AM, Lorenzo RA, Cela R, *LC-GC*, 16: 926 (1998).
23. Rodriguez I, Wasik A, Lobinski R, *Anal. Chem.*, 70: 4063 (1998).

24. Garcia Fernandez R, Montes Bayon M, Garcia Alonso JI, Sanz-Medel A, *J. Mass Spectrom.*, 35: 639 (2000).
25. Välimäki I, Perämäki P, *Mikrochim. Acta*, 137: 191 (2001).
26. Dowling TM, Uden PC, *J. Chromatogr. A*, 644: 153 (1993).
27. Tutschku S, Mothes S, Dittrich K, *J. Chromatogr. A*, 683: 269 (1994).
28. Jiang GB, Ceulemans M, Adams FC, *J. Chromatogr. A*, 727: 119 (1996).
29. Girousi S, Rosenberg E, Voulgaropoulos A, Grasserbauer M, *Fresenius' J. Anal. Chem.*, 358: 828 (1997).
30. Rodriguez Pereiro I, Wasik A, Lobinski R, *Fresenius' J. Anal. Chem.*, 363: 460 (1999).
31. Gallina A, Magno F, Tallandini L, Passaler T, Caravello GU, Pastore P, *Rapid Commun. Mass Spectrom.*, 14: 373 (2000).
32. Gui-bin J, Qun-fang Z, Bin H, *Environ. Sci. Technol.*, 34: 2697 (2000).
33. Cardellicchio N, Giandomenico S, Decataldo A, Di Leo A, *Fresenius' J. Anal. Chem.*, 369: 510 (2001).
34. Haas K, Feldmann J, Wennrich R, Stark HJ, *Fresenius' J. Anal. Chem.*, 370: 587 (2001).
35. Dirkx WMR, de la Calle MB, Ceulemans M, Adams FC, *J. Chromatogr. A*, 683: 51 (2002).
36. Lai-Feng Y, Narasaki H, *Anal. Bioanal. Chem.*, 372: 382 (2002).
37. Chong NS, Houk RS, *Appl. Spectrosc.*, 41: 66 (1987).
38. Peters GR, Beauchim D, *Anal. Chem.*, 65: 97 (1993).
39. Ruiz Encinar J, Leal Granadillo IA, Garcia Alonso JI, Sanz-Medel A, *J. Anal. At. Spectrom.*, 16: 475 (2001).
40. Hintelmann H, Falter R, Ilgen G, Evans RD, *Fresenius' J. Anal. Chem.*, 358: 363 (1997).
41. Ruiz Encinar J, Garcia Alonso JI, Sanz-Medel A, *J. Anal. At. Spectrom.*, 15: 1233 (2000).
42. Heumann KG, *Int. J. Mass Spectrom. Ion Processes*, 118/119: 575 (1992).
43. Hintelmann H, G, Evans RD, Villenueve JY, *J. Anal. At. Spectrom.*, 10: 619 (1995).
44. Gallus SM, Heumann KG, *J. Anal. At. Spectrom.*, 11: 887 (1996).
45. Garcia Alonso JI, Ruiz Encinar J, Rodridguez Gonzalez P, Sanz-Medel A, *Anal. Bioanal. Chem.*, 373: 432 (2002).
46. Ruiz Encinar J, Rodridguez Gonzalez P, Garcia Alonso JI, Sanz-Medel A, *Anal. Chem.*, 74: 270 (2002).
47. Kim AW, Foulkes ME, Ebdon L, *J. Anal. At. Spectrom.*, 7: 1147 (1992).
48. Dabek-Zlotorzynska E, Lao EPC, Timerbaev AR, *Anal. Chim. Acta*, 359: 1 (1998).
49. Quevauviller Ph, De La Calle-Guntinas MB, Maiter EA, Camara C, *Mikrochim. Acta*, 118: 131 (1995).
50. Paneli M, Rosenberg E, Grasserbauser M, Ceulemans M, Adams F, *Fresenius' J. Anal. Chem.*, 357: 756 (1997).
51. Tutschku S, Mothes S, Wennrich R, *Fresenius' J. Anal. Chem.*, 354: 587 (1996).
52. Liu Y, Lopezavila V, Alcaraz M, Beckert WF, *J. High Resolut. Chromatogr.*, 17: 527 (1994).
53. Jantzen E, Prange A, *Fresenius' J. Anal. Chem.*, 353: 28 (1995).
54. Szpunarlobinska J, Ceulemans M, Dirkx W, Witte C, Lobinski R, Adams FC, *Mikrochim. Acta*, 113: 287 (1994).
55. Tseng CM, Dediego A, Martin FM, Amouroux D, Donard OFX, *J. Anal. At. Spectrom.*, 12: 743 (1997).

56. Ceulemans M, Adams FC, *J. Anal. At. Spectrom.*, 11: 86 (1996).
57. Cai Y, Monsalud S, Jaffe R, Jones RD, *J. Chromatogr. A*, 876: 147 (2000).
58. Bulska E, Baxter DC, Frech W, *Anal. Chim. Acta*, 249: 545 (1991).
59. Qian J, Skyllberg U, Tu Q, Bleam WF, Frech W, *Fresenius' J. Anal. Chem.*, 367: 467 (2000).
60. Orellana-Velado NG, Pereiro R, Sanz-Medel A, *J. Anal. At. Spectrom.*, 13: 905 (1998).
61. Orellana-Velado NG, Pereiro R, Sanz-Medel A, *J. Anal. At. Spectrom.*, 15: 49 (2000).
62. Emteborg H, Baxter DC, Sharp M, Frech W, *Analyst*, 120: 69 (1995).
63. Mena ML, McLeod CW, Jones P, Withers A, Minganti V, Capelli R, Quevauviller P, *Fresenius' J. Anal. Chem.,* 351: 456 (1995).
64. Madson MR, Thompson RD, *J. AOAC*, 81: 808 (1998).
65. Reuther R, Jaeger L, Allard B, *Anal. Chim. Acta*, 394: 259 (1999).
66. Mothes S, Wennrich R, *J. High Resolut. Chromatogr.*, 22: 181–182 (1999).
67. Minganti V, Capelli R, De Pellegrini R, *Fresenius' J. Anal. Chem.*, 351: 471 (1995).
68. Frech W, Snell JP, Sturgeon RE, *J. Anal. At. Spectrom.*, 13: 1347 (1998).
69. Talmi Y, *Anal. Chim. Acta*, 74: 107 (1975).
70. Jimenez S, Sturgeon RE, *J. Anal. At. Spectrom.*, 12: 597 (1997).
71. Ceulemans M, Adams FC, *J. Anal. At. Spectrom.*, 11: 201 (1996).
72. Donais MK, Uden PC, Schantz MM, Wise SA, *Anal. Chem.*, 68: 3859 (1996).
73. Kim A, Foulkes ME, Ebdon L, Rowland S, *J. High Resolut. Chromatogr.*, 15: 665 (1992).
74. Prange A, Jantzen E, *J. Anal. At. Spectrom.*, 10: 105 (1995).
75. DeSmaele T, Moens L, Dams R, Sandra P, *Fresenius' J. Anal. Chem.*, 355: 778 (1996).
76. Pritzl G, Stuer-Lauridsen F, Carlsen L, Jensen AK, Thorsen TK, *Int. J. Environ. Anal. Chem.*, 62: 147 (1996).
77. Milde D, Plzak Z, Suchanek M, *Coll. Czech. Chem. Commn.*, 62: 703 (1997).
78. De La Calle-Guntinas MB, Scerbo R, Chiavarini S, Quevauviller Ph, Morabito R, *Appl. Organometal. Chem.*, 11: 693 (1997).
79. Carlierpinasseau C, Lespes G, Astruc M, *Appl. Organometal. Chem.*, 10: 505 (1996).
80. Pereiro IR, Schmitt VO, Szpunar J, Donard OEX, Lobiniski R, *Anal. Chem.*, 68: 4135 (1996).
81. Szpunar J, Ceulemans M, Schmitt VO, Adams FC, Lobinski R, *Anal. Chim. Acta*, 332: 225 (1996).
82. Ceulemans M, Adams FC, *Anal. Chim. Acta*, 317: 161 (1995).
83. Jiang GB, Xu FZ, *Appl. Organometal. Chem.*, 10: 77 (1996).
84. Lu TC, Whang CW, *J. Clin. Chem. Soc.*, 42: 515 (1995).
85. Lalere B, Szpunar J, Budzinski H, Garrigues P, Donard OFX, *Analyst*, 120: 2665 (1995).
86. Martin FM, Donard OFX, *Fresenius' J. Anal. Chem.*, 351: 230 (1885).
87. Martin FM, Tseng CM, Belin C, Quevauviller PH, Donard OFX, *Anal. Chim. Acta*, 286: 343 (1994).
88. Campillo N, Aguinaga N, Vinas P, Lopez-Garcia I, Hernandez-Cordoba M, *Anal. Chim. Acta*, 525: 273 (2004).
89. Chou CC, Lee MR, *J. Chromatogr. A*, 1064: 1 (2005).
90. Cai Y, Rapsomaniakis S, Andreae MO, *Anal. Chim. Acta*, 274: 243 (1993).

91. Ponce de Leon CA, Montes-Bayon M, Caruso JA, *J. Chromatogr. A*, 974: 1 (2002).
92. Pretorius WG, Ebdon L, Rowland SJ, *J. Chromatogr. A*, 646: 369 (1993).
93. Smaele TD, Verrept P, Moens L, Dams R, *Spectrochim. Acta*, 50B: 1409 (1995).
94. Smaele TD, Moens L, Dams R, Sandra P, *Fresenius' J. Anal. Chem.*, 355: 778 (1996).
95. Smaele TD, Moens L, Dams R, Sandra P, Vandereycken J, Vandyck J, *J. Chromatogr. A*, 793: 99 (1998).
96. Loon JCV, Alock LR, Pinchin WH, French JB, *Spectroscop. Lett.*, 19: 1125 (1986).
97. Hintleman H, Evans RD, Villeneuve JY, *J. Anal. At. Spectrochem.*, 10: 619 (1995).
98. DeSmaele T, Verrept P, Moens L, Dams R, *Spectrochim. Acta Part B*, 50: 1409 (1995).
99. Pritzl G, Stuer-Lauridsen F, Carlsen L, Jensen AK, Throsen TK, *Int. J. Environ. Anal. Chem.*, 11: 287 (1992).
100. Montes-Bayon M, Gutierrez JI, Garcia-Alonson A, Sanz-Medel J, *Anal. At. Spectrom*, 14: 1317 (1999).
101. Fernandez RG, Montes-Bayon M, Alonso JIG, Medel AS, *J. Mass Spectrom.*, 35: 639 (2000).
102. Encinar JR, Granadillo IL, Alonson JIG, Sanz-Medel A, *J. Anal. At. Spectrom.*, 16: 475 (2001).
103. Pelaez MV, Bayon MM, Alonson JIG, Medel AS, *J. Anal. At. Spectrom.*, 15: 1217 (2000).
104. Encinar JR, Villar MIM, Alonson JIG, Medel AS, *Anal. Chem.*, 73: 3174 (2001).
105. Dmuth N, Hermann KG, *Anal. Chem.*, 73: 4020 (2001).
106. Holland JF, Enke CG, Allison JT, Stults JD, Pickston JD, Newcombe B, Watson JT, *Anal. Chem.*, 55: 997A (1993).
107. Allen LA, Prang HM, Warren AR, Houk RSJ, *J. Anal. At. Spectrom.*, 10: 267 (1995).
108. Allen LA, Leach JJ, Houk RSJ, *Anal. Chem.*, 69: 2384 (1997).
109. Walder AJ, Freedman PA, *J. Anal. At. Spectrom.*, 7: 571 (1992).
110. Walder AJ, Furuta N, *Anal. Sci.*, 9: 675 (1993).
111. Liu W, Lee HK, *J. Chromatogr. A*, 834: 45 (1999).
112. Pereiro IR, Diaz AC, *Anal. Bioanal. Chem.*, 372: 74 (2002).
113. Cooke WS, *Today's Chem. Work*, 5: 16 (1996).
114. Wasik A, Rodriguez Pereiro I, Dietz C, Spunzar J, Lobinski R, *Anal. Commun.*, 35: 331 (1998).
115. B'Hymer C, Brisbin JA, Sutton KL, Caruso JA, *Am. Lab.*, Feb. issue, 17 (2000).
116. Harrison RM, Tapsomanikis S, *Environmental Analysis Using Chromatography Interfaced with Atomic Spectroscopy*, Ellis Horwood, Chichester (1989).
117. De la Calle-Guntinas MB, Brunori C, Scerbo R, Chiavarini S, Quevauviller Ph, Adams F, Morabito R, *J. Anal. At. Spectrom.*, 12: 1041 (1997).

6 Speciation of Metal Ions by Reversed-Phase High-Performance Liquid Chromatography

High-performance liquid chromatography (HPLC) is a modality of liquid chromatography involving the use of solid and liquid stationary and mobile phases, respectively. HPLC is often erroneously called high-pressure liquid chromatography, as it works at high pressure of the mobile phase. Since 1980, HPLC has emerged as a technique of choice in analytical science. It has been widely used for the analyses of almost all classes of compounds, including organic and inorganic moieties. The development of a variety of columns makes it a superb analytical technique. Moreover, the use of aqueous and nonaqueous mobile phases greatly enhanced its range of application. The applications of HPLC in metal ion speciation are significant, and many metal ions have been speciated using this technique. Generally, aqueous buffers are used as the mobile phases for the speciation of metal ions by HPLC. Therefore, the running cost of an HPLC machine is quite low, which is a plus for this technique. Before the introduction of capillary electrophoresis, HPLC was the primary technique used for the simultaneous speciation of metallic species. The new approach of successfully coupling detection devices, such as electrochemical detection, inductively coupled plasma (ICP), atomic absorption spectroscopy (AAS), mass spectrometry (MS), and so forth, with HPLC widely increased its application range. HPLC has the advantage of performing separations of nonvolatile species; thus, it has greater versatility than gas chromatography (GC), which often requires a derivatization of metallic ions. In HPLC, metal ion species are separated on the chromatographic column according to their affinity for the stationary and the mobile phases. Normal-phase HPLC is characterized by the use of an inorganic adsorbent or chemically bonded stationary phase with polar functional groups. The mobile phase used is nonaqueous, with organic solvents or sometimes a very little amount of water. The liquid–liquid partition is the main phenomenon responsible for the separation. The normal-phase HPLC has not been used for metal ion speciation due to the insolubility of metal ions in organic mobile phases. Moreover, the normal-phase columns cannot be used with aqueous mobile phases, as these are unstable under aqueous situations. The reversed-phase HPLC uses a nonpolar solid with a large surface area and with more polar aqueous solutions as the mobile phase. Most used columns in this mode of HPLC are C_2 or C_8 or C_{18}, while the mobile phases used are buffers. The nonspecific hydrophobic interactions of metal ions with stationary phases are responsible for the separation process. Reversed-phase HPLC has been used widely for the speciation of neutral or weakly charged metal ions or their complexes. The speciation of metal ions by reversed-phase HPLC has been achieved by using underivatized metallic species. However, some have derivatized metallic species and reported an improvement in speciation. A number of monographs and reviews were published on the speciation of metal ions by reversed-phase HPLC.[1–15] This chapter presents the speciation strategy for metal ions in the environmental and biological samples by HPLC using reversed-phase columns only. Attempts were also made to explain the optimization and the mechanisms of metal ion speciation.

6.1 APPLICATIONS

The selection criteria for a speciation method depend on the nature of the metal ions in the environmental and biological samples. The physicochemical properties of the analyte, that is, charge, polarity, molecular mass, and so forth, are responsible for their speciation. HPLC on reversed-phase columns has been used for the speciation of neutral or weakly charged metal complexes. In 1991, Robards et al.[13] reviewed more than 440 papers on metal ion speciation by liquid chromatography. The speciation of metal ions by reversed-phase HPLC is summarized below.

6.1.1 Arsenic (As)

Many reports are available on the speciation of arsenic due to the toxicity of arsenic species. A detailed study of the speciation of arsenobetaine, arsenocholine, and tetramethylarsonium ion was carried out by Blais et al.[16] The authors used a novel high-performance liquid chromatography–atomic absorption spectrometry (HPLC-AAS) interface based on thermochemical hydride generation for the determination of arsenobetaine [$(CH_3)_3As + CH_2COOH$], arsenocholine [$(CH_3)_3As + CH_2CH_2OH$], and tetramethylarsonium [$(CH_3)_4As+$] cations. Le and Ma[17] described arsenic speciation, namely, the separation of As(III), As(V), MMAA, and DMAA using C_{18} columns of different sizes. Prohaska et al.[18] used HPLC coupled to double-focusing sector field inductively coupled plasma–mass spectrometry (ICP-MS) as a sensitive element-specific detector for the speciation of arsenic [As(III) and As(V)] of liquid and gaseous emissions from soil samples, which were equilibrated in a microcosmos experiment. Slejkovec et al.[19] used an analytical procedure for the speciation of arsenic in urban aerosol samples using HPLC. The aerosols were collected by sequential filtration through membrane filters. Do et al.[20] used a C_{18} column in HPLC with hydride generation inductively coupled plasma–atomic emission spectrometry (HG ICP-

AES) for the speciation of four arsenic species, namely, arsenite As(III), arsenate As(V), monomethylarsonate (MMA), and dimethylarsinate (DMA). This analytical method allowed for the sensitive determination of the arsenic species in the submicrogram per liter range. Pedersen et al.[21] used a single quadrupole HPLC electrospray mass spectrometry system with a variable fragmentor voltage facility in the positive ion mode for simultaneous recording of elemental and molecular mass spectral data for arsenic compounds. The method was applicable to seven organoarsenic compounds tested — four arsenic samples containing carbohydrates (arsenosugars), a quaternary arsonium compound (arsenobetaine), dimethylarsinic acid, and dimethylarsinoylacetic acid. It was not suitable for the separation of two inorganic arsenic species — arsenite and arsenate. In the case of arsenosugars, qualifying ion data for a characteristic common fragment (m/z 237) were also simultaneously obtained. The method was used to identify and quantify the major arsenosugars in crude extracts of two brown algae. Saeki et al.[22] described the speciation of arsenic using Octadecyl (ODS) reversed-phase column. Gong et al.[23] speciated MMA and DMA by using an ODS-3 reversed-phase column with 4.7 mM tetrabutylammonium hydroxide, 2 mM malonic acid, and 4% methanol (pH 5.85) as the mobile phase. Van Elteren et al.[24] speciated arsenic species [As(III) and As(V)] in five bottled mineral waters from the Radenska and the Rogaska springs (Slovenia). Both a combination technique (HPLC-HGAFS) and a more conventional technique based on selective coprecipitation of As(III) with dibenzyldithiocarbamate prior to arsenic analysis (FI-HG-AFS) were used. The techniques yielded data that were not significantly different on the 5% level, with HPLC-HG-AFS being the least sensitive (with a detection limit of 1 μ/L) and the selective coprecipitation technique suitable for sub-picogram-per-liter (pg/L) levels (with a detection limit of 0.05 μg/L). The latter technique showed recoveries of 96.4 ± 0.4%. Recently, Ali and Aboul-Enein[25] achieved the speciation of arsenic metal ion [As(III) and As(V)] on an Econosil C_{18} (250 × 4.6 mm id, particle size 10 μm) column. The authors used water–acetonitrile (80:20, v/v) as the mobile phase. The detection was carried out using an ultraviolet (UV) detector at 410 nm and AAS, respectively, and separately.

6.1.2 Chromium (Cr)

Chromium(V) is a toxic metal ion species, while Cr(III) is nontoxic and, moreover, is a daily nutritional requirement at trace levels. Therefore, the speciation of chromium metal ion by reversed-phase HPLC was reported by many.[26–28] Ebdon et al.[29] used a reversed-phase column for the speciation of chromium metallic species in soil samples. The authors used plasma quad 2 ICP-MS as the detector. Grace et al.[30] used HPLC coupled to inductively coupled plasma–mass spectrometry (ICP-MS) for the separation and detection of chromium species [Cr(III) and Cr(VI)] in azo dyes, Acid Blue 158, and Acid Blue. Ali and Aboul-Enein[26] achieved partial separation of Cr(III) and Cr(VI) metal ions on an Econosil C_{18} (250 × 4.6 mm id, particle size 10 μm) column The mobile phases consisted of 10 m*M* ammonium acetate buffer (6 pH)-acetonitrile (10:90, v/v) for chromium speciation. The detection was carried out by using a UV-visible spectrometer at 410 nm and atomic absorption spectrometer (AAS), respectively, and separately. Posta et al.[31] studied the speciation of Cr(III) and Cr(V) on a reversed-phase (RP) C_{18} column using tetrabutylammonium acetate (TBAA), phosphoric acid, and methanol-based eluent, with the detection achieved by AAS.

6.1.3 Lead (Pb)

The speciation of lead metal ion was achieved on reversed-phase columns. Robecke and Cammann[32] evaluated an HPLC method for the determination of tetramethyllead (TTML) and tetraethyllead (TTEL) using a LiChrospher 60 column. The authors used acetonitrile–$LiClO_4$ and methanol–$LiClO_4$ as eluents. Furthermore, the authors advocated an improvement of the separation of TTML and TTEL species using methanol–chloroform–$LiClO_4$ as the eluent. Cammann et al.[33] separated dimethyl-, diethyl-, trimethyl-, and triethyllead species on an Hypersil ODS reversed-phase column using methanol–citric-acid buffer (different pHs) (40:60, v/v), containing methyl thioglycolate, as the mobile phase. The authors achieved the detection by using a UV/visible spectrometer at 235 nm. Shum et al.[34] described the speciation of lead metal ion [Pb(II), $Me_3Pb(I)$, and $Et_3Pb(I)$] using a reversed-phase C_{18} column. The authors used 5 m*M* ammonium pentasulfonate–acetonitrile (20:80, v/v) as the mobile phase. Bettmer et al.[35] described the speciation of dimethyl-, trimethyl-, dimethyl-, and triethyllead organolead species on an Hypersil ODS column. The authors used methanol–citric acid (40:60, v/v) as the mobile phase. Quartz tube atomic absorption spectrometry (QTAAS) and UV/visible detectors were used for detection purposes. Al-Rashdan et al.[36] used a Nucleosil C_{18} column for the speciation of lead metal ion. The authors used methanol in water (40 to 90%) as the mobile phase. The detection was achieved by using plasma quad ICP-MS as the detection technique. Brown et al.[37] reported the speciation of trimethyl-, triethyl-, and inorganic lead species on an Hypersil ODS column using methanol–sodium-acetate buffer (0.1 M) (10:90 and 30:70, v/v) as the eluent. The detection was carried out by isotope dilution measurements.

6.1.4 Mercury (Hg)

Reversed-phase HPLC was used for the speciation of inorganic and organic mercury species. Munaf et al.[38] used Develosil-ODS and STR-ODS-H columns for the speciation of methyl-, ethyl-, and organic species using cystein–acetic acid as the eluent. Al-Rashdan et al.[36] described an HPLC method for the separation of methyl-, ethyl-, phenyl-, and inorganic mercury complexes using an ODS RP-18 column and acetonitrile–water as the mobile phase. Shum et al.[34] described the speciation of MeHg(I), EtHg(I), and PhHg(I) species using a reversed-phase C_{18} column. The authors used 5 m*M* ammonium pentasulfonate–acetonitrile (20:80, v/v) as the mobile phase. Brown et al.[37] used a C_{18}

column for the determination of methyl-, ethyl-, and inorganic mercury using methanol–acetonitrile containing 2-mercaptoethanol as the mobile phase. The detection was achieved by using the ICP-MS techniques. Bettmer et al.[35] described the speciation of methyl- and ethylmercury species on an Hypersil-ODS column. The authors used methanol–citric acid (40:60, v/v) as the mobile phase. A variety of detectors were used for detection purposes. The most important detectors used were UV/visible spectrophotometry, electrochemical detector, graphite furnace atomic absorption spectrometry (GFAAS), microwave-induced plasma–atomic emission spectrometry (MIP-AES), IC-MS, and cold vapor atomic absorption spectrometry (CVAAS) detectors. Sarzanoni et al.[39] used an RP-18 column for the speciation of mercury species, as their pyrrolidinedithiocarbamate (PCD) complexes. The authors used UV, post-column oxidation, and cold vapor atomic absorption spectrometry (UV-PCO-CVAAS) as the detection devices. Huang et al.[40] separated methyl- and ethylmercury species on an Hypersil-ODS reversed-phase column using methanol–citric-acid buffer (different pHs) (40:60, v/v) containing methyl thioglycolate as the mobile phase. The authors achieved detection by using the UV/visible spectrometer at 235 nm. Cammann et al.[33] described a fully automated system for the direct determination of methylmercury (MeHg), ethylmercury (EtHg), phenylmercury (PhHg), and inorganic mercury [Hg(II)] at the nanogram per liter (ng/L) level using octadecylsilane column. The authors used an HPLC-CV-AAS assembly for the detection purpose. Falter and Schöler[41] speciated methyl-, ethyl-, methoxyethyl-, ethoxyethyl-, phenyl-, and inorganic mercury in water samples. The authors used a C_{18} column with detection by UV, PCO, and CVAAS. Ebdon et al.[29] used a reversed-phase column for the speciation of mercury species in contact lens solution. The authors used plasma quad 2 ICP-MS as the detector.

6.1.5 Selenium (Se)

Selenium is also toxic to humans; therefore, many papers on selenium speciation were available in the literature. Quijano et al.[42] described the speciation of selenocystine, selenomethionine, selenite, and selenate with ICP-MS detection. Olivas et al.[43] used an Hamilton RP1 reversed-phase column for the speciation of selenocystine, selenomethionine, and trimethylselenium. Vilanó et al.[44] proposed an online method for organic and inorganic selenium speciation (selenite, selenate, selenocystine, and selenomethionine), consisting of liquid chromatography UV irradiation–hydride generation–quartz cell atomic absorption spectrometry. Kotrebai et al.[45] speciated selenium metal species in selenium-enriched plants, such as hyperaccumulative phytoremediation plants (*Astragalus praleongus*, 517 μg/g Se, and *Brassica juncea*, 138 μg/g Se in dry sample), yeast (1200, 1922, and 2100 μg/g Se in dry sample), ramp (*Allium tricoccum*, 48, 77, 230, 252, 405, and 524 μg/g Se in dry sample), onion (*Allium cepa*, 96 and 140 μg/g Se in dry sample), and garlic (*Allium sativum*, 68, 112, 135, 296, and 1355 μg/g Se in dry sample) by HPLC-ICP-MS after hot water and enzymatic extractions. Raessler et al.[46] determined selenium(IV) and selenium(VI) in contaminated groundwater from

Kelheim, Germany, using HPLC-HG-AAS. The authors advocated the developed assembly as being easy to handle and as being a suitable method for the long-term monitoring of species distribution in an almost routine way, taking into account the threshold values of 10 μg/L for each element. Vonderheide et al.[47] presented an HPLC separation strategy by coupling it with ICP-MS for the speciation of seleno amino acids, selenomethionine (SeMet), selenoethionine (SeEt), and selenocystine (SeCys). Furthermore, the characterization of unidentified selenium containing peaks was attempted by employing several procedures, including electrospray mass spectrometry (ES-MS). Montes-Bayon et al.[48] determined the concentrations of selenium species in *Brassica* species (Indian mustard) at parts per million (ppm) levels using RP-HPLC-ICP-MS and ES-MS methods.

6.1.6 Tin (Sn)

Some reports were published concerning the speciation of tin metal ion using reversed-phase HPLC. Astruck et al.[49,50] developed a HPLC and GFAAS combination for the analysis of butyltin species (mono-, di-, tri-, and tetrabutyltin) in water on a Nucleosil column using 0.001% tropolone solution in toluene as the mobile phase. Dauchy et al.[51] used a C_{18} column for the speciation of mono-, di-, and tritin species. The authors used 0.1% (mass/volume) troplone methanol–water–acetic acid (80:14:6, v/v) as the mobile phase. White et al.[52] developed a reversed-phase liquid chromatographic method for the determination of dibutyltin (DBT), tributyltin (TBT), diphenyltin (DPhT), and triphenyltin (TPhT) in sediments, which was compatible with both atmospheric pressure ionization (API) mass spectrometry and inductively coupled plasma (ICP) mass spectrometry. As a result of this development, both techniques may be used for the complementary speciation of organotin compounds. The chromatographic system is comprised of a Kromasil-100 5 μm C_{18} (150 × 2.1 mm) column and a mobile phase of 0.05% triethylamine in acetonitrile–acetic acid–water (65:10:25). Chiron et al.[53] developed a liquid chromatographic method with ICP-MS for the speciation of butyl- (monobutyltin, dibutyltin, tributyltin) and phenyl- (monophenyltin, diphenyltin, triphenyltin) tin compounds in sediments. Rosenberg et al.[54] presented an HPLC method with atmospheric pressure chemical ionization–mass spectrometry (HPLC-APCI-MS) for the speciation of butyl- and phenyltin compounds. Chromatography was performed on a 30 × 2 mm, 3 μm C_{18} column, enabling the separation of mono-, di-, and trisubstituted butyl- and phenyltin compounds in less than 10 min using a water/1% trifluoroacetic acid/methanol gradient. Fairman and Wahlen[55] reported the separation of dibutyltin, tributyltin, and phenyltin on a C_{18} reversed-phase column using acetonitrile–water–acetic acid–triethylamine (65:23:12:0.05, v/v/v/v) as the mobile phase. The authors achieved the detection by ICP-MS. Yang et al.[56] developed a method for the determination of tributyltin (TBT) in sediment by isotope dilution (ID) analysis using HPLC-ICP-MS. Reverse-spike ID analysis was performed to determine the accurate concentration of a 117Sn-enriched TBT spike using a well-characterized natural abundance TBT standard.

6.1.7 Miscellaneous Metal Ions

In addition to the metal ions discussed above, speciation of other metal ions has also been carried out using reversed-phase HPLC. Datta et al.[57] studied the speciation of aluminum using HPLC with Cyclobond I and Cyclobond III columns. The mobile phase used was methanol–water (1:1, v/v) containing 0.1 *M* triethylamine (TEA) and glacial acetic acid (pH, 4). Polak et al.[58] separated aluminum citrate, aluminum maleate, and Al(III) species using a fast protein liquid chromatography–inductively coupled plasma–mass spectrometry (FPLC-ICP-MS) assembly. Cairns et al.[59] reported the speciation of platinum metal ion using a reversed-phase column. The authors detected platinum species by ICP-MS. Ebdon et al.[29] used a reversed-phase column for the speciation of cobalt and iron metallic species in soil samples. The authors used plasma quad 2 ICP-MS as the detector. Houck et al.[60] analyzed Se(IV) and Se(VI) on a C_{18} column using methanol–water–tetrabutylammonium ion as the mobile phase. The detection was achieved by ICP-MS with a detection limit of 10 to 20 ng/mL. B'Hymer and Caruso[61] analyzed different selenium species in yeast solution on a Phenomenex C_8 column. The authors used 10% methanol and 1% trifluroacetic acid as the mobile phase.

Speciation of different metal ion species by reversed-phase HPLC is summarized in Table 6.1. To make this chapter more useful and to show the nature of the chromatograms of the separated metal ion species by reversed-phase HPLC, a typical example of the chromatograms of arsenic speciation is shown in Figure 6.1. Quantitatively, separated metallic species are determined by the retention time (t), capacity (k), separation (α), and resolution factors (R). The quantitative estimation is carried out by comparing the area of the identified peak with the area of the peak obtained by the standard metallic species of the known concentration.

6.2 OPTIMIZATION OF METAL ION SPECIATION

To achieve the maximum speciation of metal ions by HPLC, optimization of the experimental parameters is of utmost importance. Optimization may be achieved by using suitable reversed-phase columns and the mobile phases. The other controlling factors for speciation of metal ions are pH and the concentration of the mobile phases. Additionally, derivatization of metal ions, detection, extraction of the environmental samples, and amount injected into the HPLC machine may be used to control speciation in reversed-phase HPLC. Therefore, the following discussion on optimization strategies may be helpful when optimizing the speciation of metal ions in environmental and biological samples.

6.2.1 Stationary Phases

Selection of a suitable column is the key issue in the speciation of metal ions by reversed-phase HPLC. In reversed-phase HPLC, the stationary phases used are low-polar in nature (derivatized silica, i.e., bonded to an organic moiety such as C_2 or C_8 or C_{18} alkyl chains with or without certain groups). The separating

TABLE 6.1
Speciation of Metal Ions by Reversed-Phase, High-Performance Liquid Chromatography

Metal Ions	Sample Matrix	Columns	Detection	Ref.
		Arsenic (As)		
Arsenobetaine [$(CH_3)_3As$ + CH_2COOH] and arsenocholine [$(CH_3)_3As$ + CH_2CH_2OH]	—	Reversed phase	THG	16
As(III), As(V), MMA, and DMA	—	C_{18} column	—	17
As(III) and As(V)	—	—	ICP-MS	18
AS(III) and As(V)	Urban aerosol	—	—	19
AS(III) and As(V)	—	C_{18} column	HG-ICP-AES	20
Arsenosugars, a quaternary arsonium compound (arsenobetaine), dimethylarsinic acid, and dimethylarsinoylacetic acid	Algae	—	—	21
Arsenic species	—	ODS reversed-phase column	—	22
MMA and DMA	—	ODS-3 reversed-phase column	—	23
As(III) and As(V)	Mineral waters	—	HG-AFS	24
As(III) and As(V)	—	Econosil C_{18}	UV/visible	25
MMA and DMA	Mineral water	C_{18} columns	ICP-MS	80
MMA and DMA	Human urine	Water Bondapak C_{18}	ICP-MS	81
MMA and DMA	Human urine	ODS column	ICP-MS	82
MMA and DMA	Wine	ODS column	ICP-MS	83
MMA and DMA	—	Vydac C_{18}	ICP-MS	84
As(III), AS(V), MMA, and DMA	—	C_{18} column	ICP-MS	85
As(III), AS(V), MMA, and DMA	—	C_{18} column	ICP-MS	86
As(III), AS(V), MMA, and DMA	—	C_{18} column	UV-HG-QF-AAS	87, 88
As(III), AS(V), MMA, and DMA	—	C_{18} column	ICP-MS	89
Arsenic species	—	C_{18} column	ICP-MS	90
Arsenic species	—	C_{18} column	—	91
Arsenic species	—	Hamilton PRP-X100	ICP-MS	92
As(III), As(V), MMA, and DMA	River sediment	Novapak C_{18}	HG-AFS	93
		Chromium (Cr)		
Cr(III) and Cr(VI)	Soil	C_{18} column	ICP-MS	29
Cr(III) and Cr(VI)	Dyes	—	—	30
Cr(III) and Cr(VI)	—	C_{18} column	AAS	31
Cr(III) and Cr(VI)	—	Econosil C_{18}	UV/visible	25

TABLE 6.1 (Continued)
Speciation of Metal Ions by Reversed-Phase, High-Performance Liquid Chromatography

Metal Ions	Sample Matrix	Columns	Detection	Ref.
	Lead (Pb)			
Tetramethyllead (TTML), 7 tetraethyllead (TTEL)	—	LiChrospher 60	—	32
Dimethyl-, diethyl-, trimethyl-, and triethyllead species	—	Hypersil ODS	UV/visible	33
Pb(II), $Me_3Pb(I)$, and $Et_3Pb(I)$	—	C_{18} column	—	34
Dimethyl-, trimethyl-, dimethyl-, and triethyllead organolead species	—	Hypersil ODS, 120-50	—	35
Lead species	—	Nucleosil C_{18}	ICP-MS	36
Trimethyl-, triethyl-, and inorganic lead	—	Hypersil ODS	Isotopic dilution	37
	Mercury (Hg)			
Methyl-, ethyl-, and organic mercury species	—	Develosil-ODS and STR-ODS	—	38
Methyl-, ethyl-, phenyl-, and inorganic mercury complexes	—	ODS RP-18 column	—	36
MeHg(I), EtHg(I), and PhHg(I)	—	C_{18} column	—	34
Methyl-, ethyl-, and inorganic mercury	—	C_{18} column	—	37
Methyl- and ethylmercury species	—	Hypersil ODS, 120-50		
Mercury species	—	C_{18} column	—	39
Methyl- and ethylmercury species	—	Hypersil ODS	—	40
MeHg, EtHg, PhHg, and Hg(II)	—	Octadecylsilane column	CV-AAS	33
Organomercury species	—	Hypersil ODS, 120-50	—	35
Organomercury species	—	C_{18} column	—	94
Methyl- and inorganic mercury	—	ODS column	ICP-MS	95
Methyl- and inorganic mercury	—	Spherisorb ODS-2	ICP-MS	96
Mercury species	Contact lens solution	C_{18} column	ICP-MS	29
HgCl, MeHgCl, EtHgCl, and PheHgCl	Seafoods	Shim-pack CLC-ODS	CV-AFS	97
	Selenium (Se)			
Selenocystine, selenomethionine, selenite, and selenate	—	—	ICP-MS	42

TABLE 6.1 (Continued)
Speciation of Metal Ions by Reversed-Phase, High-Performance Liquid Chromatography

Metal Ions	Sample Matrix	Columns	Detection	Ref.
Selenocystine, selenomethionine, and trimethylselenium	—	Hamilton RP1	—	43
Selenite, selenate, selenocystine, and selenomethionine	—	—	UV-IH	44
Se(IV) and Se(VI)	Groundwater	C_{18} column	AAS	46
Seleno-amino acids, selenomethionine (SeMet), selenoethionine (SeEt), selenocystine (SeCys), and selenomethionine	—	—	ICP-MS	47
13 selenium species	Human urine	Phenomenex RP-18	VG Elemental PQ-2	98
Seleno-amino acids	Yeast	Zorbax SB-C_8	ICP-MS	99, 100
Se(IV), Se(VI), and seleno-amino acids	—	Phenomenex RP-8	ICP-MS	60
Seleno-amino acids, trimethylselenonium, and inorganic selenium	—	Nucleosil 120 C_{18}	ICP-MS	101
Organic and inorganic selenium species	Human urine	Nucleosil 120 C_{18}	ICP-MS	102
Trimethylselenonium, selenocystine, selenomethionine, selenoethionine, selenite, and selenate	Biological samples	Spherisorb 5 ODS	—	103
MMSe and TMSe	—	C_{18} column	ICP-MS	104
Selenium species	Biological samples	C_{18} column	ICP-MS	105
Selenium species	Biological samples	C_{18} coulmn	ICP-MS	106
Selenium species	Astragalus praleongus	—	ICP-MS	45
Selenium species	*Brassica* species	C_{18} column	ICP-MS/ ES-MS	48
Se(IV) and seleno-amino acids	—	LiChroCART	HG-AFS	107
Selenium species	—	Hypercarb	MS	108
		Tin (Sn)		
Mono-, di-, tri-, and tetrabutyltin	Water	Nucleosil column	—	49, 50
Mono-, di-, and tritin	—	C_{18} column	—	51

TABLE 6.1 (Continued)
Speciation of Metal Ions by Reversed-Phase, High-Performance Liquid Chromatography

Metal Ions	Sample Matrix	Columns	Detection	Ref.
Dibutyltin (DBT), tributyltin (TBT), diphenyltin (DPhT), and triphenyltin (TPhT)	Sediments	Kromasil-100	API/ICP	52
Mono-, di-, tri-, and phenyltin	Sediment	Reversed phase	—	53
Butyl- and phenyltin	—	C_{18} column	APCI-MS	54
Dibutyltin, tributyltin, and phenyltin	—	C_{18} column	—	55
Tributyl- and triphenyltin	Water	Kromasil 100 C_{18}	ICP-MS	109
Trimethyl-, tributyl, and triphenyltin	Fish tissue	C_{18} column	ICP-MS	110
Tin species	Sediment	C_{18} column	ICP-MS	111
Tin species	—	C_{18} column	ICP-MS	56
Miscellaneous metal ions				
Aluminum species	—	Cyclobond I and II	—	57
Aluminum species	—	FPLC column	ICP-MS	58
Aluminum species	Biological samples	C_{18} column	ETAAS	112
Organo-aluminum species	Biological samples	C_{18} column	UV	113
Platinum species	—	C_{18} column	ICP-MS	114
Silicon species	Biological samples	C_{18} column	ET-AAS	109

capacities of these phases depend on the types and the structures of the alkyl chain. These are very stable stationary phases and can be used with a variety of aqueous mobile phases, including various buffers and acidic and basic solutions. The inertness of these phases makes them ideal for the speciation of metal ion species, particularly in environmental samples. Moreover, these phases have a wide range of applications due to their reversed-phase nature with good stability, efficiency, and high separation power. Silica is most suitable for the preparation of reversed phases. These columns are available from many manufacturers and go by different trade names. The most commonly used commercial columns are Nucleosil, Hamilton, Hypersil, Develosil, μ-Bondapak, Econosil, and Inertsil. The commonly available reversed phases are given in Table 6.2. Only a few reports are available in the literature that indicate the comparative speciation of the same metal ion species onto different reversed-phase columns. Le XC et al.[62] reported the speciation of arsenic metal ions on C_{18} (Figure 6.2a) and C_8 (Figure 6.2b) reversed-phase columns. The best speciation of arsenic was reported on a

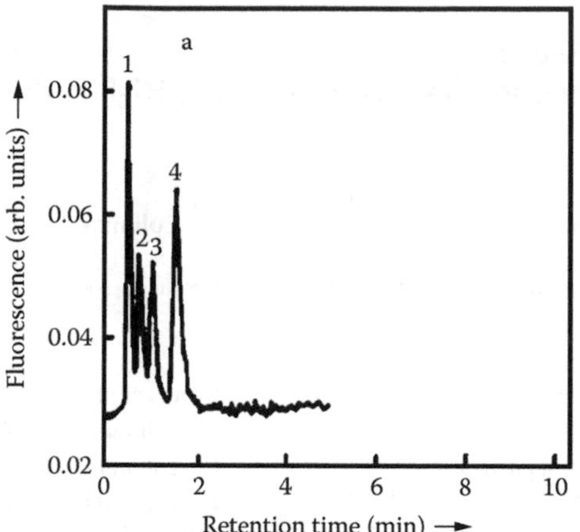

FIGURE 6.1 Chromatograms of arsenic species by reversed-phase high-performance liquid chromatography: (1) As(III), (2) As(V), (3) monomethylarsenate (MAA), and (4) Dimethylarsenate (DMA) using a Phenomenex ODS column. (From Le, X.C., and Ma, M., *Anal. Chem.*, 70: 1926 (1998). With permission.)

TABLE 6.2
Some Silica-Gel-Based Reversed Phases with Functional Groups

Type of Modification	Functional Group	Organic Si-C- Group
$-C_2$	Dimethyl-	$-(CH_3)_2$
$-C_4$	Butyl-	$-(CH_2)_3-CH_3$
$-C_8$	Octyl-	$-(CH_2)_7-CH_3$
$-C_{18}$	Octadecyl-	$-(CH_2)_{17}-CH_3$
$-C_6H_5$	Phenyl-	$-(CH_2)_3$-Phe
-CN	Cyano (nitrile)-	$-(CH_2)_3$-CN
$-NO_2$	Nitro-	$-(CH_2)_3$-Phe-NO_2
$-NH_2$	Amino-	$-(CH_2)_3-NH_2$
$-N(CH_3)_2$	Dimethylamino-	$-(CH_2)_3-N(CH_3)_2$
-OH	Diol-	$-(CH_2)_3-O-CH_2-CH(OH)-CH_2(OH)$
-SA	Sulfonic acid-	$-(CH_2)_3$-Phe-SO_3Na
-SB	Quaternary ammonium-	$-(CH_2)_3$-Phe-CH_2- $N^+(CH_3)_3Cl$

C_{18} column. Similarly, Ali and Aboul-Enein[25] attempted the speciation of arsenic and chromium metal ions on different reversed-phase columns, but the best speciation of these metal ions was achieved on an Econosil C_{18} column. Briefly, the selection of a suitable column for metal ion speciation is the first requirement.

6.2.2 Mobile Phases

After the selection of a suitable stationary phase, the choice of the mobile phase is another important factor for optimizing metal ion speciation. Because the

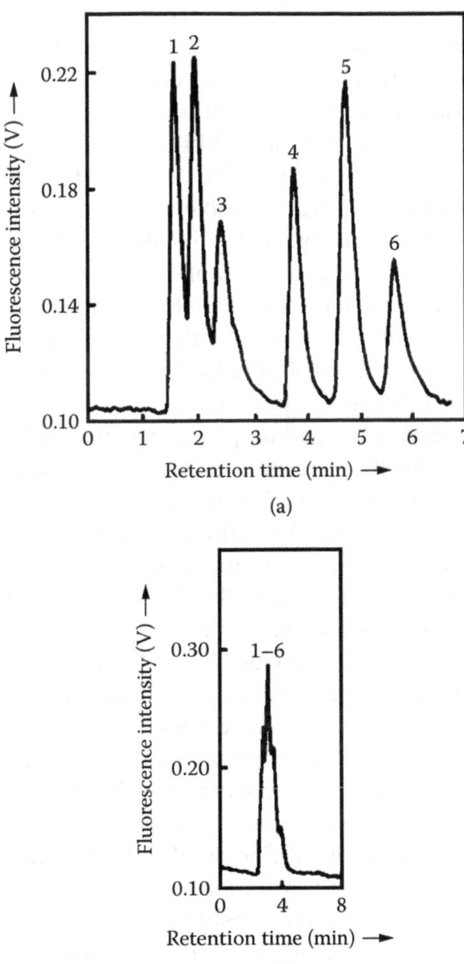

FIGURE 6.2 Effect of the reversed-phase column on the speciation of arsenic metal ion (a) on the C_{18} column and (b) on the C_8 column; (1) through (6): As(III), MMA, DMA, MMA, DMA, and As(V), respectively. (From Le, X.C., Lu, X., Ma, M., Cullen, W.R., Aposhian, H.V., and Zheng, B., *Anal. Chem.*, 72: 5172 (2000). With permission.)

columns are reversed phase in nature, the use of the polar mobile phase is required. Care must be taken to consider the detector being utilized when attempting to select the mobile phase. In the case of element-specific detectors (AAS, ICP, MS, etc.), mobile phases should be selected that may be compatible with the detectors. The samples eluted from the HPLC column are aspirated into the flame or plasma by a nebulizer, and high concentrations of nonvolatile components (in the mobile phase) may clog the nebulizer.[63] All the mobile phases are aqueous in nature, and generally, buffers of different concentrations and pHs are used for this purpose. Le et al.[17,23,64] used a variety of mobile phases for the speciation of arsenic metal ion. The authors used tetrabutylammonium chloride–malonic acid as the mobile phase. Furthermore, the same authors[62] used tetrabutylammonium hydroxide–malonic acid as the mobile phase for the speciation of arsenic metal ion. The same mobile phase was used for the speciation of arsenic metal ion by Saeki et al.[22] White et al.[52] used acetonitrile–acetic acid–triethylamine as the mobile phase for the speciation of organotin species. To optimize speciation, some organic solvents were also used as the organic modifiers. The most important organic solvents are acetonitrile,[34,52,55] methanol,[17,23,35,64] and tetrahydrofuran.[9] Methanol is the most widely used organic modifier when detection is achieved by ICP-MS, as it causes less plasma instability.[65] It is very important to mention here that the use of other organic solvents is limited due to their plasma instability. Of course, speciation is carried out by a variation of the mobile phases, but only a few research papers present the results obtained by varying the mobile phase. Shum et al.[34] varied the concentration of acetonitrile for the speciation of trimethyl- and triethyllead species. The results are shown in Figure 6.3a. The results indicate that the retention times decrease with an increase in the acetonitrile contents. Furthermore, the same authors[34] also studied the effect of acetonitrile on the speciation of MeHg(I), EtHg(I), and PhHg(I) species (Figure 6.3b). Again, the retention times decrease by increasing the concentration of acetonitrile. This response may be due to the fact that acetonitrile decreases the bonds of metal ion species with the stationary phase. Moreover, it is worthy mentioning here that the addition of organic solvent into the mobile phase may improve speciation by sharpening the peak shape. Shum et al.[34] described the peak shape of MEHg(I) organometal at different concentrations of acetonitrile. Figure 6.4 indicates that a sharp peak was obtained at 40% acetonitrile, while the peak shape broadened at a lower concentration of acetonitrile. Recently, Ali and Aboul-Enein[25] also studied the effect of acetonitrile on the speciation of arsenic, and the results are given in Figure 6.5. The high value of acetonitrile resulted in poor resolution but sharp peaks. The authors reported high values of separation and resolution factors at 20% acetonitrile. Briefly, the concentrations of the different constituents of the mobile phase should be varied for the optimization of speciation of metal ion species.

6.2.3 pH of the Mobile Phase

pH also controls speciation, as it stabilizes the bonds between metal ions and the stationary phase. Moreover, the stability of metal ion complexes with derivatizing reagent is also controlled by the pH of the mobile phase. Therefore, the optimization of speciation can be achieved by varying the pH of the mobile phase. Again, only few reports deal with the results of speciations obtained by varying the pH of the mobile phase. Bettmer et al.[35] described the effect of pH on the speciation of dimethyl-, diethyl-, and trimethyllead and reported the best speciation (maximum capacity factors) at pH 6.7 to 7 (Figure 6.6). Similarly, Cammann et al.[33] reported the effect of pH on the speciation of organolead and organomercury species. The results of these findings are shown in Figure 6.7. It

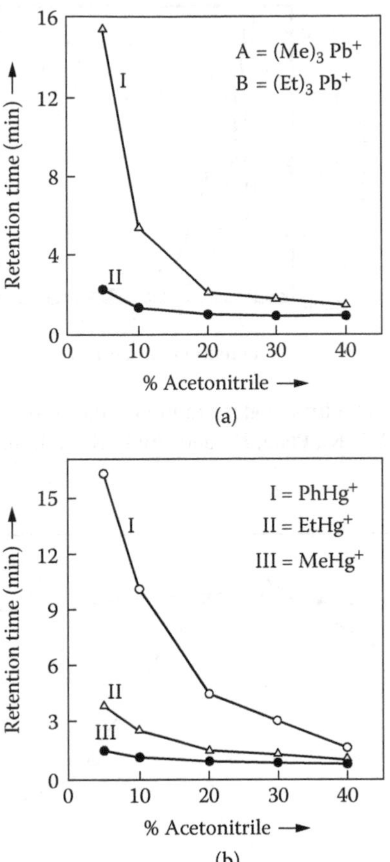

FIGURE 6.3 Effect of acetonitrile concentration on the retention of (a) $Me_3Pb(I)$ and $Et_3Pb(I)$ and (b) MeHg(I), EtHg(I), and PhHg(I) metal ion species using a reversed-phase column. (From Shum, S.C.K., Pang, H., and Houk, R.S., *Anal. Chem.*, 64: 2444 (1992). With permission.)

may be concluded from Figure 6.7 that 6.4 was the best pH for the separation of ethylmercury, methyllead, dimethyllead, diethyllead, and trimethyllead, while pH 6 was found to be suitable for the separation of triethyllead. It is also interesting to observe that the capacity factors of organollead species increased with the increase of pH, which may be due to the formation of the complexes of organollead compounds with methyl thioglycolate, a mobile-phase additive and derivatizing reagent.

6.2.4 Detection

The method of detection in reversed-phase HPLC is a crucial parameter needed to obtain optimized speciations of metal ions at trace levels, especially in envi-

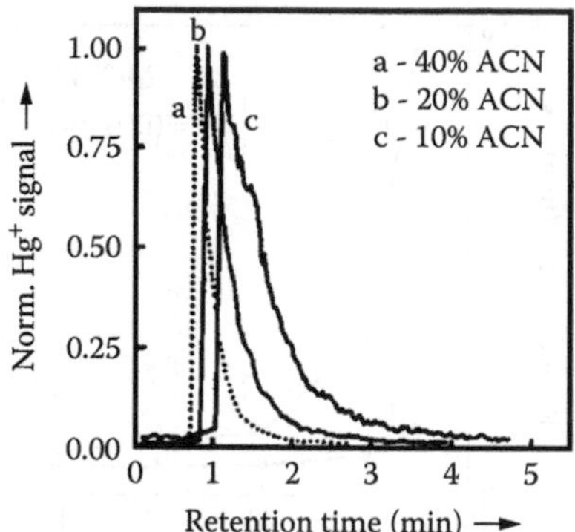

FIGURE 6.4 Effect of acetonitrile concentration on the peak shape of MeHg(I) organometal ion. (From Shum, S.C.K., Pang, H., and Houk, R.S., *Anal. Chem.*, 64: 2444 (1992). With permission.)

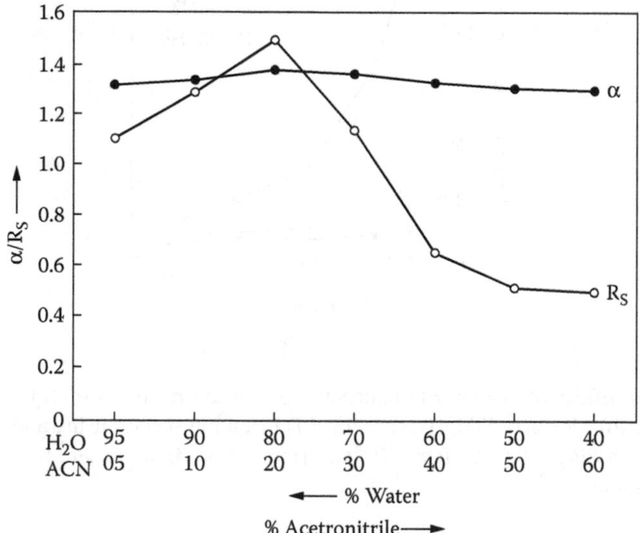

FIGURE 6.5 Effect of acetonitrile concentration on the speciation of arsenic and chromium metal ions. (From Ali, I., and Aboul-Enein, H.Y., *Chemosphere*, 48: 275 (2002). With permission.)

ronmental samples where the concentrations of metallic species are very low. The peak size and area of metallic species depend on the response of the detector. A detector with low limits of detection is supposed to be the best. As in the case of GC, two types of detectors were used in reversed-phase HPLC. The first category belongs to the nonelement-specific, which have a wide range of applications but show interference due to the presence of impurities. The second class of detectors is element specific and can be used for a particular element only

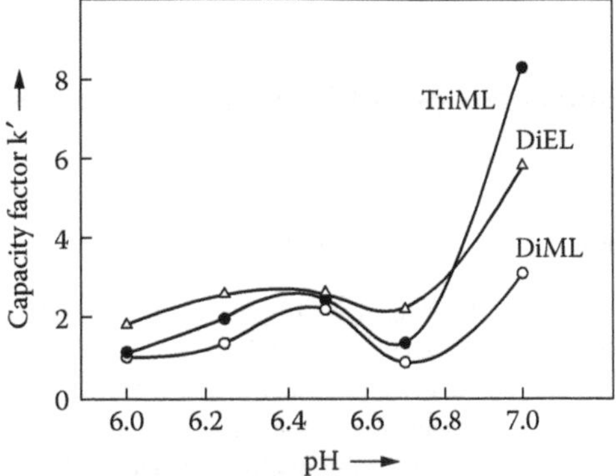

FIGURE 6.6 Effect of pH on the speciation of DiML, TriML, and DiML metal ion species. (From Bettmer, J., Camman, K., and Robecke, M., *J. Chromatogr. A*, 654: 177 (1993). With permission.)

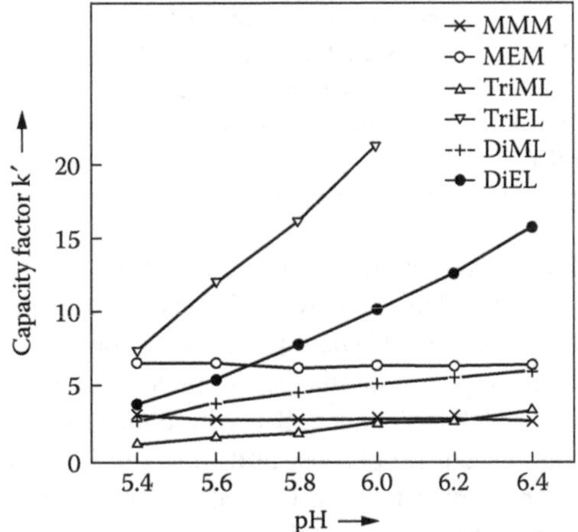

FIGURE 6.7 Effect of pH on the speciation of organomercury and organolead metal ion species. (From Cammann, K., Robecke, M., and Bettmer, J., *Fresenius' J. Anal. Chem.*, 350: 30 (1994). With permission.)

without any interference. UV/visible and electrochemical detectors have been used as the nonelement-specific detectors for the detection of many metal ions.[65] Metal ions are generally UV/visible transparent, so the application of this detector is not common. But, the pre- or postcolumn derivatization of metallic species makes this mode of detection universal and popular. To increase the sensitivity of detection, UV visualization agents (probe) were also used in the mobile phase. The important probes involve chromate, aromatic amines, and heterocyclic compounds, such as benzylamine, 4-methylbenzylamine, dimethylbenzylamine, imidazole, *p*-toludine, pyridine, creatinine, ephedrine, and anionic chromophores (benzoate and anisates). The various metallochromic reagents for the detection

TABLE 6.3
Some Organic Compounds (with Chromophores) for the Detection of Metal Ions Species

Metal Ions Species	Complexing Reagents	Detection Wavelength
Fe(II), Cu(II), Co(II), Zn(II), Pb(II), V(IV), and Ni(II)	4,7-Dimethyl-1,10-phenanthroline, 1,10-phenanthroline, α,β,γ,δ-tetrakis(4-carboxyphenyl)porphyrin, 2-nitroso-1-naphthol-4-sulfonic acid, 2-(5-bromo-2-pyridylazo)-5-diethylaminophenol, 2-(5-bromo-2-pyridylazo)-5-(*N*-propyl-*N*-sulfopropyl-amino) phenol	220–587 nm
Al(III), Ba(II), Sr(II), U(IV), La(III), and lanthanoides	1-(2,4-Dihydroxy-1-phenylazo)-8-hydroxy-3,6-naphthalenesulfonic acid, 3,6-bis-(2-sulfophenylazo-4,5-dihydroxy-2,7-naphthalenesulfonic acid, arsenazo III	450–654 nm
Fe(II), Co(II), and Ni(II)	2-(5-Nitro-2-pyridylazo)-5-(*N*-propyl-*N*-sulfopropylamino) phenol	560 nm
Fe(II, III), Cu(II), Co(II, III), Zn(II), Ni(II), and Cr(III)	4-(2-Pyridylazo) resorcinol	500 nm

of metal ions in reversed-phase HPLC are summarized in Table 6.3. To achieve high sensitivity in indirect detection by a UV/visible detector, metallic species peak widths should be minimized. The interaction of metal ions and visualization agent with the stationary phase should be suppressed. The visualization agents should exhibit a higher UV absorbance. 8-Hydroxyquinoline-5-sulfonic acid and lumogallion have fluorescent properties, and they can be used for the detection of metal ions by fluorescent detectors. Overall, fluorescence detectors have not yet received wide acceptance in metal ion speciation, though their gains in sensitivity and selectivity over photometric detectors are significant, and they are even already commercial available. In spite of this, nonelement-specific detectors are being replaced rapidly by element-specific detectors.

The different element-specific detectors used in reversed-phase HPLC are atomic absorption spectrometry (AAS), atomic emission detection (AED), atomic fluorescence spectrometry (AFS), inductively coupled plasma (ICP), microwave-induced plasma–atomic emission detection (MIP-AED), mass spectrometry–selected ion-monitoring mode (MS-SIM), inductively coupled plasma–flame photometric detection (ICP-FPD), microwave-induced plasma–atomic emission spectroscopy (MIP-AES), cold vapor atomic absorption spectrometry (CVAAS), quartz tube–atomic absorption spectrometry (QT-AAS), among others (Table 6.1).[1–13] The application ranges of the element-specific detectors can be increased by changing the energy sources and the working wavelengths. The chances of interference in these detectors due to the presence of impurities are very low.

Basically, in speciation analysis, the choice of the detector depends on the chemical forms to be determined and on the mode of the separation used. According to Byrdy et al.,[66] plasma source mass spectrometry has good potential as a method for the detection and speciation of trace elements. This is due to the highly selective nature and excellent sensitivity of the detector. In comparison to atomic emission detection, detection limits are usually two to three orders of magnitude lower for plasma MS determinations.

The interfacing of HPLC with element-specific detectors is a most important aspect to be discussed here. HPLC has the advantages of performing the separations of many metallic species by coupling with an ICP-MS detection device, due to its superior sensitivity. Sutton and Caruso[67] reviewed the application of ICP-MS as the detector for metal ion speciation in RP-HPLC. The flow of the mobile phase in HPLC is 0.5 to 2 mL/min, which is compatible with the flow of the solvent in AAS and ICP-MS systems; hence, the interfacing of these detectors with HPLC can be achieved easily and successfully. The main concern with this type of interfacing is the transfer line from HPLC to the detector, which must be of minimum length. This is done to minimize possible peak broadening from excessive postcolumn volume. Many interfacing (nebulizer) designs were developed and used for the coupling of HPLC with ICP-MS and AAS, among others, detection systems, and some of them are described herein. Boorn et al.[68] described a concentric nebulizer design that is shown in Figure 6.8a. The authors reported that conventional pneumatic nebulizers were inefficient from the liquid transfer point of view at high flow rates. However, Olesik et al.[69] presented a nebulizer with small bore and advocated for more efficient liquid transfer due to the small bore of the nebulizer. Lu and Barnes[70] reported the use of an ultrasonic nebulizer for the optimum flow of the mobile phase in ICP. The design of this nebulizer is given in Figure 6.8b. This nebulizer has a transducer plate over which sample liquid flows. The liquid sample forms a thin film and is nebulized by high-frequency mechanical vibrations from the transducer. The transport efficiencies are in the range of 10 to 30%. Pergantis et al.[71] and Olesik et al.[72] reported the use of a high-efficiency nebulizer at low flow rates (up to the range of 100 μL/min). Shum et al.[73] described another nebulizer design (Figure 6.9a) by which the sample liquid can be introduced very near the plasma inside the torch, and this nebulizer eliminates the spray chamber part of the ICP-MS detector. The authors reported good efficiency at low flow rates for this detector. This nebulizer was found to be successful up to the flow rate of 30 μL/min. Furthermore, the authors[34] reported an improvement in the detection of metallic species by a factor of 2.5. Another development in nebulizer design is that of the oscillating nebulizer, which was reported to be successful for interface with small-bore HPLC columns.[74,75] The nebulizer has two capillaries with liquid flow in the inner and gas flow in the outer capillaries, respectively (Figure 6.9b). The other approach of nebulizer design is a hydraulic high-pressure nebulizer,[76] which was shown to increase sensitivity for some elements. In addition to these nebulizers, Koropchak et al.[77] and Tomlinson et al.[78] used thermospray as an interface. This involves forcing the liquid sample through an electrically heated capillary to form the

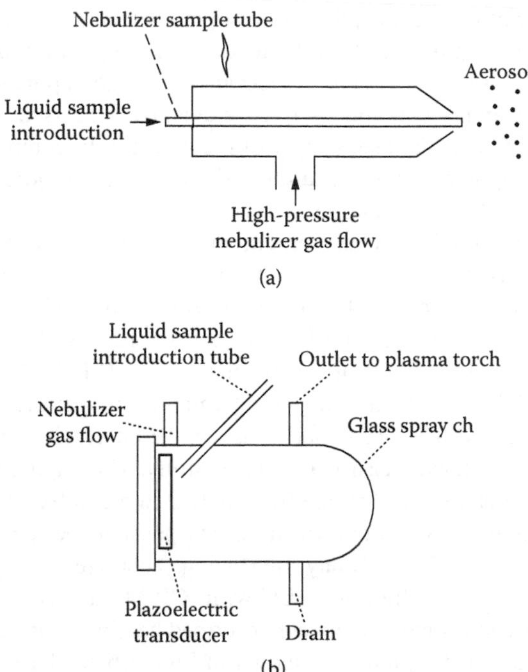

FIGURE 6.8 (a) Concentric and (b) ultrasonic nebulizers. (From B'Hymer, C., Brisbin, J.A., Sutton, K.L., and Caruso, J.A., *Am. Lab.*, 32: 17 (2000). With permission.)

aerosol in ICP. A complete schematic diagram showing the coupling of HPLC with ICP-MS, including the mounting of the nebulizer and postcolumn reactor is shown in Figure 6.10.

Various detection limits were reported for the speciation of different metal ions.[1–13] The limits of detection depend on the type of detector used. Ma and Le[64] reported the detection limits of arsenic species at the microgram level using hydride generation–atomic fluorescence spectrometry (HG-AFS) and ICP-MS detectors. The same authors[17] reported detection limits at the nanogram level for arsenic using atomic fluorescence detection. Cammann et al.[33] reported detection limits of 270 to 800 ng for organomercury and organolead species using a UV/visible detector. Recently, Ali and Aboul-Enein[26] reported detection limits at the microgram level for arsenic and chromium metal ion species using a UV/visible detector. Martinez et al.[79] used a hollow-cathode (HC) radiofrequency glow-discharge (rf-GD) optical-emission spectrometry (OES) detector for the determination of inorganic mercury species by cold vapor (CV) generation in a flow-injection (FI) system. The analytical performance characteristics of FI-CV-rf-GD-OES for mercury detection were evaluated at the 253.6 nm emission mercury line, and the detection limit obtained was 1.2 to 1.8 ng/mL. Bettmer et al.[35] compared the detection limits of organomercury and organolead species using mercaptoethanol and methyl thioglycolate as the derivatizing reagents. The results

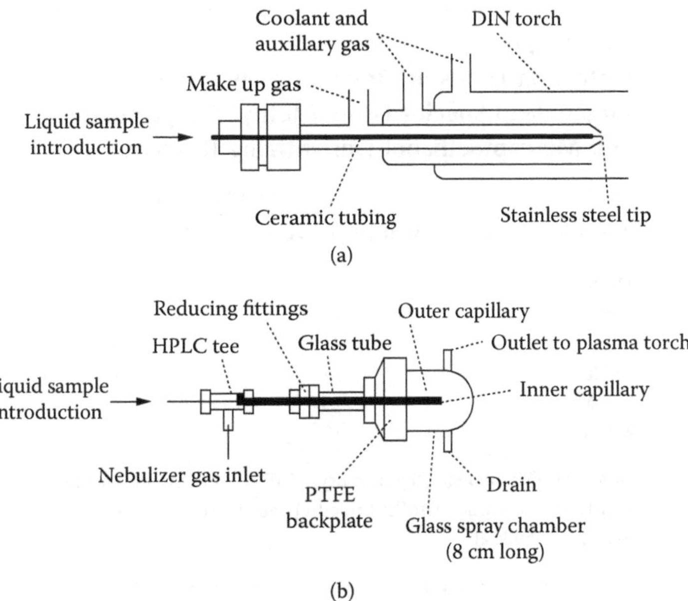

FIGURE 6.9 (a) Direct injection and (b) oscillating capillary nebulizers. (From B'Hymer, C., Brisbin, J.A., Sutton, K.L., and Caruso, J.A., *Am. Lab.*, 32: 17 (2000). With permission.)

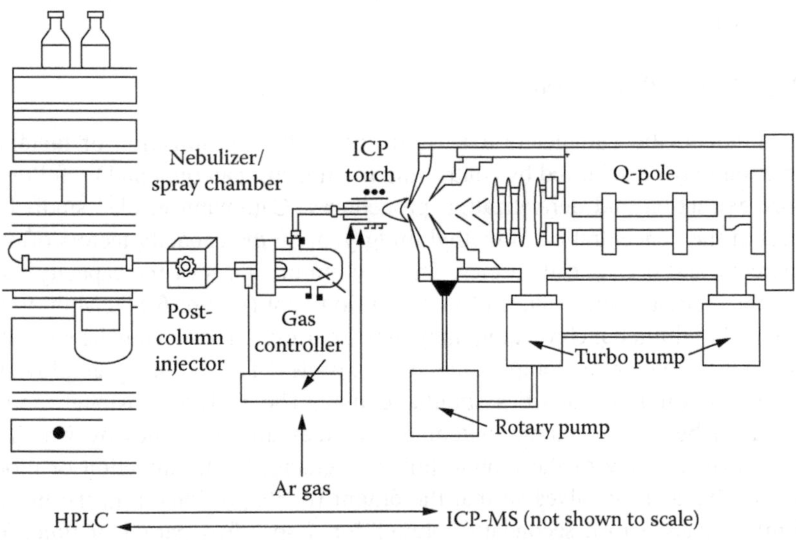

FIGURE 6.10 The coupling of reversed-phase HPLC with ICP-MS, indicating the mounting of the nebulizer and postcolumn reactor. (From Fairman, B., and Wahlen, R., *Spectroscopy Europe,* 13: 16 (2001). With permission.)

TABLE 6.4
Detection Limits of Organomercury and Organolead Species Using Methyl Thioglycolate and Mercaptoethanol Derivatizing Reagents

	Detection Limits (ng)	
Metal Ions Species	**Methylthioglycolate**	**Mercaptoethanol**
DiML	12	18
TriML	17	35
DiEL	31	30
TriEL	32	61
MMM	25	—
MEM	19	—

Note: MMM: monomethylmercury, MEM: methylethylmercury, DiML: dimethyllead, TriML: trimethyllead, DiEL: diethyllead, and TriEL: triethyllead.

Source: From Bettmer, J., Camman, K., and Robecke, M., *J. Chromatogr. A*, 654, 177, 1993. With permission.

are given in Table 6.4, and it may be concluded from this table that the lower limits of the detection were obtained by using methyl thioglycolate. Therefore, selection of a suitable derivatizing reagent is another factor in optimizing metal ion speciation.

6.2.5 Other Parameters

In addition to the parameters discussed above, the optimization of metal ion speciation may be achieved by controlling the injecting amount and by following proper extraction and derivatization procedures. Cammann et al.[33] studied the effects of the concentration of methyl thioglycol on the capacity factors of organomercury and organolead species. The maximum values of the capacity factor were obtained at higher values of methyl thioglycol (Figure 6.11). Multidimensional and multimodal chromatography may be useful in achieving the speciation of a mixture containing complex metal ion species. This modality employs two columns of different polarities simultaneously. The multidimensional method, also called heart-cut column switching, includes all techniques by which the direction of the flow of the mobile phase is changed. The direction of mobile phase is changed by valves so that the eluent (or its portion) from the primary column is passed to a secondary column for a specified period of time. This modality of chromatography is not fully developed, and only a few reports are available on metal ion speciation using the multidimensional mode of chromatography. Briefly, these parameters can be used to optimize metal ion speciation by reversed-phase HPLC. Based on our experience and the literature available,

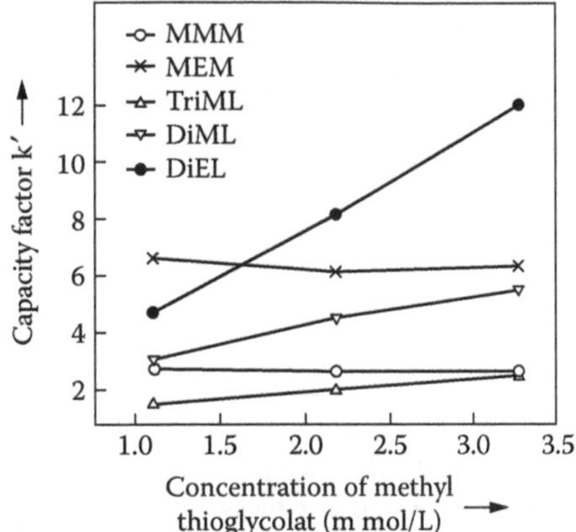

FIGURE 6.11 Effect of the concentration of methyl thioglycolate derivatizing reagent on the capacity factors of organomercury and organolead metal ion species. (From Cammann, K., Robecke, M., and Bettmer, J., *Fresenius' J. Anal. Chem.*, 350: 30 (1994). With permission.)

a scheme was developed to carry out and optimize the speciation of metal ions by reverse-phase HPLC in the environmental and biological samples. The developed protocol is given in Scheme 6.1.

6.3 DERIVATIZATION OF METAL IONS

Some metal ion species cannot be separated by reversed-phase HPLC directly; therefore, derivatization is required, which changes the physicochemical properties of metallic species. Therefore, such types of metal ion species can be separated easily after undergoing the derivatization process. Moreover, this process changes the spectroscopic properties of metal ions. Generally, metal ions are phototransparent and cannot be detected by a UV/visible detector. Therefore, derivatization changes these UV/visible transparent species into photochromatic analytes, which, in turn, can be detected by a UV/visible detector. Derivatization may also be used to separate trace elements from their matrices and to concentrate the analytes species, that is, to generate species that may be easily separated by reversed-phase HPLC. The derivatization may be carried out by two methods — pre- and postcolumn derivatization. In the first case, metal species are usually allowed to react with a derivatizing agent prior to loading onto the reversed-phase column. In the second case, metal ions, after elution from the column, are allowed to react with suitable reagents. In this case, a postcolumn reactor (reaction coil) is required to be inserted between the column and the detector. The most important properties of the derivatizing reagents are fast reaction rates at room temperature, a wide range of reactivity, a pure form, and UV absorption capability. Because most of the detection in reversed-phase HPLC was carried out by element-specific detectors, there are only a few reports on the derivatization of metal ions in reversed-phase HPLC. However, the most commonly used derivatizing reagents

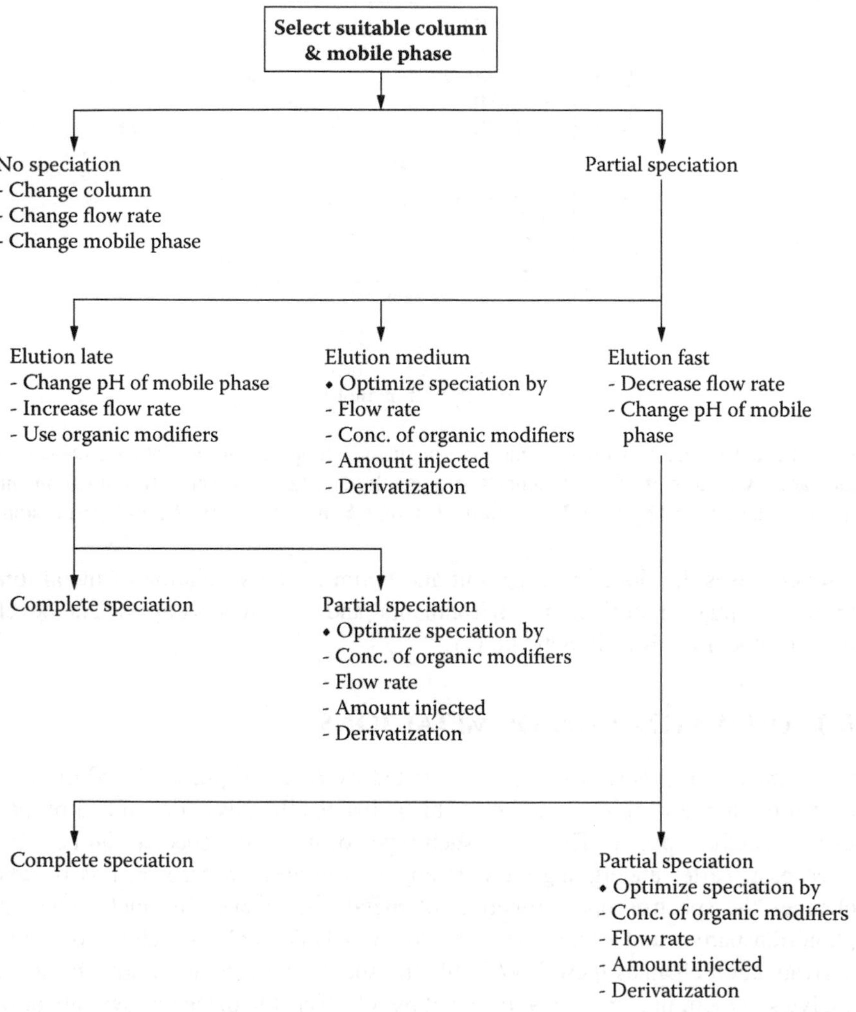

SCHEME 6.1

are dithiocarbamates, as the sulfur atom of these reagents has a very strong tendency to form complexes with all metal ions except alkali and alkaline earth metals. Other important derivatizing reagents are azo dyes and 8-hydroxyquinol for the successful separation of metal ions on reversed-phase columns.[7]

Bettmer et al.[35] used online column derivatization with mercaptoethanol for the separation of lead compounds. Based on the optimized chromatographic conditions, an online enrichment system was developed with recoveries between 75 and 88%. The use of methyl thioglycolate as the second complexing reagent made it possible to determine all analytes, organolead, and organomercury compounds, simultaneously. Camman et al.[33] reported the pre- and postcolumn derivatizations of organomercury and organolead species with methyl thioglycolate

TABLE 6.5
Some Sulfur-Containing Derivatizing Reagents for the Speciation of Mercury and Lead Metal Ions

Metal Ions Species	Derivatizing Reagents	Detectors	Ref.
Pb(II), DiML, DiEL, TriML, and TriEL	Dithiozone	QT-AAS	115, 116
Hg(II), MMM, and MEM	Dithiozone	UV-VIS	117, 118
Pb(II), DiML, DiEL, TriML, and TriEL	Alkyldithiocarbamate	QT-AAS	115, 116
Hg(II), MMM, and MEM	Alkyldithiocarbamate	UV-VIS	119
Hg(II), MMM, and MEM	Mercaptoethanol	ED	120
Hg(II), MMM, and MEM	Mercaptoethanol	GF-AAS	121
Hg(II), MMM, and MEM	Mercaptoethanol	UV-VIS	122, 123
Hg(II), MMM, and MEM	Mercaptoethanol	MIP-AES	124
Hg(II), MMM, and MEM	Mercaptoethanol	ICP-MS	124
MMM, MEM, DiML, DiEL, and TriEL	Mercaptoethanol	UV	35
Hg(II), MMM, and MEM	Cysteine	CVAAS	125

Note: MMM: monomethylmercury, MEM: methylethylmercury, DiML: dimethyllead, TriML: trimethyllead, DiEL: diethyllead, and TriEL: triethyllead.

QT-AAS: quartz tube–atomic absorption spectrometer, UV-VIS: ultraviolet visible, ED: electrochemical detector, GF-AAS: graphite furnace–atomic absorption spectrometer, MIP-AES: microwave-induced plasma–atomic emission spectrometer, ICP-MS: inductively coupled plasma–mass spectrometer, and CVAAS: cold vapor atomic absorption spectrometer.

(Figure 6.11). The concentration of methyl thioglycolate was varied from 1.1 to 3.3 m*M*. According to the authors, a significant difference in the capacity factors of lead species was observed, while very little effect was observed on the retention of mercury species. Bettmer et al.[35] reported the use of mercaptoethanol as a derivatizing reagent for the speciation of lead and mercury metal ions. Sulfur compounds have generally been used as the derivatizing reagents (online or offline) for the complexation of lead and mercury metal ions in speciation analysis. Therefore, attempts were made to compile information on these sulfur-containing reagents and metal ion speciations, and the results are summarized in Table 6.5. Inorganic and organic lead and mercury compounds show a high affinity for sulfur compounds, and mercaptoethanol was used for the speciation of lead and mercury metallic species. The primary reaction during pre- and postcolumn derivatization is presented in Figure 6.12. It is also important to mention here that the mercaptoethanol complexes of lead are sensitive toward UV and visible radiations and, hence, decomposed during the course of time. Bettmer et al.[35] studied the degradation of mercaptoethanol complexes of dimethyllead, trimethyllead, diethyllead, and triethyllead in daylight, and the results are given in Figure 6.13. It may be concluded from this figure that the complexes of trimethyllead and triethyllead were stable, while the complexes of dimethyllead and trimethyllead decomposed with time. Therefore, these complexes should be kept in dark bottles and exposure

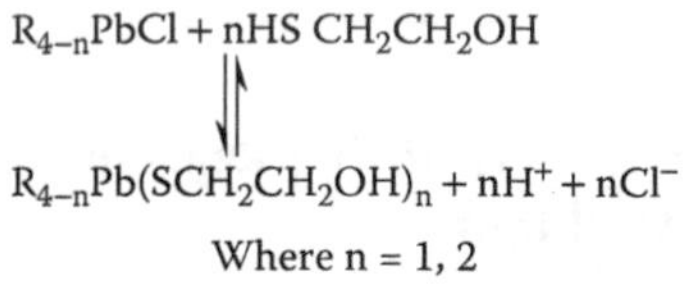

$$R_{4-n}PbCl + nHS\ CH_2CH_2OH$$

$$\rightleftharpoons$$

$$R_{4-n}Pb(SCH_2CH_2OH)_n + nH^+ + nCl^-$$

Where n = 1, 2

FIGURE 6.12 The reaction scheme of mercaptoethanol with lead metal ions. (From Bettmer, J., Camman, K., and Robecke, M., *J. Chromatogr. A*, 654: 177 (1993). With permission.)

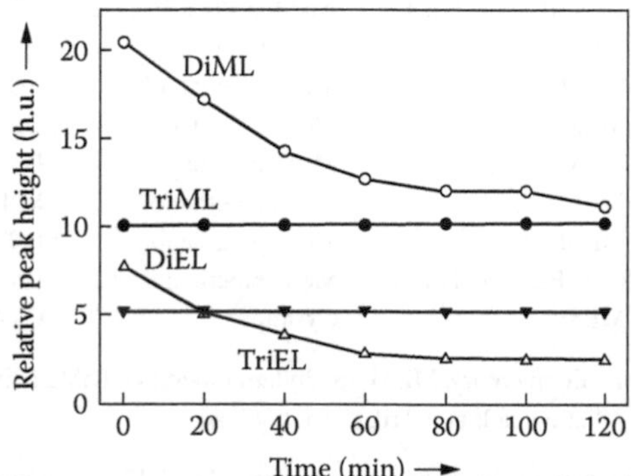

FIGURE 6.13 The effect of daylight on the stability of DiML, TriML, DiEL, and TriEl complexes with mercaptoethanol. (From Bettmer, J., Camman, K., and Robecke, M., *J. Chromatogr. A*, 654: 177 (1993). With permission.)

to UV and visible radiation should be avoided. Generally, the peaks of lead and mercury complexes of mercaptoethanol overlap and create some confusion in the determination of metallic species of these metal ions. This problem can be overcome by using methyl thioglycolate as a derivatizing agent. Moreover, metal ion complexes of methyl thioglycolate can be detected photometrically at 235 nm.

6.4 MECHANISMS OF METAL IONS SPECIATION

Many reviews and books describe the mechanisms of the separation of analytes in reversed-phase HPLC. But, the exact mechanisms of the separation on reversed-phase columns remain unknown. However, a brief discussion on the separation of different metallic species of the same metal ion on RP columns is presented herein. As known, different species of the same metal ion have different physicochemical properties. Moreover, the derivatization of metallic species results in the differentiation of their physical and chemical properties. Therefore, these species distributed at different rates between the mobile and the stationary phases. Basically, adsorption and the partition phenomena of metallic species between the stationary and the mobile phases control their distribution, which depends on the natures of the stationary phase, the mobile phases, and the metal ion species.

Generally, the stationary phase tends to bind metallic species (to different extents) by the adsorption phenomenon. At the same time, the mobile phase tries to carry forward these metallic species (due to partition process), and as a result, different metallic species are eluted at different elution times and the separation occurs. The forces responsible for the separation of metallic species include hydrogen bonding and van der Waals forces, especially in the case of derivatized metal ion species. Steric effects also play a crucial role in the separation of different organometallic species. Ionic interactions are also responsible for the separation phenomenon. Electrostatic interactions, dipole forces, and dispersion forces are also important for the separation of metal ion species on reversed-phase columns. Generally, reversed-phase columns are silica based; hence, the different types of interactions mentioned above may occur between the hydroxyl groups of silica gel and metallic species.

6.5 CONCLUSION

In the literature, there are many reports that indicate the utility and superiority of reversed-phase HPLC in metal ion speciation. Since 1980, it has been thought of and used as the best technique for the qualitative and quantitative speciation of metal ions. It replaced GC in the field of metal ion speciation, due to the fact that, as in GC, derivatization is not necessary in reversed-phase HPLC. In addition, the reversed-phase nature of the columns makes this technique popular in the field of metallic speciation. Reversed-phase columns are now being replaced by ion-exchange columns, which have good separating capacities for metallic species. HPLC with an ion-exchange column is called ion chromatography, which will be discussed in detail in Chapter 7. There is no serious drawback to reversed-phase HPLC in metal ion speciation, but the separation powers of reversed-phase columns for metal ion species are not very high; hence, some metal ion species cannot be speciated satisfactorily using this modality of liquid chromatography. And, the coupling with element-specific detectors requires some special attention due to the fact that the sample from the HPLC column is introduced into the detectors in the form of liquid. Ion-exchange columns replaced reversed-phase HPLC through ion-exchange HPLC (ion chromatography).

REFERENCES

1. Nielen MWF, Frei RW, Brinkmann UATh, *Selective Sample Handling and Detection in High Performance Liquid Chromatography*, Elsevier, Amsterdam (1988).
2. Robarts K, Starr P, Patsalides E, *Analyst*, 116: 1247 (1991).
3. Cagniant D, *Complexation Chromatography*, Marcel Dekker, New York (1992).
4. Quevauviller Ph, *J. Chromatogr. A*, 750: 25–33 (1996).
5. Sarzanini C, Mentasti E, *J. Chromatogr. A*, 789: 301–321 (1997).
6. Liu W, Lee HK, *J. Chromatogr. A*, 834: 45–63 (1999).
7. Sarzanini C, Mentasti E, *J. Chromatogr. A*, 850: 213–228 (1999).

8. Benramdane L, Bressolle F, Vallon JJ, *J. Chromatogr. Sci.*, 37: 330 (1999).
9. B'Hymer C, Brisbin JA, Sutton KL, Caruso JA, *Am. Lab.*, Feb. issue, 17 (2000).
10. Brown PR, Grushka E (Eds.), *Advan. Chromatogr.*, 41: 250 (2001).
11. Fytianos K, *J. AOAC Int.*, 84: 1763 (2001).
12. Ponce de Leon C, Montes-Bayon M, Caruso JA, *J. Chromatogr. A.*, 974: 1–21 (2002).
13. Robards K, Starr P, Patsalides E, *Analyst*, 116: 1247 (1991).
14. Ferrarello CN, Fernandez de la Campa MR, Sanz-Medel A, *Anal. Bioanal. Chem.*, 373: 412 (2002).
15. B'Hymer C, Caruso JA, *J. Chromatogr. A*, 1045: 1 (2004).
16. Blais JS, Momplaisir GM, Marshall WD, *Anal. Chem.*, 62: 1161 (1990).
17. Le XC, Ma M, *Anal. Chem.*, 70: 1926 (1998).
18. Prohaska T, Pfeffer M, Tulipan M, Stingeder G, Mentler A, Wenzel WW, *Fresenius' J. Anal. Chem.*, 364: 467 (1999).
19. Slejkovec Z, Salma I, van Elteren JT, Zemplen-Papp E, *Fresenius' J. Anal. Chem.*, 366: 830 (2000).
20. Do B, Alet P, Pradeau D, Poupon J, Guilley-Gaillot M, Guyon F, *J. Chromatogr. B, Biomed. Sci. Appl.*, 740: 179 (2000).
21. Pedersen SN, Francesconi KA, *Rapid Commun. Mass Spectrom.*, 14: 641 (2000).
22. Saeki K, Sakakibara H, Sakai H, Kunito T, Tanabe S, *Bio. Metals*, 13: 241 (2000).
23. Gong Z, Lu X, Cullen WR, Le XC, *J. Anal. At. Spectrom.*, 16: 1409 (2001).
24. van Elteren JT, Stibilj V, Slejkovec Z, *Water Res.*, 36: 2967 (2002).
25. Ali I, Aboul-Enein HY, *Chemosphere*, 48: 275 (2002).
26. Yao X, Liu J, Cheng J, Zheng Y, *Fresenius' J. Anal. Chem.*, 342: 702 (1992).
27. Andrle CM, Broekaert JAC, *Fresenius' J. Anal. Chem.*, 346: 653 (1993).
28. Martinez-Bravo Y, Roig-Navarro AF, Lopez FJ, Hernandez F, *J. Chromatogra. A*, 926: 265 (2001).
29. Ebdon L, Evans EH, Pretorius WG, Rowlands SJ, *J. Anal. At. Spectrom.*, 9: 939 (1994).
30. Grace K. Zoorob GK, Caruso JA, *J. Chromatogr. A*, 773: 157 (1997).
31. Posta J, Berndt H, Luo SK, Schaldach G, *Anal. Chem.*, 65: 2590 (1993).
32. Robecke M, Cammann K, *Fresenius' J. Anal. Chem.*, 341: 555 (1991).
33. Cammann K, Robecke M, Bettmer J, *Fresenius' J. Anal. Chem.*, 350: 30 (1994).
34. Shum SCK, Pang H, Houk RS, *Anal. Chem.*, 64: 2444 (1992).
35. Bettmer J, Camman K, Robecke M, *J. Chromatogr. A*, 654: 177 (1993).
36. Al-Rashdan A, Vela NP, Caruso JA, Heitkemper DT, *J. Anal. At. Spectrom.*, 7: 551 (1992).
37. Brown AA, Ebdon L, Hill SJ, *Anal. Chim. Acta*, 286: 391 (1994).
38. Munaf E, Haraguchi H, Ishii D, Takeuchi T, Goto M, *Anal. Chim. Acta*, 235: 399 (1990).
39. Sarzanoni C, Sacchero G, Aceto M, Abollino O, Mentasi E, *J. Chromatogr. A*, 626: 151 (1992).
40. Huang CW, Jiang SJ, *J. Anal. At. Spectrom.*, 8: 681 (1993).
41. Falter R, Schöler HF, *J. Chromatogr. A*, 645: 253 (1994).
42. Quijano MA, Gutierrez AM, Perez-Conde MC, Camara C, *J. Anal. At. Spectrom.*, 11: 407 (1996).
43. Olivas RM, Donard OFX, Gilon N, Poyin-Gautier M, *J. Anal. At. Spectrom.*, 11: 1171 (1996).
44. Vilanó M, Padró A, Rubio R, Rauret G, *J. Chromatogr. A*, 819: 211 (1998).

45. Kotrebai M, Birringer M, Tyson JF, Block E, Uden PC, *Analyst*, 125: 71 (2000).
46. Raessler M, Michalke B, Schulte-Hostede S, Kettrup A, *Sci. Total Environ.*, 258: 171 (2000).
47. Vonderheide AP, Wrobel K, Kannamkumarath SS, B'Hymer C, Montes-Bayon M, Ponce De Leon C, Caruso JA, *J. Agric. Food Chem.*, 25: 5722 (2002).
48. Montes-Bayon M, Yanes EG, Ponce D, Jayasimhulu K, Stalcup A, Shann J, Caruso JA, *Anal. Chem.*, 74: 107 (2002).
49. Astruck A, Astruck M, Pinel R, *Anal. Chim. Acta*, 228: 129 (1990).
50. Astruck A, Astruck M, Pinel R, *Mikrochim. Acta*, 109: 83 (1992).
51. Dauchy X, Cottier R, Batel A, Jeannot R, Borsier M, Astruc A, Astruc M, *J. Chromatogr. Sci.*, 31: 416 (1993).
52. White S, Catterick T, Fairman B, Webb K, *J. Chromatogr. A*, 794: 211 (1998).
53. Chiron S, Roy S, Cottier R, Jeannot R, *J. Chromatogr. A*, 879: 137 (2000).
54. Rosenberg E, Kmetov V, Grasserbauer M, *Fresenius' J. Anal. Chem.*, 366: 400 (2000).
55. Fairman B, Wahlen R, *Spectroscopy Europe,* 13: 16 (2001).
56. Yang L, Mester Z, Sturgeon RE, *Anal. Chem.*, 74: 2968 (2002).
57. Datta AK, Wedlund PJ, Yokel RA, *J. Trace Elem. Electrolytes Health Dis.*, 4: 107 (1990).
58. Polak TB, Milacic R, Pihlar B, Mitrovic B, *Phytochemistry,* 57: 189 (2001).
59. Cairns WRL, Ebdon L, Hill SJ, *Fresenius' J. Anal. Chem.*, 355: 202 (1996).
60. Houck RS, Shum SCK, Wiederin DR, *Anal. Chim. Acta*, 250: 61 (1991).
61. B'Hymer C, Caruso JA, *J. Anal. At. Spectrom.*, 15: 1531 (2000).
62. Le XC, Lu X, Ma M, Cullen WR, Aposhian HV, Zheng B, *Anal. Chem.*, 72: 5172 (2000).
63. Zoorob GK, McKiernan JW, Caruso JA, *Mikrochim. Acta*, 128: 145 (1998).
64. Ma M, Le XC, *Clin. Chem.*, 44: 3 (1998).
65. Ellis LA, Roberts DJ, *J. Chromatogr. A*, 774: 3 (1997).
66. Byrdy FA, Caruso JA, *Environ. Health Perspect.*, 1: 21 (1995).
67. Sutton KL, Caruso JA, *J. Chromatogr. A*, 856: 243 (1999).
68. Boorn AW, Browner RF, *Anal. Chem.*, 54: 1402 (1982).
69. Olesik JW, Stewart I, Hartshone J, Kinzer JA, Winter conference on plasma spectrochemistry, January 5, 1998, paper #M1, Scotland (1998).
70. Lu Q, Barnes RM, *J. Microchem.*, 54: 129 (1996).
71. Pergantis SA, Heithmar EM, Hinners TA, *Anal. Chem.*, 67: 4530 (1995).
72. Olesik JW, Dinzer JA, Harkleroad B, *Anal. Chem.*, 66: 2022 (1994).
73. Shum SCK, Neddersen R, Houk RS, *Analyst*, 117: 577 (1992).
74. B'Hymer C, Sutton KL, Caruso JA, *J. Anal. At. Spectrom.*, 13: 855 (1998).
75. Wang L, May SW, Browner RF, Pollock SH, *J. Anal. At. Spectrom.*, 11: 1137 (1996).
76. Jakubowski N, Jepkens B, Stuewer D, Berndt H, *J. Anal. At. Spectrom.*, 9: 193 (1994).
77. Koropchak JA, Winn DH, *Trends Anal. Chem.*, 6: 171 (1987).
78. Tomlinson MJ, Caruso JA, *Anal. Chim. Acta*, 322: 1 (1996).
79. Martinez R, Pereiro R, Sanz-Medel A, Bordel N, *Fresenius' J. Anal. Chem.*, 371: 746 (2001).
80. Thomas P, Sniatecki K, *J. Anal. At. Spectrom.*, 10: 615 (1995).
81. Le XC, Cullen WR Reimer KJ, *Talanta*, 41: 495 (1994).
82. Chaterji A, *Sci. Total Environ.*, 228: 25 (1999).

83. Wangkarn S, Pergantis S, *J. Anal. At. Spectrom.*, 15: 627 (2000)
84. Wang J, Jiang SJ, *Anal. Chim. Acta*, 289: 213 (1994).
85. Ding H, Wang J, Dorsey LG, Caruso J, *J. Chromatogr. A*, 694: 425 (1995).
86. Pergantis SA, Heithmar EM, Hinners TA, *Analyst*, 122: 1063 (1997).
87. Zhang X, Cornelis R, DeKimpe J, Mees L, Vanderbiesen V, DeCubbber A, Vanholder R, *Clin. Chem.*, 42: 1231 (1996).
88. Zhang X, Cornelis R, DeKimpe J, Mees L, *Anal. Chim. Acta*, 319: 177 (1996).
89. Le XC, Mingsheng M, Norris AW, *Anal. Chem.*, 68: 4501 (1996).
90. Dean JR, Ebdon L, Foulkes ME, Crew HM, Massey RC, *J. Anal. At. Spectrom.*, 4: 285 (1989).
91. Sanz-Medel A, Aizpun B, Marchante JM, Segovia E, Fernandez ML, Blanco E, *J. Chromatogr. A*, 683: 233 (2002).
92. Iserte LO, Roig-Navarro AF, Hernandez F, *Anal. Chim. Acta*, 527: 97 (2004).
93. Huerga A, Lavilla I, Bendicho C, *Anal. Chim. Acta*, 534: 121 (2005).
94. Evans O, Mckee GD, *Analyst*, 112: 983 (1987).
95. Harrington CF, Catterick T, *J. Anal. At. Spectrom.*, 12: 1053 (1997).
96. Wan CC, Chen CS, Jiang SJ, *J. Anal. At. Spectrom.*, 12: 683 (1997).
97. Liang LN, Jiang GB, Liu JF, Hu JT, *Anal. Chim. Acta*, 477: 131 (2003).
98. Le XC, Li XF, Lai V, Ma M, Yalcin S, Feldmann J, *Spectrochim. Acta* 53: 899 (1998).
99. Kotrebai M, Tyson JF, Block E, Uden PC, *J. Chromatogr. A*, 866: 51 (2000).
100. Bird SM, Ge HH, Uden P, Tyson JF, Block E, Denoyer E, *J. Anal. At. Spectrom.*, 12: 785 (1997).
101. Gayon JMM, Thomas C, Feldmann I, Jakuboswski N, *J. Anal. At. Spectrom.*, 15: 1093 (2000).
102. Gayon JMM, Thomas C, Feldmann I, Jakuboswski N, *J. Anal. At. Spectrom.*, 16: 457 (2001).
103. Quijano MA, Moreno P, Gutierrez AM, Perez-Conde MC, Camara C, *J. Mass Spectrom.*, 35: 878 (2000).
104. Shiobara Y, Ogra Y, Suzuki KT, *Analyst*, 124: 1237 (1999).
105. Suzuki KT, *Tohoku J. Exp. Med.*, 178: 27 (1996).
106. Suzuki KT, Itoh M, Ohmichi M, *J. Chromatogr. B, Biomed. Appl.*, 666: 13 (1995).
107. Ipolyi I, Stefanka Z, Fodor P, *Anal. Chim. Acta*, 435: 367 (2001).
108. Lindemann T, Hintelmann H, *Anal. Bioanal. Chem.* 372: 486 (2002).
109. Fairman B, Catterick T, Wheals B, Polina E, *J. Chromatogr. A*, 758: 85 (1997).
110. Kumar UT, Dorsey JG, Caruso JA, Evans EH, *J. Chromatogr. A*, 654: 261 (1993).
111. Kim A, Hill S, Ebdon L, Rowland S, *J. High Resolut. Chromatogr.*, 15: 665 (1992).
112. Wrobel K, Gonzalez EB, Wrobel K, Sanz-Medel A, *Analyst*, 120: 809 (1995).
113. Canteros-Picotto A, Fernandez-Martin JL, Cannata-Andia JB, *Am. J. Kidney Dis.*, 36: 969 (2000).
114. Lustig S, Michalke B, Beck W, Schramel P, *Fresenius' J. Anal. Chem.*, 360: 18 (1998).
115. Blais JS, Marshall WD, Adams FC, in Brockaert JAC, Gucer S, Adams FC (Eds.), *Metal Speciation in the Environment*, Springer Verlag, Berlin (1990).
116. Blais JS, Marshall WD, *J. Anal. At. Spectrom.*, 4: 271 (1989).
117. Cammann K, Robecke M, *Chem. Indus.*, 6: 32 (1990).
118. Langseth W, *Anal. Chim. Acta*, 185: 249 (1986).
119. Langseth W, *Fresenius' J. Anal. Chem.*, 325: 267 (1986).
120. MacCrehan WA, *Anal. Chem.*, 53: 74 (1981).

121. Brinckman FE, Blair WR, Jewett KL, Iverson WP, *J. Chromatogr. Sci.*, 15: 393 (1977).
122. Wilken RD, *Fresenius' J. Anal. Chem.*, 342: 795 (1992).
123. Kollotzek D, Oechsle D, Kaiser G, Tschöpel P, Tölg G, *Fresenius' J. Anal. Chem.*, 318: 485 (1984).
124. Bushee DS, *Analyst*, 113: 1167 (1988).
125. Fujita M, Takabatake E, *Anal. Chem.*, 55: 454 (1983).

7 Speciation of Metal Ions by Ion Chromatography

Basically, ion chromatography (IC) is an advanced version of reversed-phase, high-performance liquid chromatography (HPLC), where the reversed-phase (RP) column of RP-HPLC is replaced by an ion-exchange column, and the other parts are similar to those used with RP-HPLC. IC was earlier known by the name of ion-exchange chromatography, but with the development of various ion-exchange columns, its name was replaced by ion chromatography. IC has been used widely for the analysis of all inorganic and organic metal ions, including ionized species. The development of a variety of ion-exchange columns makes it the best technique in metal ion speciation. In IC, anion- and cation-exchange columns are used, and mixed (anion and cation) columns are also available that improve separation efficiency. In cation-exchange chromatography, the stationary phase is usually composed of resins containing sulfonic acid groups or carboxylic acid groups of negative charges, and cation metallic species are attracted to the stationary phase by electrostatic interactions. In anion-exchange chromatography, the stationary phase is a resin generally containing primary or quaternary amine functional groups of positive charge, and these stationary-phase groups pull solutes of negative charge. Pure cation-exchange columns are useful for alkali and alkaline earth metal ions, while anion-exchange columns separate transition and heavy metal ions successfully. Again, as in the case of RP-HPLC, the use of aqueous and nonaqueous mobile phases enhanced its application. Generally, aqueous buffers are used as the mobile phases for the speciation of metal ions by IC, making the running cost low. In IC, metal ion species are separated on the chromatographic column according to their reversible interchange of ions between a solution and a solid, inorganic, or polymeric insoluble material containing fixed ions and exchangeable counterions. Generally, the chromatographic separation of metal ions on anion-exchange columns requires the presence of negatively charged complexes. Thus, the complexation can be obtained by offline or online methods using a suitable ligand. The combination of IC with advanced detectors in this technique is similar to RP-HPLC. Many detection devices such as electrochemical detection, conductometric detection, ICP, MS, AAS, and so forth, are used for the detection of metallic species. The speciation of derivatized and underivatized metal ions has been achieved using these columns. A number of monographs and reviews were published on the speciation of metal ions by IC.[1–14] We describe here the speciation strategy for metal ions in environmental and biological samples by using IC and its related modalities.

Attempts were also made to explain the optimization of the speciation and the mechanisms of metal ion speciation by IC.

7.1 APPLICATIONS

IC is recognized as the best separation technique for different metal ion species due to its higher capability to separate metal ions using ion-exchange columns and a variety of mobile phases. The physicochemical properties of the analyte (i.e., charge, polarity, molecular mass, etc.) also govern their speciation. IC has been used for the speciation of many cationic, anionic, and organometallic species. Some examples of metal ion speciation by IC are summarized below.

7.1.1 Aluminum (Al)

Aluminum is an important element in an environmental point of view; hence, many researchers reported its speciation using IC. Hils et al.[15] described a procedure for the speciation of aluminum in the percolating water of forest soil by online coupling with ICP-MS. Inorganic and organic aluminum species were fractionated on a cation-exchange column, IonPac CG122. Phytotoxic polymeric aluminum hydroxides, for example, Al_{13} $[AlO_4Al_{12}(OH)_{24}(H_2O)_{12}{}^{7+}]$, were determined using pyrocatechol violet (PCV) as a species-dependent complexing reagent prior to the cation-exchange step. Yamada et al.[16] used flow injection analysis (FIA) with fluorescence detection of aluminum species in soil extracts on ionic exchangers. Aluminum complexes with organic substances (anion species) were also separated by using a strongly acidic cation exchanger in the pH range of 3 to 5. Furthermore, aluminum was separated into three categories, namely, the $Al_3{}^+$ and AlOH complex; aluminum organic complexes (cation species); and its anion species, by using a strongly acidic and a weakly acidic cation exchanger at around pH 5. Hara et al.[17] reported using IC for the speciation of aluminum in rainwater into three categories — free aluminum ($Al_3{}^+$), fluoride complexes (sum of $AlF_2{}^+$ and $AlF_2{}^+$), and other forms of aluminum. The free form of the aluminum ion ($Al_3{}^+$) was directly obtained from the separation data of aluminum species according to their charges using gradient elution cation-exchange chromatography. Bantan et al.[18] used cation (Mono S) and anion (Mono Q) exchange columns for the speciation of aluminum organic acid complexes in plant sap using fast protein liquid chromatography (FPLC). The authors used inductively coupled plasma–atomic emission spectroscopy (ICP-AES) or electrothermal atomic absorption spectrometry (ET-AAS) as the detectors. Ammonium nitrate, 8 m*M*, was used as the eluent. Mitrovic and Milacic[19] reported the speciation of aluminum metal ion, for example, $Al(OH)_2{}^+$ and $Al(OH)_3$, on a cation-exchange Mono-S-type column using NH_4NO_3 as the eluent in a linear gradient manner.

7.1.2 Arsenic (As)

Revealed in the literature are many reports of arsenic speciation by IC. Anion-exchange columns were generally used for the speciation of arsenic metal ion.

However, some reports are available on arsenic speciation using cation-exchange and mixed exchangers columns. Martin et al.[20] developed an online ion chromatography–microwave-assisted oxidation–hydride generation–atomic absorption spectrometry system (using columns of different kinds) for the determination of arsenite, arsenate, dimethylarsinate (DMA), monomethylarsonate (MMA), arsenobetaine (AsB), and arsenocholine (AsC) in environmental samples. The anion-exchange Hamilton PRP-X-100 anionic column was proposed for the determination of six species. The proposed method was applied to water and sediment samples, and the potential of the developed method for future applications was demonstrated. Tsalev et al.[21] coupled IC with continuous-flow hydride generation–atomic absorption spectrometry (HG-AAS) for the speciation of As(III), monomethylarsonate, dimethylarsinate, arsenobetaine, arsenocholine, trimethylarsine oxide, and tetramethylarsonium ion. The authors used anion-exchange columns (i.e., Hamilton PRP X-100 and Supelcosil SAX 1) and gradient elution with phosphate buffers containing KH_2PO_4-K_2HPO_4. Mattusch and Wennrich[22] reported the separation of MMA, DMA, AsC, AsB, trimethylarsine oxide (TMAO), and tetramethylarsonium bromide (TMA) using an anion-exchange (IonoPac AS4-SC) column. The authors used 5 m*M* sodium carbonate, 40 m*M* sodium hydroxide, and 4% methanol as the mobile phase with the detection by ICP-MS. Zheng et al.[23] speciated six arsenic compounds including arsenocholine, arsenobetaine, dimethylarsinic acid, methylarsonic acid, arsenous acid, and arsenic acid on a Hamilton PRP-X100 anion-exchange column using isocratic elution and detected by ICP-MS. This analytical procedure was applied to the speciation of arsenic compounds in human urine. Slejkovec et al.[24] developed an ion chromatography–hydride generation–atomic fluorescence spectrometry (IC-HG-AFS) system for the speciation of arsenite, arsenate, MAA, and DMA in aerosol samples using an anion-exchange column (i.e., Adsorbosphere SAX). The authors used 10 m*M* KH_2PO_4 (pH 6) as the mobile phase. Miller et al.[25] used anion exchange for the separation of arsenate and arsenite species. Ion-exchange separations were performed for four arsenic species common in drinking water sources. Gettar et al.[26] developed an analytical methodology for the specific determination of arsenite, arsenate, MMA, and DMA using ion chromatography. Different types and sizes of anion-exchange columns, silica and polymeric, were tested using ethylenediaminetetraacetic acid (EDTA) as the eluent. The method is based on an ion chromatographic separation, coupled online to postcolumn generation of the gaseous hydrides by reaction with sodium tetrahydroborate in acidic medium. Bissen et al.[27] speciated two inorganic arsenic species, As(III) and As(V), and two organic arsenic species, MMA and DMA, by IC. The separation of arsenic species was achieved on the anionic exchange column IonPac AS11 (Dionex) with NaOH as mobile phase with the detection by ICP-MS. The technique was successfully applied to analyze extracts of two contaminated soils, sampled at a former tannery site and a former paint production site. Lindemann et al.[28] performed speciation analysis of As(III), As(V), arsenobetaine, MMA, DMA, and phenylarsonic acid (PAA) by online coupling of the anion-exchange IC with ICP-MS. He et al.[29] speciated arsenic metal ions using an online method capable of the separation of arsenic species of arsenite As(III), arsenate As(V), MMA, and DMA. The method was based on the combination of IC with UV photooxidation for sample digestion and hydride generation–atomic fluorescence spectrometry (HG-AFS) for sensitive detection. The best separations were obtained with an anion exchange AS11 column protected by an AG11 guard column, and gradient elution with NaH_2PO_4 and water as the mobile phase. Schmidt et al.[30] reported the speciation of As(III), As(V), MMA, DMA, AsC, AsB, TMA, and TMAO using an anion-exchange column. The eluents used were 0.4 m*M* HNO_3 containing 0.05 m*M* benzene-1,2-disulfonic acid with the detection achieved by ICP-MS. Roig-Navarro et al.[31] reported the analysis of As(III), As(V), MMA, and DMA using Ion 12 column with 4 m*M* HNO3-2% methanol and 0.3 *M* NH_4HCO_3-2% methanol as the mobile phases. The detection was carried out by ICP-MS. Martinez-Bravo et al.[32] described a new method for the simultaneous chromatographic separation and determination of arsenite, arsenate, MMA, DMA, selenite, and selenate in water. Speciation was achieved by the online coupling of anion-exchange IC and ICP-MS. The mobile phase used was NH_4NO_3 20 m*M*, pH 8.7-NH_4NO_3 60 m*M*, pH 8.7 with gradient elution. Day et al.[33] used an anion-exchange column coupled to ICP-MS for the detection of arsenic species — As(III), As(V), MMA, and DMA. The mobile phase containing 2 m*M* NaH_2PO_4 and 0.2 m*M* EDTA at pH 6 allowed for adequate separation of these arsenic species. Schlegel et al.[34] reported the separation method for As(III), As(V), and DMA species on an anion-exchange Hamilton PRP X-100 column using inductively coupled plasma–atomic emission spectrometry (ICP-AES) as an atomic spectrometric detection technique. Van Hulle et al.[35] determine arsenic species in three common Chinese edible seaweeds, one brown (*Laminaria japonica*) and two red (*Porphyra crispata* and *Eucheuma denticulatum*) using anion- and cation-exchange columns with detection by ICP-MS.

Zhang et al.[36] reported the speciation of As(III), As(V), MMA, and DMA on cation-exchange column (Dionex Ionopac CS 10) with the detection by UV photooxidation spectrometer. The mobile phase used was 100 m*M* HCl and 50 m*M* NaH_2PO_4. Larsen et al.[37] speciated anionic arsenic species — As(III), As(V), MMA, DMA — and cationic species — TMAO, AsB, AsC, TMAs, 2-dimethylarsinylethanol (DMAE), glycerylphosphorylarsenocholine (GPAC), and trimethylarsonium ion (TMSe) — in 11 shrimp, crab, fish, and lobster samples. The authors used a cation-exchange column (Ionosphere C) with 20 m*M* pyridinium ion adjusted to pH 2.65 with HCO_2H and an anion-exchange column (Ion 120) with 100 m*M* NH_4HCO_3 adjusted to pH 10.3 with NH_4OH as the mobile phases, separately, and respectively, using ICP-MS detector. Johnson and Aldstad[38] described an improved method for the determination of inorganic arsenic in drinking water. The method is based on comprehensive optimization of the anion-exchange IC separation of arsenite and arsenate with postcolumn generation and detection of the arsenate molybdate heteropoly acid (AMHPA) complex ion. The arsenite capacity factor was improved from 0.081 to 0.13 by using a mobile phase composed of 2.5 m*M* Na_2CO_3 and 0.91 m*M* $NaHCO_3$ (pH 10.5, 2 mL/min). A postcolumn photooxidation reactor (2.5 m × 0.7 mm) was optimized using 0.37 μ*M* potassium persulfate at 0.50 mL/min such that arsenite was converted to arsenate with 99.8 ± 4.2% efficiency. Multivariate optimization of the complexation reaction conditions yielded 1.3 m*M* ammonium molybdate, 7.7 m*M* ascorbic acid, 0.48 *M* nitric acid, 0.17 m*M* potassium antimony tartrate, and 1% (v/v) glycerol. A long path length flow cell (Teflon AF, 100 cm) was used to measure the absorption of the AMHPA complex (818 ± 2 nm). Figures of merit for arsenite/arsenate include limit of detection (1.6/0.40 μL^{-1}), standard error in absorbance ($5.1 \times 10^{-3}/3.5 \times 10^{-3}$), and sensitivity ($2.9 \times 10^{-3}/2.2 \times 10^{-3}$) with absorbance units per parts per billion (ppb). Successful application of the method to fortified surface water and groundwater was also described. Cornelis et al.[39] speciated DMA and AsB in three candidate lyophilized urine reference materials. The measurements were based on cation-exchange liquid chromatography coupled to HG-AAS with online digestion of the organic. Sheppard et al.[40] developed an IC method for the speciation of As(III), As(V), DMA, and MMA in urine, club soda, and wine samples with the detection by ICP-MS. van Elteren et al.[41] described a novel separation for cationic arsenic compounds on a polymer-based cation-exchange column using 3-carboxy-4-hydroxybenzenesulfonic acid in the mobile phase. By combining both separation techniques, eight environmentally important arsenic compounds can be determined using an online UV decomposition reactor prior to HG and AFS. The method was applied to test the stability of arsenic compounds (in aqueous media) related to food treatment procedures. Pantsar-Kallio and Manninen[42] described an IC method in conjunction with ICP-MS for the speciation of As(III), As(V), DMA, MMA, AsB, and AsC species using KNO_3 at pH 9.8 as the eluent. Chausseau et al.[43] reported the separation of As(III), As(V), DMA, and MMA using ion-exchange chromatography coupled to an axially viewed sequential ICP-AES. Vela et al.[44] presented an IC-ICP-MS method for the speciation of As(III), As(V), MMA, DMA, and AsB. Jackson et al.[45] used IC-ICP-MS for the speciation of arsenic species — As(III), As(V), DMA, and AsB — using an AS7 column and HNO_3 gradient elution in liver and gill extracts of large mouth bass (*Micropterus salmoides*) samples. Bohari et al.[46] developed a method for the determination of arsenic species (arsenite, arsenate, MMA, and DMA) in plants. The method was comprised of ion-exchange liquid chromatography coupled online to AFS through continuous hydride generation. This method was applied to cultivated plant parts. Arsenate appears to predominate in soils, roots, and leaves; unidentified species (probably arsenosugars) play an important role (60%) in rice fruits. The carrot was found to be the most contaminated edible plant part, containing 1 $mgkg^{-1}$, essentially as arsenate species. MMA was detected in all soils and some plant parts, especially shallots at low levels, whereas DMA was found only in one soil sample and in hot pepper leaves. Kohlmeyer et al.[47] reported IC-ICP-MS as the effective method for the speciation of arsenite, arsenate, MMA, DMA, AsB, AsC, trimethylarsine oxide, and tetramethylarsonium ion in fish, mussel, oyster, and marine algae samples.

7.1.3 Chromium (Cr)

Chromium(VI) is a toxic metal ion, even at trace levels. Scancar and Milacic[48] presented a new analytical procedure using an anion-exchange separation support based on convective interaction media (CIM) for the speciation of chromium. The separation of Cr(VI) was performed on a weak anion-exchange CIM diethylamine (DEAE) fast monolithic chromatographic disc. Buffer A (0.005 *M* Tris-HCl, pH 8) and buffer B (buffer A plus 3 *M* NH_4NO_3) were employed in the separation procedure. The separated chromium species were determined off-line by ET-AAS in 0.5 mL fractions. The applicability of the CIM DEAE-ET-AAS procedure was investigated for the determination of airborne Cr(VI) at a plasma cutting workplace. Byrdy et al.[49] speciated Cr(III) and Cr(VI) on an anion-exchange column with online detection by ICP-AES. A mobile phase consisting of ammonium sulfate and ammonium hydroxide was used, and a simple chelation procedure with EDTA followed to stabilize the Cr(III) species in standard solutions. Saverwyns et al.[50] used a new commercially available microbore anion-exchange column (Cetac ANX-3202) for the speciation of chromium metal ion. Nitric acid was used as the mobile phase, and the developed methodology was reported as ideal by coupling with ICP-MS. Vanhaecke et al.[51] speciated Cr(III) and Cr(VI) on an anion-exchange column with detection by ICP-MS. The authors used 60 m*M* HNO_3 as the mobile phase. Milacic et al.[52] reported the speciation of chromium metal ion in aerosol samples on an anion-exchange column (Mono Q HR 5/5) with detection by ET-AAS. Inoue et al.[53] combined IC with ICP-MS for the speciation of Cr(III) and Cr(VI) species. The authors used Excelpak ICS-A23 packed with hydrophilic polymer-based anion-exchange resin (ion-exchange capacity: 0.05 mequivg^{-1} dry weight) as the column with EDTA-NH_4-oxalic acid (pH 7) [1×10^{-3} *M* 0.01 *M*] as a mobile phase. Pobozy et al.[54] used IC procedures for the speciation of different oxidation states of chromium using an anion-exchange Hamilton PRP-X-100 column with phthalate as the mobile phase and spectrophotometry as the detection. Due to element-specific detection, determination was carried out with retention of Cr(VI) and elution of Cr(III) in the void peak. In the second procedure developed, Cr(III) was complexed with DCTA, and the obtained complex was also retained on an anion-exchange column. Collins et al.[55] described the speciation of Cr(VI) and Cr(III) (as the hexaaquo species) using a cation-exchange column. The detection was achieved by spectrophotometric or radiometric technique. These results were compared with those obtained with open-column cation-exchange chromatography using radiometric detection. Furthermore, the same authors[56] described the speciation of Cr(VI) and Cr(III) on a cation-exchange column with detection by radiometry. The mobile phase used was perchloric acid of different concentrations. Pantsar-Kallio and Manninen[57] used IC-ICP-MS for the speciation of chromium in wastewaters. By coupling an anion column with a cation column, both the anionic Cr(VI) and cationic Cr(III) species can be analyzed with detection limits below 0.5 μg/L. Sikovec et al.[58] described the speciation of Cr(III) and Cr(VI) using a HPIC CS 5A column by precolumn derivatization of Cr(III) with pyridine-2,6-dicarboxylic acid and postcolumn derivatization of Cr(VI) by diphenylcarbazide. The authors used a mixture of 2 m*M* pyridine-2,6-dicarboxylic acid (PDCA), 2 m*M* $NaHPO_4$, 1 m*M* NaI, 5 m*M* CH_3COONH_4, and 2.8 m*M* LiOH as the mobile phase with the detection by an online thermal lens spectrometer. Sikovec et al.[59] presented the speciation of Cr(III) and Cr(VI) using an HPIC CS 5A column using 2 m*M* PDCA, 2 m*M* Na_2HPO_4, 1 m*M* NaI, 5 m*M* CH_3COONH_4, and 2.8 m*M* LiOH as a mobile phase containing various organic modifiers. The authors used online thermal spectroscopic detection for chromium species.

7.1.4 Selenium (Se)

Shum et al.[60] used an anion-exchange column for the speciation of Se(IV) and Se(VI) using a sodium carbonate-bicarbonate mobile phase. The authors used ICP-MS as the detection technique. Laborda et al.[61] used a Nucleosil 100-SB anion-exchange column for the analysis of Se(IV), Se(VI), and TMSe species with ammonium citrate as the eluent. Detection was achieved by ET-AAs. Pitts et al.[62] described a two-step eluent switching procedure for the speciation of inorganic selenium species in aqueous solution using an anion-exchange column. The best separation was obtained with 25 m*M* K_2SO_4 eluent, switching to 200 m*M* after 200 sec at a flow rate of 2 mL/min. The detection was obtained with an hydride generation atomic fluorescence detector. Jackson et al.[45] used IC-ICP-MS for the speciation of selenium species — Se(IV), Se(VI), selenomethionine, or selenocystine — using an AS7 column and HNO_3 gradient elution in liver and gill extract samples. Kannamkumarath et al.[63] speciated selenium metal ion using IC with the detection by ICP-MS. Lindemann et al.[28] speciated Se(IV), Se(VI), and selenomethionine using IC, and the detection was achieved by ICP-MS. Bird et al.[64] reported the separation of selenite and selenate using an anion-exchange column. Detection was achieved by ICP-MS. Pedersen et al.[65] presented the speciation of selenomethionine (SeMet), selenocystine (SeCys), selenite, and selenate by IC. An organic polymeric strong anion-exchange column was used as the stationary phase in combination with an aqueous solution of 6 m*M* of salicylate ion at pH 8.5 as the mobile phase, which allowed for the isocratic separation of the four selenium analytes within 8 min. The separated selenium species were detected online by flame atomic absorption spectrometry (FAAS) or ICP-MS. Vilano et al.[66] reported the speciation of selenium compounds — selenate, selenite, SeCys, and SeMet — using an anion-exchange column of Hamilton PRP X-100 with the detection by UV irridation and an hydride generation quartz cell AAS (HG-QC-AAS) assembly. The authors used phosphate buffer (pH 7) as the mobile phase. Zhang et al.[67] evaluated the IC method for the speciation of selenium in plants, soil, and sediment samples. The authors used an anion-exchange resin (Dowex 1-10X) to separate selenium species — trimethylselenonium ion ($TMSe^+$), dimethylselenoxide (DMSeO), SeMet), Se(IV), and Se(VII). HG-AAS was used to determine concentrations of these selenium species. Gomez-Ariza et al.[68] speciated SeCys, SeMet, selenoethionine (SeET), selenite [Se(IV)], and selenate [Se(VI)] using anion-exchange columns (SAX and PRP X-100). Online microwave-assisted digestion and hydride generation steps were performed prior to the atomic fluorescence detection. Michalke[69] speciated selenium metal ion [SeU, SeMet, SeET, Se(IV), SeC and Se(VI)] in human serum by using a strong anion-exchange column with detection by ICP-MS.

7.1.5 Miscellaneous Metal Ions

Cardellicchio et al.[70] used an ion-exchange mixed-bead column for the speciation of Fe(II) and Fe(III) metal ion species using online complexation with pyridine-2,6-dicarboxylic acid. The detection was achieved by UV/visible spectrometer after postcolumn derivatization with 4-(2-pyridylazo)resorcinol or 2-(5-bromo-2-2-pyridylazo)-5-diethylaminophenol. Weber et al.[71] achieved the speciation of iron metal using anion-exchange column with NaOH gradient elution with the detection by AAS. Up to five different iron species were separated with the detection limit in the range of micromole concentration. The method was applied to root washings of iron-deficient wheat and barley plants and to a xylem exudate of nondeficient maize. Ulrich et al.[72] developed an IC method for the separation and quantification of Sb(III) and Sb(V) using a PRP-X100 column with ICP-MS detection. The optimum conditions for the separation of the antimony species were established with 15 m*M* nitric acid at pH 6 as eluent. Lindemann et al.[28] reported the speciation of Sb(III) and Sb(V) by IC-ICP-MS assembly. Miravet et al.[73] reported the speciation of inorganic [Sb(III) and Sb(V)] and organic (Me_3SbCl_2) antimony species on a polystyrene–divinylbenzene-based anion-exchange column (Hamilton PRP-X100). The detection was achieved by coupling HG-AFS with IC. The different mobile phases with different concentrations and pHs were tested, but the best efficiency and resolution were achieved by using a gradient elution between diammonium tartrate 250 mML^{-1} pH 5.5 and KOH 20 mML^{-1} pH 12. The detection limits obtained were 0.06 μgL^{-1} [Sb(V)], 0.09 μgL^{-1} (Me_3SbCl_2), and 0.04 μgL^{-1} [Sb(III)]. Takaya and Sawatari[74] described the separation of vanadium(IV) and vanadium(V) species using an anion-exchange column with ICP-AES detection. In this method, 1 m*M* HNO_3 solution and 100 m*M* HNO_3 solution were applied in sequence as eluents. A vanadium(IV) and vanadium(V) mixture was injected onto an anion-exchange column; and vanadium(IV) cation was then eluted by 1 m*M* HNO_3, while vanadium(V) oxoacid anion was trapped on the column. After this separation, vanadium(V) was eluted as a cation from the column by 0.1 *M* HNO_3. In this separation, about 15% of vanadium(V) interfered with vanadium(IV), and trace vanadium(IV) interfered with vanadium(V). This interference was estimated by a simple calculation based on standard observations, and the speciation of vanadium(IV) and vanadium(V) was performed. Nachtigall et al.[75] determined tetrachloroplatinate ($PtCl_4^2$) and hexachloroplatinate ($PtCl_6^2$) species by applying UV absorption detection. Independent and specific platinum quantification were obtained by off-line ICP-MS. Qualitatively, a number of Pt(II) and Pt(IV) complexes, formed by hydrolysis from tetra- and hexachloroplatinate, were also detected. Al-Rashdan et al.[76] speciated inorganic lead (Pb_2^+) and several trialkyllead species (trimethyllead chloride [TML], triethyllead chloride [TEL], and triphenyllead chloride [TPhL]) using an ion-exchange column. The detection was carried out by ICP-MS, and an isocratic separation with a 30% methanol mobile phase was found to be the best compromise between plasma stability and chromatographic resolution. The cation-exchange chromatography of tributyltin and triphenyltin with ammonium acetate eluents was optimized by adding benzyltrimethylammonium chloride.[77] The method was also applied for the analysis of trimethyltin and triethyltin. Svete et al.[78] developed an analytical procedure for the speciation of zinc metal ion using fast protein liquid chromatography (FPLC) and convective interaction media (CIM) fast monolithic chromatography with FAAS and electrospray (ES)-MS-MS detection. The speciation of zinc (Zn) was also carried out on a weak anion-exchange CIM DEAE fast monolithic disc with aqueous 0.4 *M* NH_4NO_3 linear gradient elution (7.5 min, at a flow rate of 2 mL/min) with detection by off-line AAS. Zinc-binding ligands in separated fractions were also characterized by electrospray ES-MS. The CIM DEAE disc was found to be more efficient in the separation of negatively charged zinc complexes than the Mono Q FPLC column. On the CIM DEAE disc, Zn-citrate was separated from both Zn-oxalate and Zn-EDTA. All these species were also separated from hydrated Zn^{2+}, which was eluted with the solvent front. This method is supposed to have an advantage over commonly used analytical techniques for the speciation of zinc, which were only able to distinguish between labile and strong zinc complexes. Good repeatability of the measurements (relative standard deviation: 2 to 4%), tested for six parallel determinations (2 μg of zinc) of Zn-EDTA, Zn-citrate, and Zn-oxalate was achieved at a pH of 6.4 on a CIM DAEA disc. The limit of detection for the separated zinc species was 10 ng/L. The proposed analytical procedure was applied for the speciation of zinc in aqueous soil extracts and industrial wastewater from a lead and zinc mining area. Wang et al.[79] speciated rare earth elements (REE) on a cation-exchange column combined with a post-column derivatization and online ICP-MS detection.

For easy reference, the speciations of different metal ion species by IC are summarized in Table 7.1. As a typical example of metal ion speciation by IC, Figure 7.1 indicates the separation of arsenic species using an anion-exchange column. As in the case of RP-HPLC, the quantitative determination of separated metallic species is calculated by the retention times (t), capacity (k), separation (α), and resolution factors (R). The quantitative estimation is carried out by comparing the area of identified peak with the area of the peak obtained by the standard metallic species of the known concentration.

7.2 OPTIMIZATION OF METAL ION SPECIATION

As in the case of RP-HPLC, the optimization of metal ion species can be achieved by varying the stationary phase, mobile-phase concentrations, and pH. In addition, derivatization of metal ions, detection, extraction of the collected samples, flow rate, and amount injected onto ion-exchange columns may be used to achieve the maximum speciation of metal ions. In view of this, the optimization of these parameters is discussed herein.

7.2.1 Stationary Phases

The use of a suitable column is a critical issue in IC. With the increased use of IC,[126] ion-exchange columns were developed, and generally, two types of columns

TABLE 7.1
The Speciation of Metal Ions by Ion Chromatography

Metal Ions	Sample Matrix	Columns	Detection	Ref.
		Aluminum (Al)		
Al_{13} and ($AlO_4Al_{12}(OH)_{24}(H_2O)_{12}^{7+}$)	Water	IonPac CG12	ICP-MS	15
Aluminum species	Soil extract	Cation exchange	FIA fluorescence	16
AlF_2^+ and AlF_2^+	Rain	Cation exchange	Ion-selective electrode	17
Aluminum species	—	Mono S and Mono Q	ICP-AES	18
$Al(OH)_2^+$ and $Al(OH)_3$	—	Mono S	—	19
		Arsenic (As)		
Arsenite, arsenate, dimethylarsinate (DMA), monomethylarsonate (MMA), arsenobetaine (AsB), and arsenocholine (AsC)	Environmental samples	Hamilton PRP-X-100	HG-AAS	20
As(III), monomethylarsonate, dimethylarsinate, arsenobetaine, arsenocholine, trimethylarsine oxide, and tetramethylarsonium ion	—	Hamilton PRP X-100 and Supelcosil SAX 1	HG-AAS	21
MMA, DMA, AsC, AsB, trimethylarsine oxide (TMAO), and tetramethylarsonium bromide (TAMB)	—	IonoPac AS4-SC	ICP-MS	22
Arsenocholine, arsenobetaine, dimethylarsinic acid, methylarsonic acid, arsenous acid, and arsenic acid	—	Hamilton PRP-X100	ICP-MS	23
Arsenite, arsenate, MAA, and DMA	Aerosol	Adsorbosphere SAX	HG-AFS	24
Arsenate and arsenite	Water	Anion exchange	—	25
Arsenite, arsenate, MMA, and DMA	—	Anion exchange	—	26
As(III), As(V), MMA, and DMA	—	IonPac AS11	ICP-MS	27
As(III), As(V), MMA, DMA, and phenylarsonic acid (PAA)	—	Anion exchange	ICP-MS	28
As(III), As(V), MMA, and DMA	—	AS11	HG-AFS	29
As(III), As(V), MMA, DMA, AsC, AsB, TMA, and TMAO	—	Anion exchange	ICP-MS	30

TABLE 7.1 (Continued)
The Speciation of Metal Ions by Ion Chromatography

Metal Ions	Sample Matrix	Columns	Detection	Ref.
As(III), As(V), MMA, and DMA	—	Ion 120	ICP-MS	31
Arsenite, arsenate, MMA, and DMA	Water	Anion exchange	ICP-MS	32
As(III), As(V), MMA, and DMA	—	Anion exchange	ICP-MS	33
As(III), As(V), and DMA	—	Hamilton PRP X-100	ICP-AES	34
As(III) and As(V)	Seaweeds	Cation exchange	ICP-MS	35
As(III), As(V), MMA, and DMA	—	Ionopac CS 10	UV-AAS	36
As(III), As(V), MMA, DMA, TMAO, AsB, AsC, TMAs, 2-dimethylarsinylethanol (DMAE), glycerylphosphoryl-arsenocholine (GPAC), and trimethylarsonium ion (TMSe)	Fish	Ionosphere C and Ion 120	ICP-MS	37
Arsenite and arsenate	Water	Anion exchange	—	38
DMA and AsB	Urine	Cation exchange	HG-AAS	39
As(III), As(V), MMA, and DMA	Cold drinks	—	ICP-MS	40
Arsenic species	—	Cation exchange	HG-AFS	41
As(III), As(V), dimethylarsinate, monomethylarsonate, arsenobetaine, and arsenocholine	—	—	ICP-MS	42
As(III), As(V), DMA, and MMA	—	—	ICP-AES	43
As(III), As(V), and MMA	—	—	ICP-MS	44
As(III), As(V), dimethylarsenate, and arsenobetaine	—	AS7	ICP-MS	45
Arsenite, arsenate, MMA, and DMA	Plants	—	HG-AFS	46
Arsenite, arsenate, MMA, DMA, AsB, AsC, TMAO, and TMA	Fish	—	ICP-MS	47
As(III), As(V), MMA, and DMA	Water	Hamilton PRP X-100	HG-QF-AAS	80
As(III), As(V), MMA, and DMA	Biological samples	Hamilton PRP X-100	HG-QF-AAS	81

TABLE 7.1 (Continued)
The Speciation of Metal Ions by Ion Chromatography

Metal Ions	Sample Matrix	Columns	Detection	Ref.
As(III), As(V), MMA, and DMA	—	SAX and PRP-X	—	82
As(III), As(V), MMA, and DMA	—	Hamilton PRP X-100	ICP-AES	83
As(III), As(V), MMA, and DMA	—	—	ICP-AES	84
As(III), As(V), MMA, and DMA	Urine	Adsosphere-NH_2	ICP-MS	85
MMA and DMA	—	AG 50 W-X8	HG-AAS	86
As(III), As(V), MMA, DMA, and AsB	—	Anion exchange	HG-AAS	87
As(III), As(V), MMA, DMA, AsB, and AsC	—	Anion and cation exchangers	AAS	88
As(III), As(V), MMA, DMA, AsB, TMAs, and AsC	—	Anion and cation exchangers	ICP-MS	89
As(III), As(V), MMA, DMA, and AsB	—	CAS1	ICP-MS	90
As(III), As(V), MMA, DMA, and AsB	—	BAX-10	—	91
As(III), As(V), MMA, DMA, AsB, TMAs, and AsC	—	PRP-X100	ICP-MS	92
As(III), As(V), MMA, and DMA	Biological samples	AS11	HG-AFS	93
As(III), As(V), DMA, and PhAs	—	ASGA	ICP-AES	94
As(III), As(V), DMA, and As-Bet	Seaweed	Nucleosil-*N*-(CH_3)-10	ICP-AES	95
As(III), As(V), DMA, As-Bet, and As-Col	Biological samples	Hamilton PRP X-100	ICP-AES	100
As(III), As(V), DMA, As-Bet, and As-Chol	River water	Anion exchange	ICP-MS	97
As(III), As(V), DMA, As-Bet, and As-Chol	Biological samples	Hamilton PRP X-100	ICP-MS	98
As(III), As(V), DMA, As-Bet, and As-Chol	Water	Hamilton PRP X-100	ICP-MS	99
As(III), As(V), DMA, and As-Bet	Urine	Adsosphere-NH_2	ICP-MS	100
As(III), As(V), As-Bet, As-sugar, and Me_2As-ethanol	Urine	Asaphipak	ICP-MS	101
As(III), As(V), DMA, As-Bet, and As-Chol	Urine	Hamilton PRP X-100	ICP-MS	102

TABLE 7.1 (Continued)
The Speciation of Metal Ions by Ion Chromatography

Metal Ions	Sample Matrix	Columns	Detection	Ref.
Selenous acid, selenic acid, selonocystine, selenohomocystine, SeThr, and $SeMe_3$	—	Supelcosil LC-SCX	ICP-MS	103
Arsenate, arsenite, MMA, DMA, AsB, and AsC	—	—	HG-AS	104
AsB, AsC, and TMAs	—	Cation exchange	HG-AAS	105
Arsenic species	—	Cation exchange	ICP-MS	106
		Chromium (Cr)		
Chromium species	—	Anion exchange	DEAE-ET-AAS	48
Cr(III) and Cr(VI)	—	Anion exchange	ICP-AES	49
Chromium species	—	Cetac ANX-3202	ICP-AES	50
Cr(III) and Cr(VI)	—	Anion exchange	ICP-MS	51
Cr(III) and Cr(VI)	Aerosol	Mono Q HR 5/5	ET-AAS	52
Cr(III) and Cr(VI)	—	Excelpak ICS-A23	ICP-MS	53
Cr(III) and Cr(VI)	—	Hamilton PRP-X-100	—	5
Cr(III) and Cr(VI)	—	Cation exchange	—	55
Cr(III) and Cr(VI)	—	Cation exchange	—	56
Cr(III) and Cr(VI)	—	Pak A	ICP-MS	57
Cr(III) and Cr(VI)	—	HPIC CS 5A	—	58
Cr(III) and Cr(VI)	—	HPIC CS 5A	Thermal spectrometry	59
Cr(III) and Cr(VI)	—	Chelex 100 and AGMP-1	AAS	107
Cr(III) and Cr(VI)	—	—	ICP-AES	108
Cr(III) and Cr(VI)	Wastewater	Cation exchange	ICP-MS	109
Cr(III) and Cr(VI)	—	IonPac AG5	ICP-MS	109
Cr(III) and Cr(VI)	—	ASGA	ICP-AES	94
Cr(III) and Cr(VI)	—	Mixed ion exchanger	Chemiluminescence	110
Cr(III) and Cr(VI)	—	IonPac AS4A	Chemiluminescence	111
Cr(III) and Cr(VI)	—	Pak A	—	109
Cr(III) and Cr(VI)	—	IonPac AS4A	ICP-MS	112
Cr(III) and Cr(VI)	—	Anion exchange	ICP-MS	113
Chromium complexes	—	AG50W	Radioactivity	114
Chromium complexes	—	AG50W	UV/visible	115
Cr(VI) and hydroxy complexes	—	AG50W	Radioactivity	116
Cr(VI) and hydroxy complexes	—	Dowex 50-X8	Radioactivity	117
Cr complexes	—	Dowex 50-X8	UV/visible	118

TABLE 7.1 (Continued)
The Speciation of Metal Ions by Ion Chromatography

Metal Ions	Sample Matrix	Columns	Detection	Ref.
		Selenium (Se)		
Se(IV) and Se(VI)	—	Anion exchange	ICP-MS	60
Se(IV) and Se(VI)	—	Nucleosil 100-SB	AAS	61
Se(IV) and Se(VI)	—	Anion exchange	HG-AFS	62
Se(IV) and Se(VI)	—	Anion exchange	ICP-MS	28
Selenite and selenate	—	Anion exchange	ICP-MS	64
Selenomethionine, selenocystine, selenite, and selenate	—	Anion exchange	AAS	65
Selenate, selenite, SeCys, and SeMet	—	Hamilton PRP X-100	HG-QC-AAS	66
Se(IV) and Se(VI)	Soil, sediment, and plants	Dowex 1-10X	HG-AAS	67
Selenocystine (SeCys), selenomethionine (SeMet), selenoethionine (SeET), selenite (Se(IV)), and selenate (Se(VI))	—	SAX and PRP X-100	HG-AFS	68
Selenoniocholine and trimethylselonium cation	—	Cation exchange	AAS	119
SeU, SeM, SeE, Se(IV), SeC, and Se(VI)	Human serum	Strong anion exchange	ICP-MS	69
		Miscellaneous metal ions		
Aluminum species	—	—	Fluorescence	120
Inorganic [Sb(III) and Sb(V)] and organic (Me_3SbCl_2) antimony species	Fresh water	Polystyrene-divinylbenzene-based anion-exchange column (Hamilton PRP-X100)	HG-AFS	73
Fe(II) and Fe(III)	—	Mixed bead	UV/visible	70
Iron species	—	Anion exchange	AAS	71
Fe(II) and Fe(III) oxides	Geological samples	CS-5	UV/visible	121
Iron species	—	—	ICP-MS	122
Sb(III) and Sb(V)	—	Hamilton PRP-X-100	ICP-MS	72
Sb(III) and Sb(V)	—	Anion exchange	ICP-MS	28
Sb(III) and Sb(V)	—	Hamilton PRP-X-100	ICP-MS	123
V(IV) and V(V)	—	Anion exchange	ICP-AES	74
Manganese species	—	—	ICP-MS	122
Platinum species	—	—	ICP-MS	122

TABLE 7.1 (Continued)
The Speciation of Metal Ions by Ion Chromatography

Metal Ions	Sample Matrix	Columns	Detection	Ref.
($PtCl_4^2$) and hexachloroplatinate ($PtCl_6^2$)	—	—	ICP-MS	75
Pb_2^+, trimethyllead chloride (TML), triethyllead chloride (TEL), and triphenyllead chloride (TPhL)	—	—	ICP-MS	76
Trimethyltin and triethyltin	—	Cation exchange	—	77
V(IV) and V(V)	—	—	AES	124
V(IV) and V(V)	—	CS5	ICP-MS	125
Zinc species	—	Anion exchange	AAS	78
Rare earth element species	—	—	ICP-MS	79

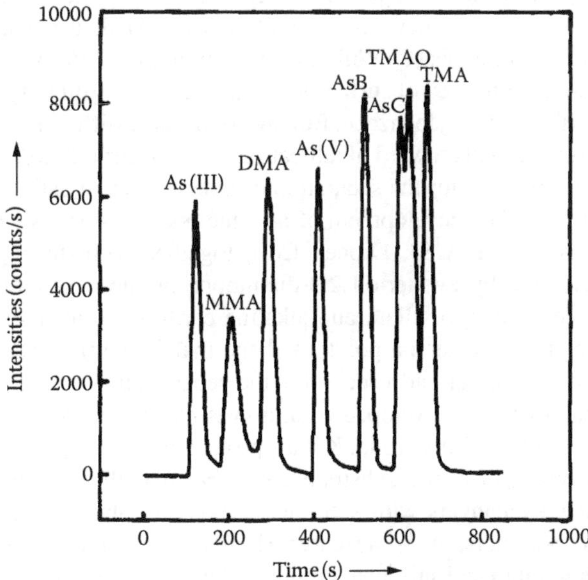

FIGURE 7.1 Chromatograms of arsenic species on an anion-exchange column (IonPac AS7/AG7) and detection with inductively coupled plasma–mass spectrometry (ICP-MS). (From Schmidt, A.C., Reisser, W., Mattusch, J., Popp, P., and Wennrich, R., *J. Chromatogr. A*, 889: 83 (2000). With permission.)

have been used in ion chromatography, those that are cation and anion exchange based. However, some authors used the mixed columns based on cation- and anion-exchange materials.[12] The separation of simple mixtures of metal ions was carried out using ion-exchange columns; hence, it is possible to use these columns for the speciation purpose. Normally, cation-exchange columns are used for the speciation of alkali and alkaline earth elements, while the transition and heavy metal ions are allowed to react with an anion of a weak acid to reduce their charges (derivatization) before entering the cation-exchange column. These derivatized metal ions are separated due to their respective affinities toward the active sites of the separating resin. On the opposite pole, anion-exchange columns are used for the speciation of negatively charged metal ion species, which can be obtained through off-line complexation (before loading onto an IC machine) or online complexation (in the chromatographic system). These can be achieved by adding some suitable ligands into the mobile phase.

Cation-exchange columns are based on the material having the capacities to exchange cations, and they can be prepared by bonding or coating cation-exchange groups onto a silica backbone. The chemically bonded cation-exchange materials are more stable than coated exchange materials, over a wide range of mobile phases. The most preferred cation-exchange compound is poly(butadiene-maleic acid). The most serious drawback of using a silica-based cation-exchange column is the limitation of the pH of the mobile phase — both higher and lower pH values of the mobile phases should be avoided. Below pH 2, a covalent bond of cation exchanger with silica gel may break, while above pH 8, a silica matrix may dissolve in the mobile phase. With the evolution of IC,[126] polymeric cation exchangers mainly consisted of surface sulfonated 20 to 25 μm polystyrene beads cross-linked with 2% divinylbenzene. But due to the large differences in selectivity of alkaline earth elements toward alkali cations, it was difficult to provide simple isocratic elution to allow for the separation of both classes of cations in a reasonable period of time. The development of new methods of synthesis for IC (latex-coated columns, IonPac CS3, Dionex Co.), together with the replacement of *m*-phenylenediamine by zwitterion 2,3-diaminopropionic acid monochloride,[127] made it possible to analyze alkali and alkaline earth metal ions simultaneously. In 1987, Kolla et al.[128] coated a poly(butadiene-maleic acid), containing carboxylate groups, on silica gel and used it for the separation of alkali, alkaline, and transition metal ions. The developed column was named the Schomburg column, after the senior worker of this group. Kolla's product was characterized by solvent-compatible ethylvinylbenzene substrate cross-linked with 55% divinylbenzene, resulting in new exchangers with sulfonic, carboxylic, phosphonic, crown ether groups. These polymers, in some cases (IonPac CS5 and CS5A), have both quaternary ammonium and sulfonate groups and have been used for the separation of many of the metal ionic species (Table 7.1). Röllin et al.[129] used latex agglomerated cation exchangers (IonPac CS10), in the presence of chelating agents, for the determination of lanthanides. In this way, the affinity of the lanthanides for cation-exchange resin was lowered, and the elements were eluted according to the stabilities of the complexes formed. Later, Nesternko et al.[130,131] used new polymer-based columns containing carboxylic (IonPac CS12, CS14), carboxylic/phosphonic/crown ether (IonPac CS15) functional groups for the separation of alkali and alkaline earth elements. Rey and Pohl[132] reported the increased efficiency of IonPac CS15 by increasing the carboxylic groups. The new column (IonPac CS16) did not require the mobile phase based on organic solvents.

Anion-exchange-based columns offer the potential of better selectivities and reduced problems of metal ion hydrolysis and can be used for the separation of complex sample matrices. As discussed above, some ligands are used to convert positive metal ions into negative species, and hence, the order of the elution becomes reversed with respect to the cation-exchange columns. Moreover, cations and anions can be analyzed in a single run using anion-based columns. Hajos et al.[133] reported the separation of transition metal ions on an anion-exchange column (As 9, Dionex). Bruzzoniti et al.[134] described the separation of lanthanides on the mixed-bed column (IonPac CS5, CS5A). Otha and Tanaka[135] used TSKguradgel QAE-SW anion-exchange column for the separation of the cation and anion mixture. The AG 50W-X4 anion-exchange column was used for the separation of rare earth elements by Strelow and Victor.[136] The mixed-bed columns containing cation and anion exchangers are useful for the separation of a wide variety of metallic species. Sun et al.[137] proposed a mixed-bed column (cation and anion exchangers based) for the separation of transition metal ions and further reported more sophisticated separations. Besides, many research papers were published on the separation of metal ions using mixed-exchangers-based columns (IonPac CS5 and CS5A).[134,138–140] Tanaka et al.[141] used TSKgel OA-PAK-A ion-exclusion column for the separation of sodium, ammonium, magnesium, and calcium metal ions. The most important group of ion exchangers is comprised of the organic materials based on synthetic resins. A copolymer of styrene and divinylbenzene is frequently employed as the support. The cation exchangers are obtained by subsequent sulfonation of these styrene and divinylbenzene resin and anion exchangers by chloromethylation followed by amination.

In view of these points, the optimization of metal ions can be achieved using cation- or anion- or mixed–exchanger-based columns. There are only a few reports dealing with the comparative studies of metal ions speciation on cation- and anion-exchange columns. Larsen et al.[37] compared the speciation of inorganic and organic arsenicals using cation (Ionosphere-C) and anion (ION 120) exchange columns. The authors reported the clear separation of AsB, AsC, TMSe, and TMAO on cation-exchange columns and As(III), As(V), MMA, and DMA species on anion-exchange columns (Figure 7.2). Furthermore, the same group[65] used anion-exchange (Polyspher IC-AN-2) and cation-exchange (Ionosphere-C) columns for the comparative study of the speciation of selenium metal ion species (Figure 7.3). It may be observed from this figure that the best separation with greater peak areas was achieved with use of the Polyspher IC-AN-2 column. An interesting study on the speciation of arsenic species was carried out by Roig-Navarro et al.[31] using two anion-exchange columns — ION 120 and Hamilton PRP-X-100 columns. The authors reported the best speciation on the latter column (Figure 7.4). It may be seen from Figure 7.4 that the best separation of As(III),

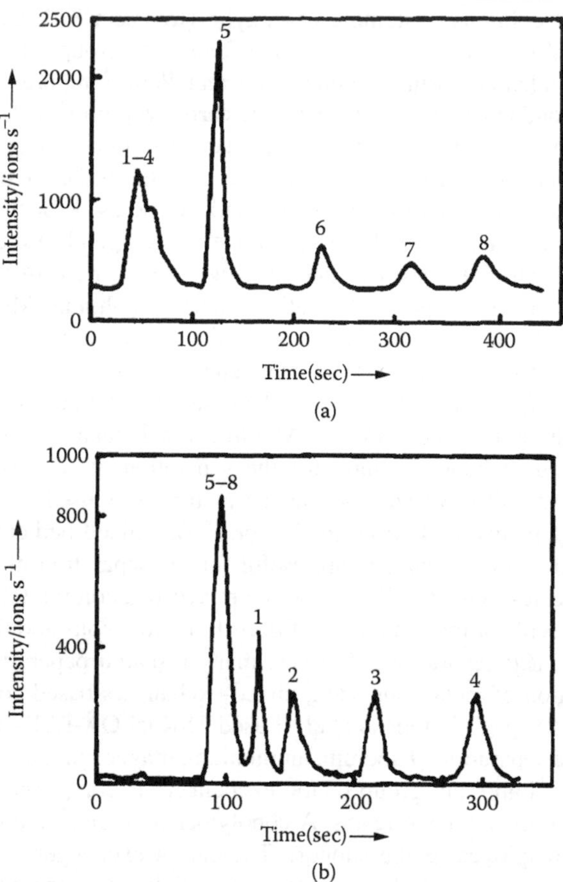

FIGURE 7.2 A comparison of arsenic metal ion speciation using (a) cation (Ionosphere-C) and (b) anion (ION-120) exchange columns: (1) DMA, (2) As(III), (3) MMA, (4) As(V), (6) TMAO, (7) AsC, and (8) TMAs. (From Larsen, E.H., Pritzl, G., and Hansen, S.H., *J. Anal. Atom. Spectrom.*, 8: 1075 (1993). With permission.)

As(V), MMA, and DMA species was achieved on the Hamilton PRP-X-100 column with better separation factors. Some commercially used cation and anion columns are summarized in Table 7.2. These findings indicate that the maximum and best speciation can be achieved by selecting a suitable column.

7.2.2 Mobile Phases

The selection of the mobile phase for the speciation of metal ions in IC is a very important issue. Because the columns are based on ion-exchange materials, the use of the polar mobile phase is required. Usually, buffers of different concentrations and pHs are used as the mobile phases. The most commonly used buffers are phosphate, formate, perchlorate, carbonate, phthalate, and acetate. However,

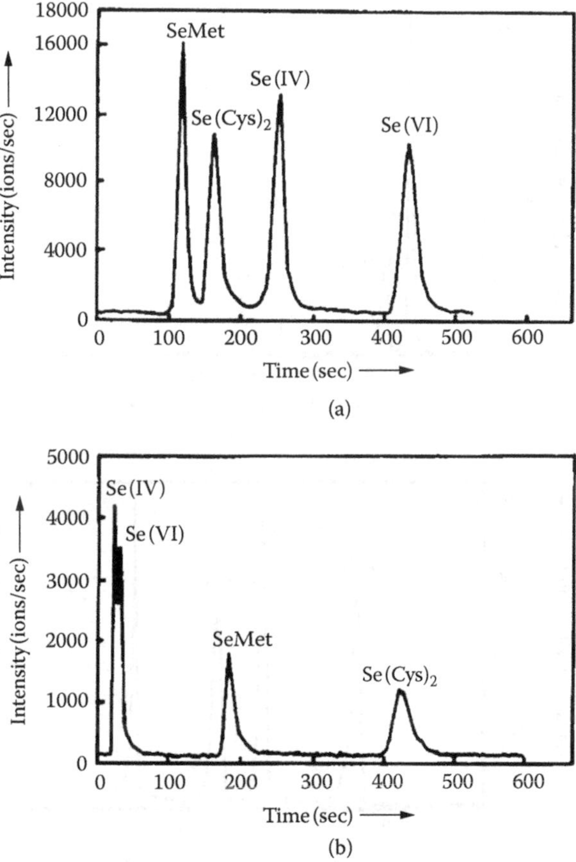

FIGURE 7.3 A comparison of selenium metal ion speciation using (a) anion (Polyspher IC-AN-2) and (b) cation (Ionosphere-C) exchange columns. (From Pedersen, G.A., and Larsen, E.H., *Fresenius' J. Anal. Chem.*, 358: 591 (1997). With permission.)

pure acidic and basic aqueous solutions were also used as the mobile phases. Other reagents were also used as the mobile phases. Larsen et al.[37] used 20 m*M* pyridinium ion, adjusted to pH 2.65 with formic acid, as the mobile phase for the speciation of arsenic. A solution of 0.1 m*M* HCl and 50 m*M* NaH_2PO_4 was used for the speciation of arsenic by Zhang et al.[36] Pedersen and Larsen[65] used 6 m*M* salicylate ion in a 3% methanol–97% water solution, adjusted to pH 8.5 with TRIS, as the mobile phase for the speciation of selenium metal ion.

For optimization purposes, variation in the concentrations of mobile-phase constituents is essential. Of course, all the mobile phases are optimized by selecting suitable concentrations of the mobile-phase components, but few reports are available that deal with the results of these variations. Martinez-Bravo et al.[32] studied the effect of ammonium nitrate on the speciation of arsenic, selenium, and chromium metal ions. The authors used two mobile phases — 20 m*M*

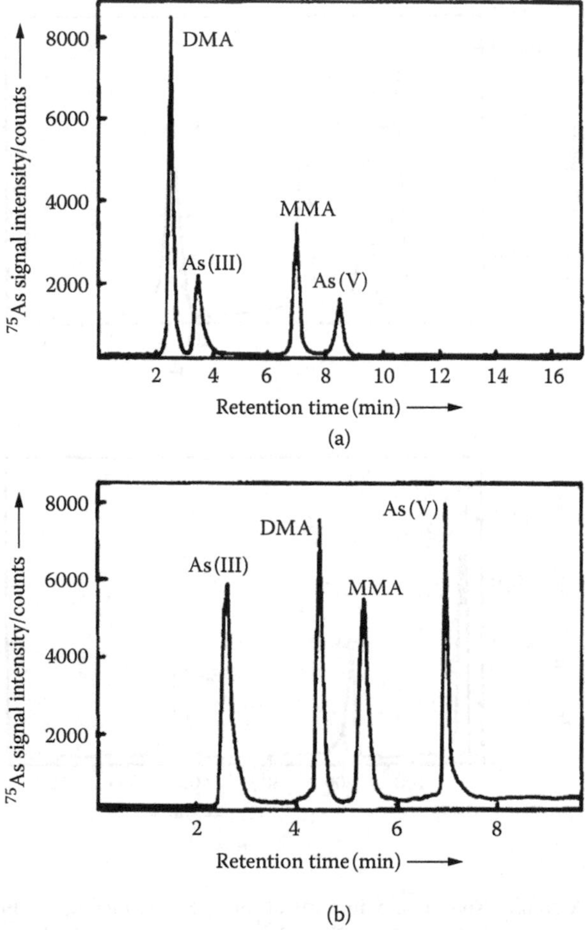

FIGURE 7.4 A comparison of arsenic metal ions speciation using different anion-exchange columns: (a) ION-120 and (b) Hamilton PRP-X-100. (From Roig-Navarro, A.F., Martinez-Bravo, Y., Lopez, F.J., and Hernandez, F., *J. Chromatogr. A*, 912: 319 (2001). With permission.)

ammonium nitrate (pH 8.2) and 60 m*M* ammonium nitrate (pH 8.2). Furthermore, the authors reported a decrease in the elution times of As(V), Cr(VI), and Se(VI) metal ions at high concentrations of ammonium acetate. These authors also studied the effect of sodium chloride (effect of ionic strength) on the chromatographic behavior of arsenic, chromium, and selenium metal ions with a slight variation in the retention times of these metal ions at high concentrations of sodium chloride. Similarly, Pobozy et al.[54] studied the effect of nitrate, sulfate, and chloride on the speciation of chromium metal ion. The authors reported a pronounced effect of these salts on the speciation of chromium metal ion (Figure 7.5). These authors also studied the effect of phthalate concentration on the

TABLE 7.2
Commercial Cation- and Anion-Exchange Columns Used for the Speciation of Metal Ions

Name of Columns	Metal Ions Speciated	Ref.
	Anion-Exchange Columns	
Hamilton PRP-X-100	Arsenic, selenium	32, 94, 142
IonPac AS14	Arsenic	143
IC-Pak	Chromium	144
IonPac AG5	Chromium	112
IonPac AS4A-SC	Arsenic	22
IonPac AS7/AG7	Arsenic	30
IonPac AHC	Chromium	42
IonPac AS5	Platinum	75
IonPac AS7	Chromium	145
ION 120	Arsenic	37
Waters IC-PAK	Arsenic	134
Polyspher IC AN-2	Selenium	65
AS11 and copper	Selenium, molybdenum	146, 147
Nucleosil-$N(CH_3)_3$-10	Arsenic	83, 95
Elite AS-3	Arsenic	87
Adsorbsphere-NH_2	Arsenic	101
Mono Q HR 5/5	Chromium	52
Polyspher IC AN-2	Selenium	65
ANX-3202	Chromium	49
	Cation-Exchange Columns	
IonPac CS10	Arsenic	36
IonPac CS5	Selenium	148
IonPac CS5A	Lanthanides	149
Ionosphere-C	Arsenic, selenium	37, 65
AG50W-X8	Chromium	55

speciation of chromium using phthalate eluent at pH 3.5 (Figure 7.6). It may be seen from Figure 7.6 that the separation factor for Cr(III) and Cr(VI) species decreases at higher concentrations of phthalate ion. Vilano et al.[66] varied the concentrations of phosphate buffer (pH 7) for the speciation of selenium metal ion. The authors varied the concentrations from 40 to 100 m*M* and reported the best separation using 40 m*M* concentration (Figure 7.7). Zhang et al.[36] studied the effect of HCl concentration on the speciation of organoarsenicals using HCl-NaH_2PO_4 as the eluent. The authors used 10, 100, and 200 m*M* HCl and reported the best separation using a 100 m*M* concentration of HCl (Figure 7.8a). Byrdy et al.[49] studied the effect of ammonium sulfate on the speciation of chromium metal ion (Figure 7.8b). The authors reported that the capacity factors for Cr(III) and Cr(VI) decrease rapidly with an increase in the concentration of ammonium

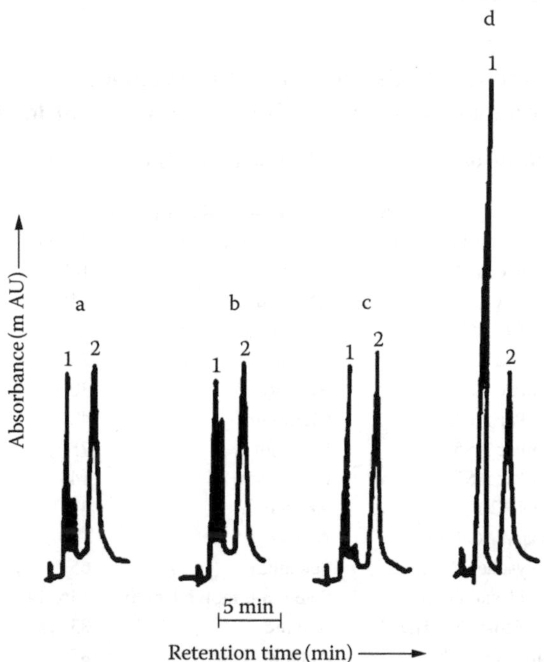

FIGURE 7.5 Effect of different anion concentrations on the speciation of chromium metal ion using (a) 20 mg/L nitrate, (b) 50 mg/L nitrate, (c) 2 m*M* sulfate, and (d) 100 m*M* chloride in the mobile phase. (From Pobozy, E., Wojasinska, E., and Trojanowicz, M., *J. Chromatogr. A*, 736: 141 (1996). With permission.)

sulfate. It is interesting to observe that the effect of ammonium sulfate was pronounced on Cr(VI) species, while Cr(III) bears little concentration effect. Pantsar-Kallio and Manninen[57] studied the effect of the interfering ions (chloride, chlorate, perchlorate, sulfate, sulfite, sulfide, thiosulfate, carbonate, cyanide, and some organic compounds) on the speciation of cationic Cr(III) and anionic Cr(VI) species using ICP-MS detection.

Some organic solvents were also used as the organic modifiers for the optimization of metal speciation by IC.[11,14] The most commonly used organic modifiers are methanol, acetonitrile, ethanol, acetone, and formic acid.[14,59,150] Methanol causes less plasma instability in ICP-MS and has been used frequently as the organic modifier in IC. The use of other organic solvents is limited due to their plasma instability. Sikovec et al.[59] studied the effect of methanol, acetonitrile, and acetone on the speciation of Cr(III) and Cr(VI), and the results are shown in Figure 7.9. It may be observed from Figure 7.9 that the speciation is affected greatly by changing the organic modifier. Similarly, Vilano et al.[66] studied the effect of the concentration of methanol for the separation of SeCys and SeMet species. The authors used 0 to 5% methanol and reported the best results using 0% of methanol. These results are given in Table 7.3, and it may be observed

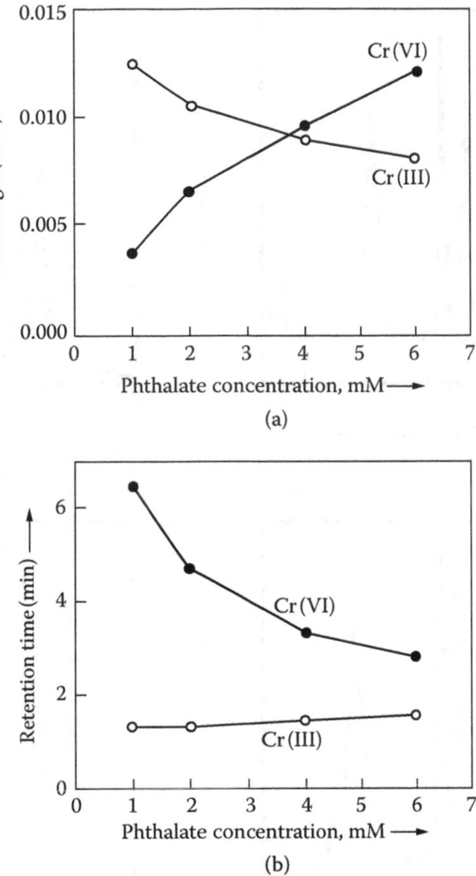

FIGURE 7.6 Effect of phthalate ion concentrations on the speciation of chromium metal ion: (a) concentration versus peak height and (b) concentration versus retention time. (From Pobozy, E., Wojasinska, E., and Trojanowicz, M., *J. Chromatogr. A*, 736: 141 (1996). With permission.)

from this table that the separation decreases by increasing the methanol concentration. It is interesting to observe that the peak area was greater using 1% methanol. The use of organic modifiers decreases the interactions between metal ion species and the column, and hence, the speciation improved in terms of sharp peaks. Briefly, the use of an organic modifier is important for the optimization of metal ion speciation.

7.2.3 pH of the Mobile Phase

The pH of the mobile phase governs the interactions between metal ion species and the ion-exchange column, and, therefore, it may be used to control speciation in IC. In addition, the pH of the mobile phase determines the charge on the metal

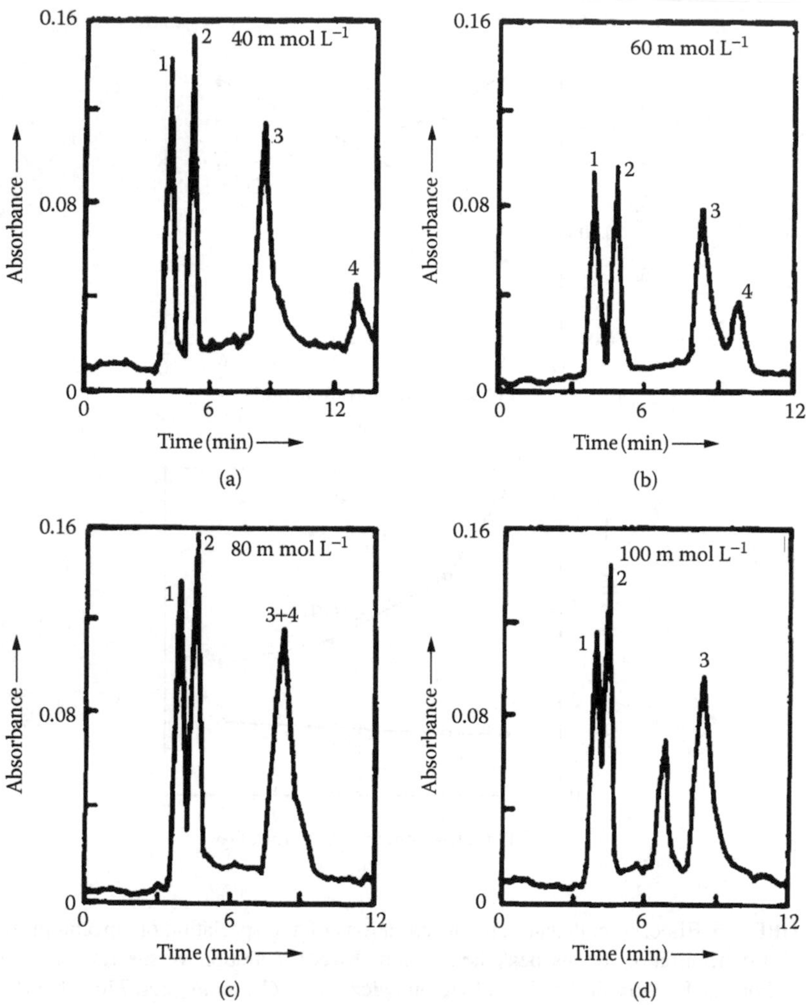

FIGURE 7.7 Effect of mobile phase concentration on the speciation of selenocystine (SeCys) and selenomethionine (SeMet) using (a) 40 m*M*/L, (b) 60 m*M*/L, (c) 80 m*M*/L, and (d) 100 m*M*/L of phosphate buffer. (From Vilano, M., Padro, A., Rubio, R., and Rauret, G., *J. Chromatogr. A*, 819: 211 (1998). With permission.)

ion species, which is useful for deciding the type of column to be used. Schlegel et al.[34] presented a schematic diagram (Figure 7.10) of the charge dependence of arsenic and selenium compound on the pH values of the solutions. With the help of this diagram, a suitable pH of the mobile phase can be selected for a particular ion-exchange column, that is, the cation- or anion-exchange column. Therefore, the pH of the mobile phase is an important aspect in metal ion speciation. Only few reports deal with the variation of pH and its effect on metal ion speciation. Gettar et al.[26] studied the speciation of arsenic species using a pH range of 4.4

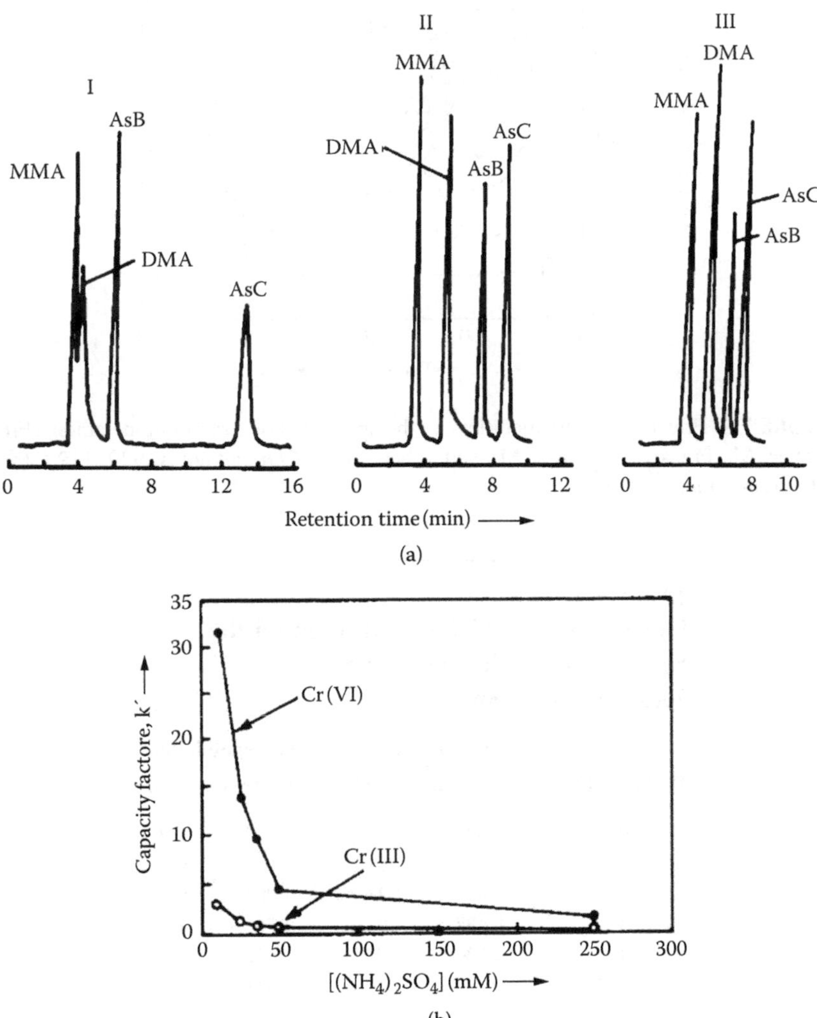

FIGURE 7.8 Effect of mobile phase concentration on the speciation of (a) organoarsenicals using cation-exchange column with (I) 10 m*M* HCl–50 m*M* NaH_2PO_4, (II) 100 m*M* HCl–50 m*M* NaH_2PO_4, and (III) 200 m*M* HCl–50 m*M* NaH_2PO_4 and (b) chromium metal ion using different concentrations of ammonium sulfate. (Part a from Zhang, X., Cornelis, R., Kimpe, J.D., and Mees, L., *Anal. Chim. Acta*, 319: 177 (1996); Part b from Byrdy, F.A., Olson, L.K., Vela, N.P., and Caruso, J.A., *J. Chromatogr. A*, 712: 311 (1995). With permission.)

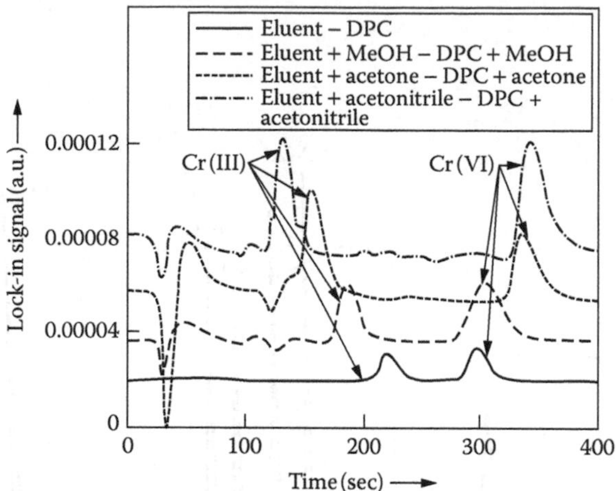

FIGURE 7.9 Effect of organic modifiers on the speciation of chromium metal ion. (From Sikovec, M., Franko, M., Novic, M., and Veber, M., *J. Chromatogr. A*, 920: 119 (2001). With permission.)

TABLE 7.3
Effect of Methanol Concentration on the Separation of Selenocystine and Selomethionine Species

	Selenocystine		Selenomethionine	
Methanol (%)	**Peak Area**	**tr(s)**	**Peak Area**	**t_r(s)**
0	2.436	293.6	1.877	581.3
1	2.595	291.7	1.955	530.5
2	2.439	288.7	1.704	541.5
3	2.487	288.7	1.565	491.7
4	2.524	292.7	1.652	465.8
5	0.726	286.7	1.531	449.9

Source: From Vilano, M., Padro, A., Rubio, R., and Rauret, G., *J. Chromatogr. A*, 819, 211, 1998. With permission.

to 6. The results are given in Figure 7.11a. Note that the retention times of As(III) were constant over the entire range of pH, while the retention times of As(V), MMA, and DMA decreased with the increase in pH. Similarly, Pobozy et al.[54] studied the speciation of chromium metal ion using 2 to 5 as pH values (Figure 7.11b). The authors reported that peak heights decreased by increasing the pH value. Briefly, the pH of the mobile phase is a very important parameter and should not be selected arbitrarily.

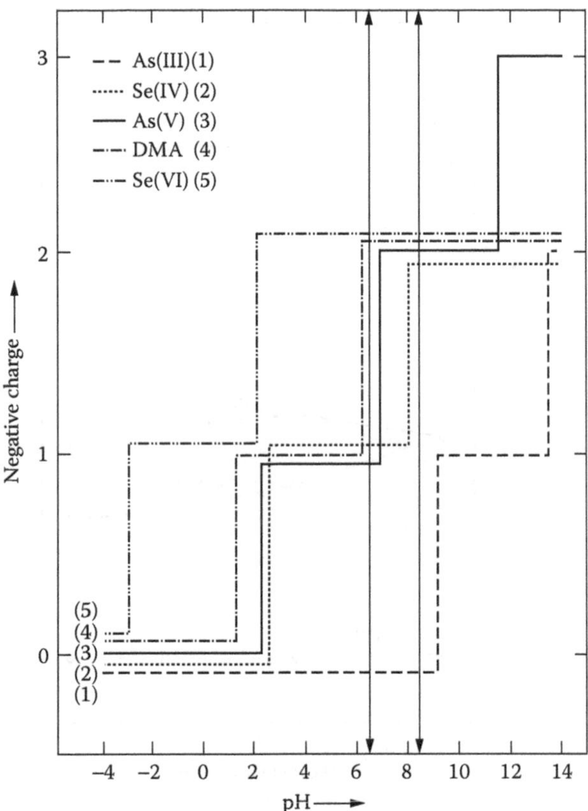

FIGURE 7.10 The charge dependence of selenium and arsenic metal ions on pH values. (From Schlegel, D., Mattusch, J., and Dittrich, K., *J. Chromatogr. A*, 683: 261 (2002). With permission.)

7.2.4 Chelating Reagents

In anion-exchange IC, the chelation of metal ions is carried out to convert them into negatively charged species. However, the chelation is very important, as the chelated species may have different physicochemical properties and can be separated easily on IC columns. Also, UV/visible detection can be used for the chelated metal ion species, which is the best alternative in comparison to the high-cost element-specific detectors. The complexation of metal ions can be carried out online or off-line as per the requirement and the facilities available. In the case of off-line complexation, metal ions are allowed to react with a suitable ligand and are then loaded on the IC machine, while online complexation of metal ions is achieved by using a suitable chelating reagent in the mobile phase. The formed complexes in the mobile phase, before entering the column, can be separated easily on an ion-exchange column. The most commonly used chelating agents/ligands used in IC are mono-, di-, and tricarboxylic acids, -hydroxyisobu-

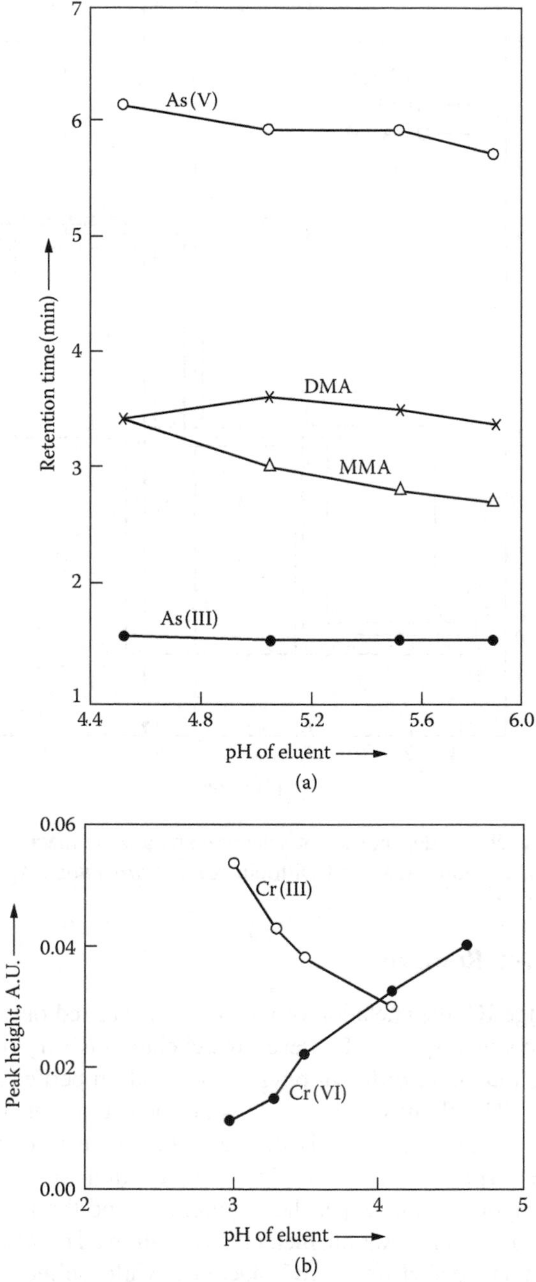

FIGURE 7.11 Effect of pH on the speciation of (a) arsenic and (b) chromium metal ions. (Part a from Gettar, R.T., Garavaglia, R.N., Gautier, E.A., and Batistoni, D.A., *J. Chromatogr. A*, 884: 211 (2000); Part b from Pobozy, E., Wojasinska, E., and Trojanowicz, M., *J. Chromatogr. A*, 736: 141 (1996). With permission.)

tyric acid, tartaric acid, 1,2-diaminocyclohexanetetraacetic acid, diethylenetriminopentaacetic acid, pyridine-2,6-dicarboxylic acid (PDCA), ethylenediaminetetraacetic acid (ESTA), and so forth.[11,12] These ligands form anionic species depending on pH values of the mobile phase and can be separated easily using an anion-exchange column.

Optimization of metal ions can be achieved by controlling the concentrations of the chelating reagents. Gettar et al.[26] studied the effect of EDTA on the speciation of arsenicals. The authors used 0 to 0.6 m*M* as the concentration of EDTA (Figure 7.12a) and reported that the retention times decreased with an increase in EDTA concentration. Furthermore, it may be seen from Figure 7.12a that the chromatographic retention of As(III) was independent of the concentration of EDTA, while the chromatographic behavior of As(V), MMA, and DMA was affected by an increasing concentration of EDTA. Similarly, Larsen et al.[37] used different concentrations of pyridinium ion and studied the separation of arsenic metal ion species (Figure 7.12b). It may be observed from Figure 7.12b that the values of the separation factor decrease by increasing the concentration of pyridinium ion. Therefore, it may be concluded from this discussion that optimization of metal ions may be controlled by using a suitable concentration of chelating reagents.

7.2.5 Detection

As in the case of RP-HPLC, detection in IC can be carried out by both nonelement- and element-specific detectors. The nonelement-specific detectors are UV/visible, fluorescent, electrochemical, conductometric, and radiometric, while element-specific detectors include AAS, AES, and ICP-MS (Table 7.1). The peak size and area of metallic species depend on the sensitivity of the detector. A detector with low limits of detection is deemed the best choice. The use of UV/visible absorbing chelates in IC makes the UV/visible detector ideal for the detection of metallic species. The most important ligands for complexation were discussed in Section 7.2.4. Sometimes a postcolumn derivatization of the separated metal ions is also required for detection purposes; hence, a special device known as a postcolumn reactor is assembled between the column and the detector. Pobozy et al.[54] described a postcolumn reactor design (Figure 7.13) for the speciation of chromium metal ion. The authors used Ce(IV) sulfate (0.5 g/L) in 0.8 *M* sulfuric acid and 0.05% aqueous solution of diphenylcarbazide (DPC) as the postcolumn reagent. The other reagents required for postcolumn reaction are similar, as discussed in Chapter 6.

The element-specific detectors have been used frequently for detection purposes in IC. The combination of IC with a suitable detector is a very important issue for the speciation of metal ions, and the coupling of AAS, AES, and ICP-MS is similar to RP-HPLC (Chapter 6). The main concern with this type of interfacing is the transfer line from the IC to the detector, which must be of the minimum length. Many interfacing (nebulizer) designs used in IC are similar, as in the case of RP-HPLC (Chapter 6). However, Saverwyns et al.[50] used a com-

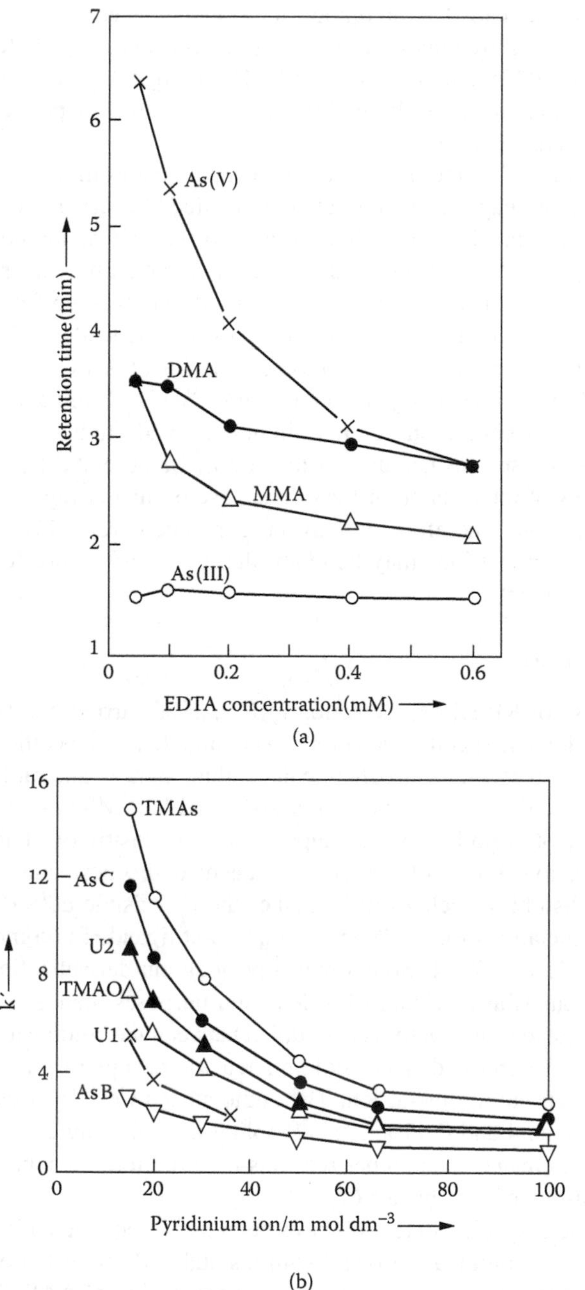

FIGURE 7.12 Effect of chelating agents on the speciation of arsenic using (a) EDTA and (b) pyridinium ion. (Part a from Gettar, R.T., Garavaglia, R.N., Gautier, E.A., and Batistoni, D.A., *J. Chromatogr. A*, 884: 211 (2000); Part b from Larsen, E.H., Pritzl, G., and Hansen, S.H., *J. Anal. Atom. Spectrom.*, 8: 1075 (1993). With permission.)

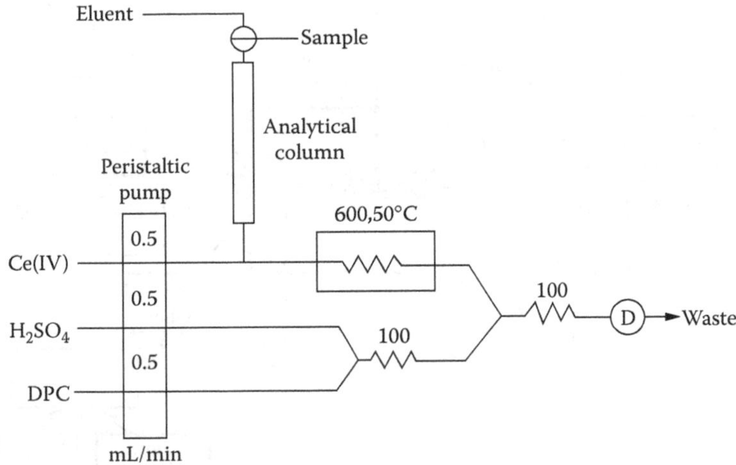

FIGURE 7.13 An ion chromatography setup with a postcolumn reactor. The length of mixing and reaction coils is shown in centimeters; D = spectrometer detector. (From Pobozy, E., Wojasinska, E., and Trojanowicz, M., *J. Chromatogr. A*, 736: 141 (1996). With permission.)

mercially available microbore anion-exchange column (Cetac ANX-3202) for the speciation of chromium metal ion. Nitric acid was used as the mobile phase, making the technique ideal for coupling with ICP-MS. Because of the low operating flow rate of the column, a microconcentric nebulizer was used as an interface between the microbore column and the ICP-mass spectrometer. For reference, a complete setup of IC-ICP-MS with nebulizer is shown in Figure 7.14.[151] Gjerde et al.[108] used a new type of nebulizer that introduces the sample directly into ICP-MS. ICP-MS is used commonly for the detection of metal ions, and, therefore, it is important to describe the operating conditions of this detector. The operating conditions of a typical ICP-MS system are presented in Table 7.4.

The limits of detection depend on the sensitivity of the detector as well as the proper combination of IC with the detector. The different columns, detectors, nebulizers, flow rates, temperatures of the detector, and so forth, determine the limits of detection. Various detection limits were reported for the speciation of different metal ions using different detectors.[1–14] Sheppard et al.[40] reported the detection limits of 0.063 ng/L of As(III), 0.037 ng/L of As(V), 0.032 ng/L of DMA, and 0.080 ng/L of MMA in club soda. The authors reported an improvement in sensitivity by using a mixture of He–Ar gases in ICP as the ionization source. Takaya and Sawatari[74] described the separation of vanadium(IV) and vanadium(V) species using anion-exchange column with ICP-AES detection. The lower determination limit was 1 μg/L, which is insufficient to speciate vanadium sampled by conventional sampling methods in a working environment. Byrdy et al.[49] calculated the absolute detection limits based on peak height calculations and reported them to be in the range of 40 pg/L for Cr(III) and 100 pg/L for Cr(VI) in aqueous media by ICP-MS. By ICP-AES, detection limits were 100 ng/L for

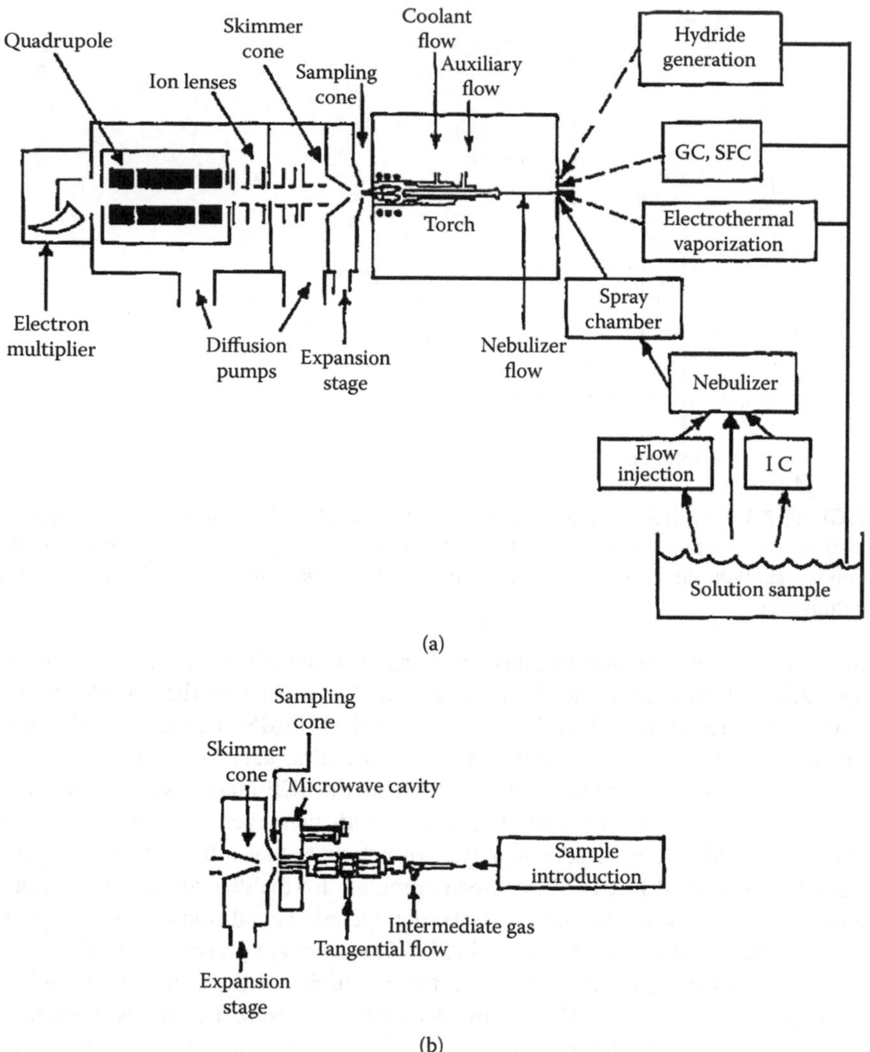

FIGURE 7.14 (a) An ion chromatography–inductively coupled plasma–mass spectrometry (IC-ICP-MS) setup and (b) a nebulizer with interface. (From Vela, N.P., Olson, L.K., and Caruso, J.A., *Anal. Chem.*, 65: 585A (1993). With permission.)

Cr(III) and 200 ng/L for Cr(VI). Inoue et al.[53] combined IC with ICP-MS for the speciation of Cr(III) and Cr(VI) species. The authors used Excelpak ICS-A23 packed with hydrophilic polymer-based anion-exchange resin. The detection limits for Cr(III) and Cr(VI) were 8.1×10^{-5} and 8.8×10^{-5} mg/L, respectively. Pobozy et al.[54] used IC procedures for the speciation of different oxidation states of chromium with the use of an anion exchange Hamilton PRP-X100 column and reported 2.5 and 1.8 ng/mL as the detection limits for Cr(III) and Cr(VI). Pedersen

TABLE 7.4
Operating Conditions of a Typical Inductively Coupled Plasma–Mass Spectrometry Detector

Parameters	Instrument Setting
General Description	
Radio frequency forward power (W)	1350
Radio frequency reflected power (W)	<5
Plasma gas flow (L/min)	0.8–1.1
Nebulizer gas flow (L/min)	0.7–1
Liquid sdample flow rate (mL/min)	1
Sample Introduction System	
Nebulizer	Concentric
Spray chamber	Scott-type double pass
Pump Setting	
Expension stage pressure	$<2 \times 10^0$
Intermediate stage pressure	4×10^6
Analyzer stage pressure	0×10^4

Source: From Ponce de Leon, C., Montes-Bayon, M., and Caruso, J.A., *J. Chromatogr. A*, 974, 1, 2002. With permission.

et al.[65] presented the detection limit of 1 μg/L for SeMet, SeCys, selenite, and selenate species using ICP-MS. He et al.[29] reported that the limits of detection of arsenicals were 0.11 to 0.15 ng/L. Chausseau et al.[43] reported a 10 μg/L detection limit for As(III), DMA, MMA, and 20 μg/L for As(V). Gomez-Ariza et al.[68] speciated SeCys, SeMet, SeET, selenite (Se(IV)), and selenate (Se(VI)) using an anion-exchange column. Online microwave-assisted digestion and hydride generation steps were performed prior to the atomic fluorescence detection. The detection limits obtained were in the range of 0.6 to 0.9 μg/L. Martinez-Bravo et al.[32] reported that the detection limits of chromium speciation were in the range of 40 to 60 ng/L with good accuracy and repeatability of the method. Day et al.[33] used an anion-exchange column coupled to ICP-MS for the detection of As(III), As(V), MMA, and DMA and reported a detection limit at 100 ng/L.

The suppression technology may be used to avoid the interferences and increase the detection sensitivity in the conductometric detection. This may be achieved by reducing the conductivity of the background mobile phase and by converting the sample counterions into a single ionic species with a higher equivalent conductivity. In anion speciation, normally, carbonate/hydrogen carbonate electrolytes are suppressor exchanged as the mobile phase. The chemical suppression is based on the use of the salt of weakly dissociating acids ($NaHCO_3$) as the mobile phase. These eluents can be eliminated to a large extent by cation exchange in a postcolumn reaction according to the following equation:

$$Na^+ + HCO_3^- \xrightarrow[-Na+]{+H+} H_2CO_3 \tag{7.1}$$

The carbonic acid formed as a result of cation exchange is very weakly dissociated, showing a very low conductivity; hence, the sample anions undergo a corresponding reaction:

$$Na^+ + Cl^- \xrightarrow[-Na+]{+H+} HCl \tag{7.2}$$

In this way, sodium chloride is converted into HCl by the suppression, which has a higher conductivity than organic acids. Thus, the signal to be measured is the sum of the conductivities due to Cl and H^+ ions against low background conductivity, and there is good detection sensitivity. The limits of detection in IC using different detectors are summarized in Table 7.5. It may be seen from Table 7.5 that the limits of detection vary from one detector to another.

7.2.6 Other Parameters

In addition to the above parameters discussed, some other factors are important in controlling metal ion speciation by IC. These include the dimension of the column, amount injected onto the column, flow rate, temperature, and other operating conditions of the detectors, especially in element-specific detectors.

Gettar et al.[26] tried to optimize the speciation of arsenicals using columns that were 100 mm and 250 mm in diameter. The authors reported the best separation on the 100 mm diameter column (Figure 7.15). Similarly, Busch and Seubert[152] studied the effect of the geometry of the IC column on the speciation of aluminum fluoride species. The authors used a standard bore and a microbore column. The results showed that the disintegration of the higher coordinated Al fluoride species (AlFn with $n > 2$) could be dramatically reduced when utilizing the microbore technique. The particle sizes of the column material may also affect metal ion speciation. The same group[152] also studied the effect of the temperature (–5 and 50°C) of the column on the speciation of aluminum metal ion. The authors reported that with the standard bore of the column, temperature was of minor importance. The agreement between speciation data experimentally determined by microbore chromatography and that calculated using stability constants was quite good. The standard bore column showed bigger differences between calculated and experimentally determined species distributions. Gettar et al.[26] optimized the separation of arsenite, arsenate, MMA, and DMA by using different flow rates of the mobile phase. The complete elution and resolution of the four species was achieved in about 6 min. Similarly, Slejkovec et al.[24] tried the optimization of arsenite and arsenate using 0.5, 0.75, and 1.2 mL/min as the flow rate (Figure 7.16). The authors reported the best separation using 1.2 mL/min as the flow rate. The peak shapes were sharp with the 1.2 mL/min flow rate, while the peaks were broad at lower values of flow rate.

TABLE 7.5
Detection Limits in Ion Chromatography Using Different Detectors

Metal Ions Species	Detectors	Detection Limits	Ref.
As	ICP-MS	0.032–0.80 ng/L	40
		0.11–0.15 ng/L	29
		100 ng/L	33
		20–300 pg	40, 82, 85
	ICP-AES	0.56 mg/L	34
	HG-AAS	1–1.4 μg/L	36
	UV-HG-ICP-AES	0.23–0.30 μg/L	96
	HG-QF-AAS	1.3–4.8 μg/L	90
	HG-QF-AAS	2–3 ng/L	81
	HG-GF-AAS	1.6–1.9 μg/L	87
Cr	ICP-MS	40–100 ng/L	49
		8.1×10^5 and 8.8×10^5 mg/L	53
		1.8–2.56 ng/L	54
		40–60 ng/L	32
		2 ppb	153
	Diphenylcarcazide spectroscopic detection — thermal lens	300 ng/L	58
	Diphenylcarcazide spectroscopic detection — thermal lens	1–10 μg/L	59
	ICP-AES	100–200 ng/L	49
Se	ICP-MS	1 μg/L	65
	HG-atomic fluorescence	0.6–0.9 μg/L	68
Sn	—	400–1000 pg/L	154
V	ICP-AES	1 μg/L	72
Rare earth	—	1–5 ppt	155

Larsen et al.[37] studied the effect of loading amount on the separation of arsenic metal ion species by injecting 10 and 100 mL samples (Figure 7.17). The authors reported the best separation using 10 mL as the loading amount. Column switching has also been proven as a separation-improving technique, particularly in separating mono- and divalent ions simultaneously. The column-switching valves allow for two or more configurations of the system. In this direction, Gómez-Ariza et al.[68] optimized the separation of SeCys, SeMet, SeET, selenite (Se(IV)), and selenate (Se(VI)) by coupling an anion exchange column and a reversed-phase column, both connected through a six-port switching valve. Online microwave-assisted digestion and hydride generation steps were performed prior to the atomic fluorescence detection. The elution of the seleno amino acids was accomplished in the reversed-phased column using water as the mobile phase. Selenite

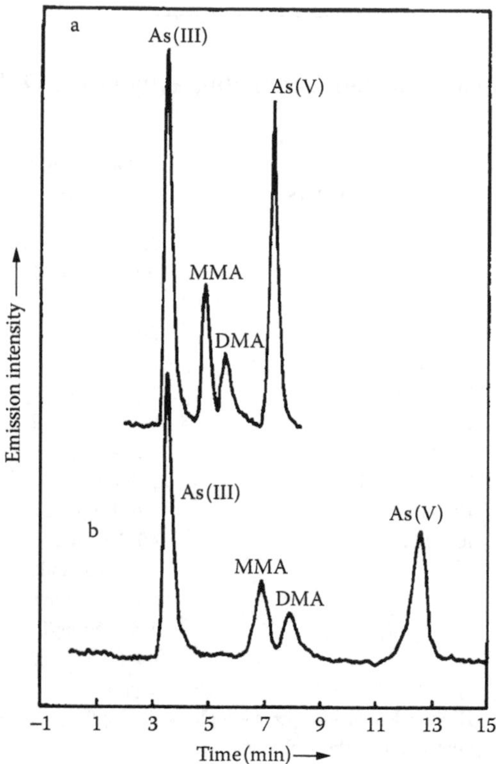

FIGURE 7.15 Effect of column length on the speciation of arsenic metallic species using (a) 100 and (b) 250 mm lengths, respectively. (From Gettar, R.T., Garavaglia, R.N., Gautier, E.A., and Batistoni, D.A., *J. Chromatogr. A*, 884: 211 (2000). With permission.)

and selenate were separated in the anion-exchange column, using gradient elution with an acetate buffer.

Optimization can also be achieved by controlling the experimental conditions of element-specific detectors. The most important experimental conditions are the purity of the gas, flow of gas, temperature of the detector, and concentration of the reducing agents, as in a hydride generation unit. Zhang et al.[36] studied the effect of the flow of argon gas in HG-AAS on the speciation of arsenic compounds. The authors used continuous and argon-segmented flows for the detection of arsenic species (Figure 7.18). It may be observed from Figure 7.18 that the best separation was achieved using the argon-segmented flow. Moreover, this flow resulted in an increase of the intensity and sharpness of the peaks. Schlegel et al.[34] studied the effect of $NaBH_4$ concentration on the signal/background ratio of As(III), As(V), and DMA species (Figure 7.19). The authors reported that the signal/background ratio increased with an increasing amount of $NaBH_4$, which resulted in good detection and sensitivity of these arsenic species. Briefly, these parameters are also important and can be used to control speciation by IC. Based

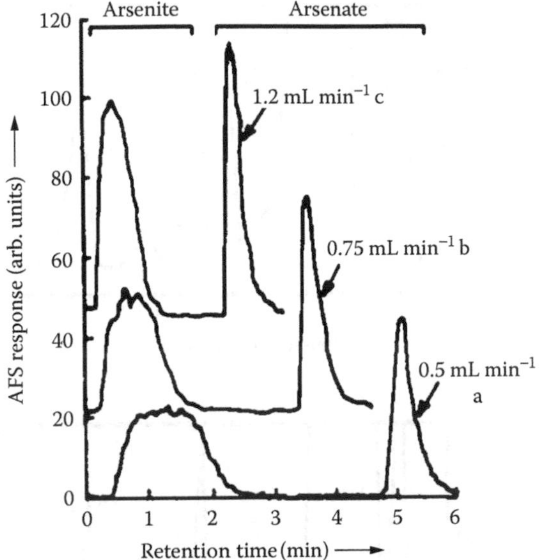

FIGURE 7.16 Effect of flow rate on the separation of arsenite and arsenate using (a) 0.5, (b) 0.75, and (c) 1.2 mL/min flow rates, respectively. (From Slejkovec, Z., Salma, I., and van Elteren, J.T., *Fresenius' J. Anal. Chem.*, 366: 830 (2000). With permission.)

on our experience and the literature available, a scheme was developed to carry out and optimize the speciation of metal ions by IC into environmental and biological samples. The developed protocol is given in Scheme 7.1.

7.3 MECHANISMS OF METAL IONS SPECIATION

Various researchers attempted to describe the mechanisms of the speciation of metal ions on ion-exchange columns.[11,13] Basically, the separation of metallic species on ion-exchange columns occurs due to the ion-exchange process, that is, the exchange of metallic species by other species of the same charge present onto the column. Metallic cation and anion are exchanged by different cation and anion exchangers, respectively, that are present in the column. Therefore, cations and anions are separated by cation- and anion-exchange columns, respectively. Therefore, metallic cations are converted into anions before their separation by anion-exchange column. The corresponding counterions are located in the vicinity of the functional groups and can be exchanged with other ions of the same charge in the mobile phase. For every ion, the exchange process is characterized by a corresponding ion-exchange equilibrium that determines the distribution between the mobile and stationary phases. For example, in the case of an anion (A), an exchange process with a dynamic equilibrium is as follows:

$$E^-_{sta} + A^-_{mob} \leftrightarrow A^-_{sta} + E^-_{mob} \quad (7.3)$$

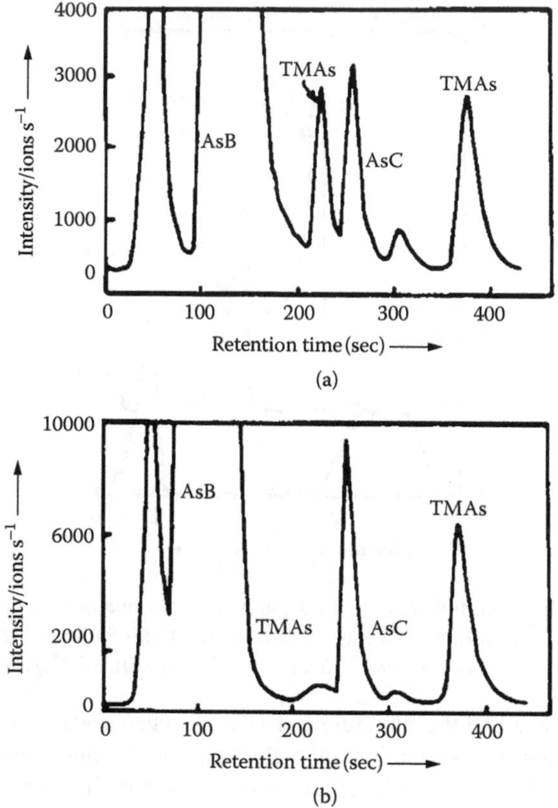

FIGURE 7.17 Effect of loading amount on the speciation of arsenic by loading (a) 25 and (b) 100 mL of extract. (From Larsen, E.H., Pritzl, G., and Hansen, S.H., *J. Anal. Atom. Spectrom.*, 8: 1075 (1993). With permission.)

$$K = [A^-]_{sta}[E^-]_{mob} / [A^-]_{mob}[E^-]_{sta} \tag{7.4}$$

where A^- and E^- are sample and mobile phase anions, respectively, while K is the equilibrium constant. Thus, different anions can be separated on the basis of their different affinities toward the stationary phase (different values of K). Similarly, the speciation of cations can be explained on the basis of Equation 7.3 and Equation 7.4. These exchange processes are governed by certain interactions, such as electrostatic and hydrophobic interactions. Hydrogen bonding, dipole-induced-dipole interactions, steric effect, and van der Waals, ionic interactions may also play a crucial role in the speciation of different organometallic species on ion-exchange columns.

In 1991, Hajos et al.[156] described ion-exchange reactions, the secondary chemical equilibria (SCE), as playing a crucial role in the separation of metallic ions on ion-exchange columns, and this hypothesis can be understood from Figure

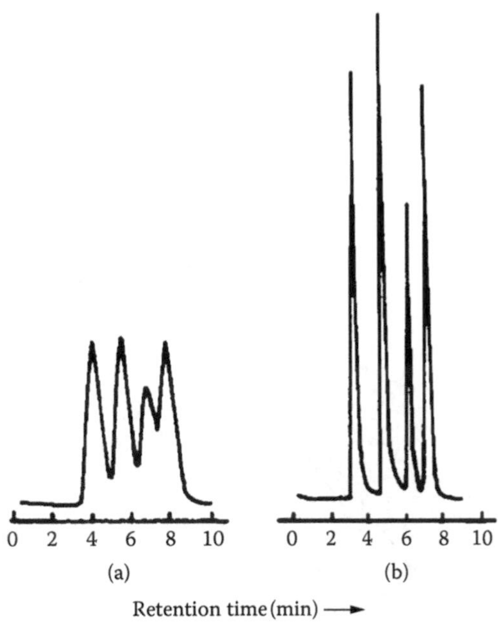

FIGURE 7.18 Effect of argon flow on the speciation of arsenic metal ions using (a) continuous and (b) argon-segmented flow rates, respectively. (From Zhang, X., Cornelis, R., Kimpe, J.D., and Mees, L., *Anal. Chim. Acta*, 319: 177 (1996). With permission.)

7.20a through Figure 7.20c. This figure indicates the equilibria existing on a cation or anion exchanger between a cation (M^{2+}), an added ligand (H_2L), and an eluent cation (enH^+). The complexation due to the protonated ligand facilitates cation exchange through a pulling effect (Figure 7.20a), while as shown in Figure 7.20b, the equilibria involves the retention of an anionic complex where the excess of the ligand (deprotonated) or another anion added to the eluent competes for the fixed sites. Figure 7.20c represents both the pushing and pulling coupled effects of the ligand and of the competing cation, where complexation reduces the availability of the free M^{2+} for the exchange. Furthermore, the general separation mechanisms of metal ion separation on cation- and anion-exchange columns are shown in Figure 7.21. Figure 7.21 clearly shows the pattern of the separation based on the exchange process. In this way, ions loaded on an IC machine get separated on cation- or anion-exchange columns due to their different affinities toward cation- and anion-exchange columns separately. It is worth mentioning here that in all cases, the pH of the mobile phase is a determining parameter for the exchange process, as it acts on both ligand dissociation and cation protonation.

In SEC, the stationary phase consists of a network of pores where ionic species larger than the pore sizes are excluded, thus eluting rapidly. In the same way, species smaller than pores permeate the pores of the packing material and are retained longer during the chromatographic run. The species of the medium sizes are functionated by their sizes.

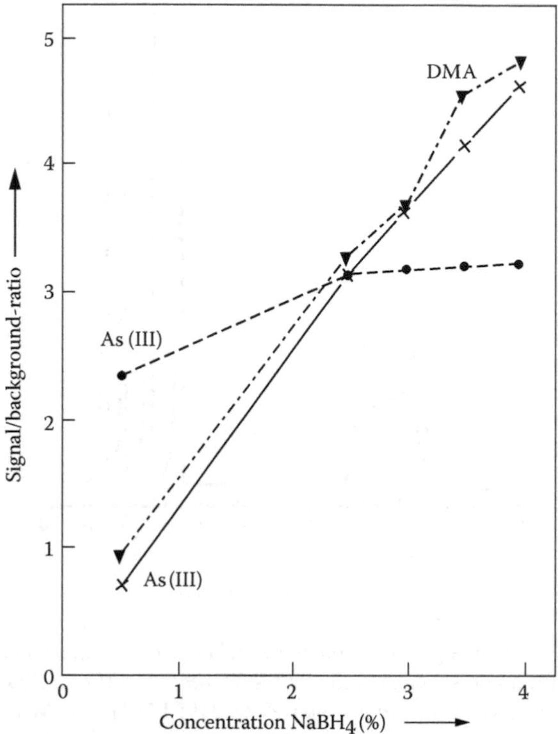

FIGURE 7.19 Effect of $NaBH_4$ concentrations on net intensity signals of arsenic metallic species. (From Schlegel, D., Mattusch, J., and Dittrich, K., *J. Chromatogr. A*, 683: 261 (2002). With permission.)

7.4 CONCLUSION

Ion chromatography, as a technique of metal ion speciation, is gaining importance since its development in 1975.[126] The utility of this technique is based on the development of cation-, anion-, and mixed-exchanger-based columns by which a wide range of metallic species can be speciated. Unlike on reversed-phase columns, the speciation on ion-exchange columns is easy to achieve, is sensitive, and is reproducible due to the effectiveness of the ion-exchange-process-based separation of metallic species. The use of the aqueous buffers as mobile phases decreased the running cost of the technique. The combination of IC with element-specific detectors greatly increased its application due to its lower limits of detection. There is no serious drawback of IC in metal ion speciation, but the coupling with element-specific detectors requires additional attention due to the fact that the sample from the IC column is introduced into the detectors in liquid form. And, the development of new ion-exchange columns, mobile phases, and chelating reagents is required in order to separate more complex mixtures of metal ion species.

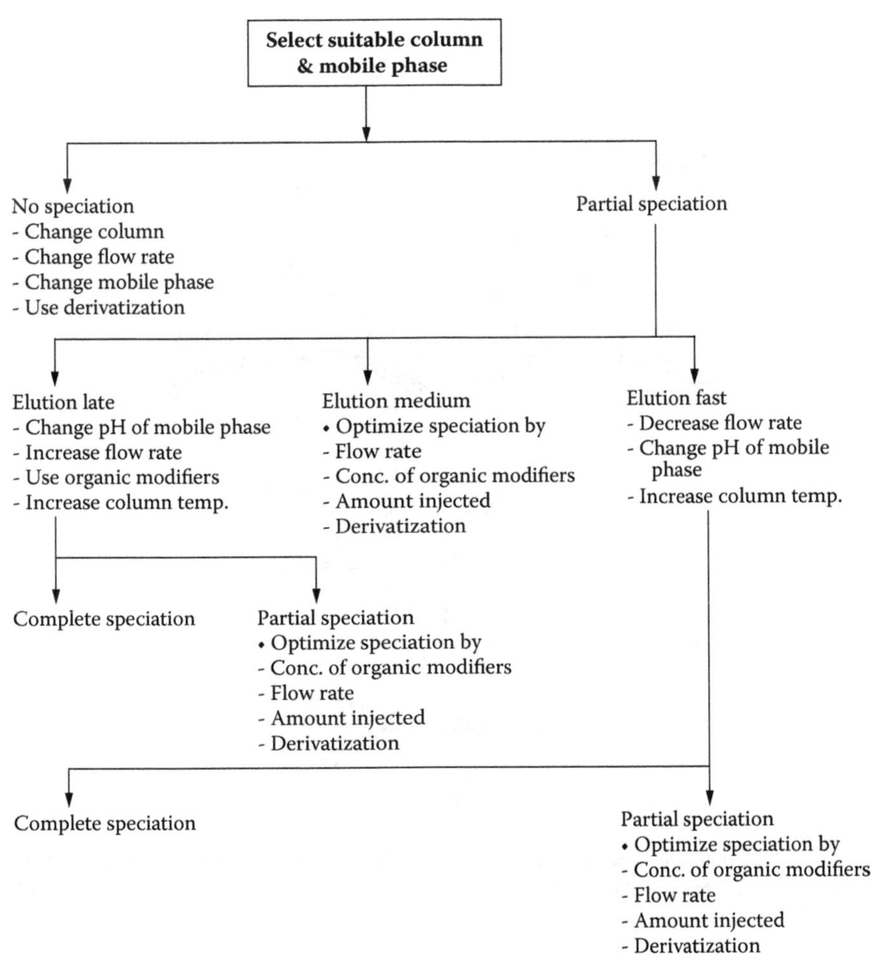
Select suitable column
& mobile phase
No speciation
- Change column
- Change flow rate
- Change mobile phase
- Use derivatization
Partial speciation
Elution late
- Change pH of mobile phase
- Increase flow rate
- Use organic modifiers
- Increase column temp.
Elution medium
• Optimize speciation by
- Flow rate
- Conc. of organic modifiers
- Amount injected
- Derivatization
Elution fast
- Decrease flow rate
- Change pH of mobile phase
- Increase column temp.
Complete speciation
Partial speciation
• Optimize speciation by
- Conc. of organic modifiers
- Flow rate
- Amount injected
- Derivatization
Complete speciation
Partial speciation
• Optimize speciation by
- Conc. of organic modifiers
- Flow rate
- Amount injected
- Derivatization

SCHEME 7.1

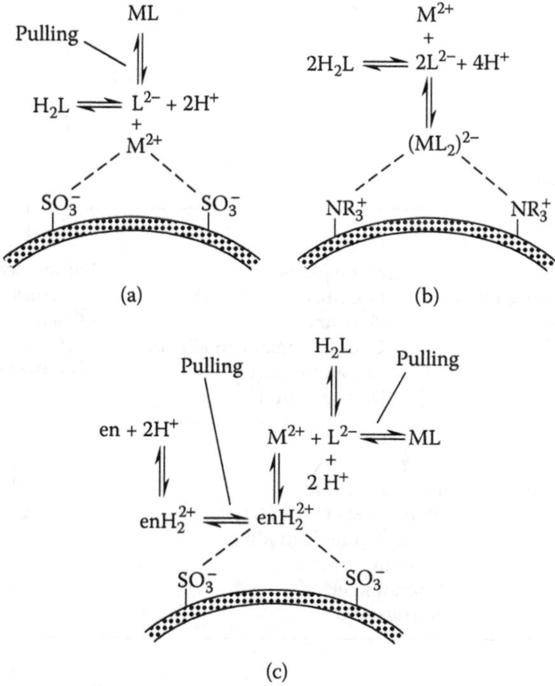

FIGURE 7.20 Chemical equilibria in (a and c) cation- and (b) anion-exchange columns. (From Brown, P.R., and Grushka, E., Eds., *Advan. Chromatogr.*, 41: 250 (2001). With permission.)

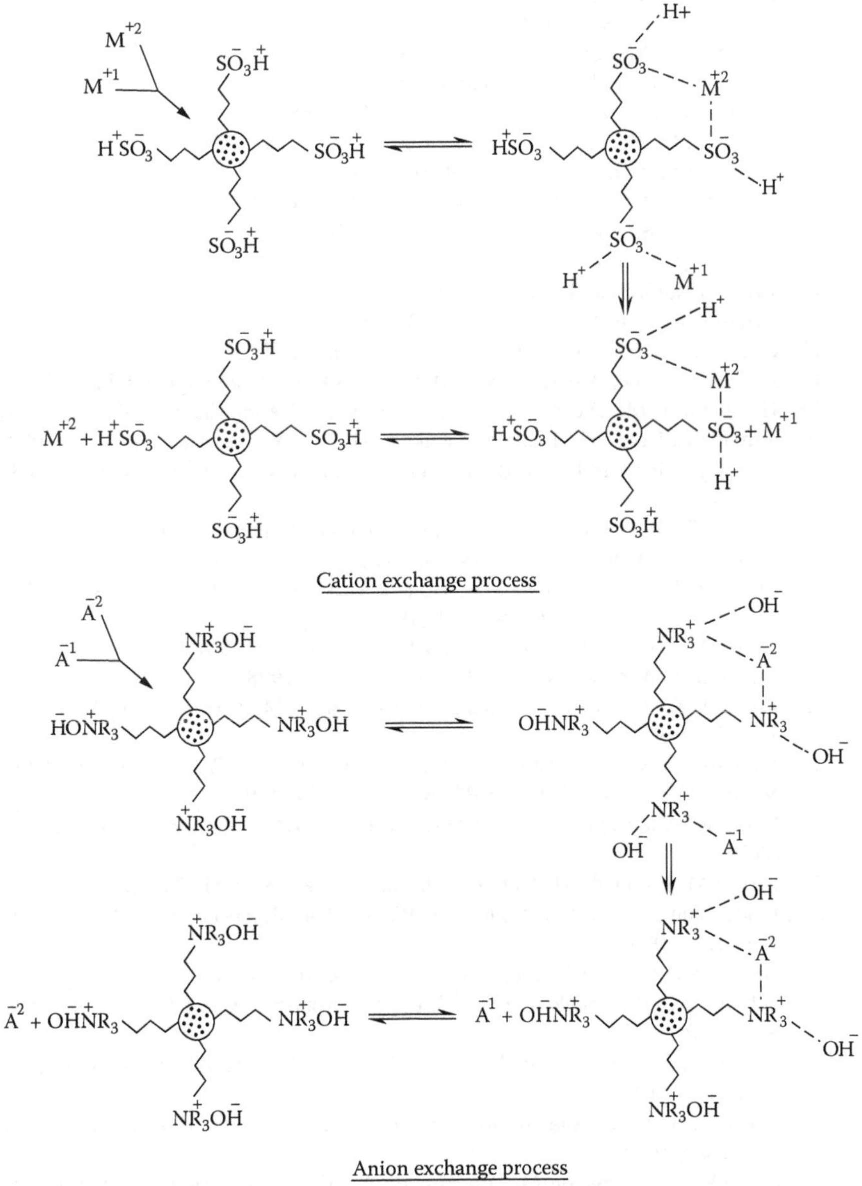

Cation exchange process

Anion exchange process

FIGURE 7.21 Metal ion speciation by ion-exchange chromatography.

REFERENCES

1. Quevauviller Ph, *J. Chromatogr. A*, 750: 25 (1996).
2. Pyrzynska K, *Analyst*, 121: 77R (1996).
3. Sarzanini C, Mentasti E, *J. Chromatogr. A*, 789: 301 (1997).
4. Ellis LA, Roberts DJ, *J. Chromatogr. A*, 774: 3 (1997).
5. Liu W, Lee HK, *J. Chromatogr. A*, 834: 45 (1999).
6. Sarzanini C, Mentasti E, *J. Chromatogr. A*, 850: 213 (1999).
7. Benramdane L, Bressolle F, Vallon JJ, *J. Chromatogr. Sci.*, 37: 330 (1999).
8. B'Hymer C, Brisbin JA, Sutton KL, Caruso JA, *Am. Lab.*, Feb. issue, 17 (2000).
9. Lopez-Ruiz B, *J. Chromatogr. A*, 881: 607 (2000).
10. Buchberger W, *J. Chromatogr. A*, 884: 3 (2000).
11. Brown PR, Grushka E (Eds.), *Advan. Chromatogr.*, 41: 250 (2001).
12. Fytianos K, *J. AOAC Int.*, 84: 1763 (2001).
13. Sarzanini C, Bruzzoniti MC, *Trends Anal. Chem.*, 20: 304 (2001).
14. Ponce de Leon C, Montes-Bayon M, Caruso JA, *J. Chromatogr. A*, 974: 1 (2002).
15. Hils A, Grote M, Janeen E, Eichhorn J, *Fresenius' J. Anal. Chem.*, 364: 457 (1999).
16. Yamada E, Hiwada T, Inaba T, Tokukura M, Fuse Y, *Anal. Sci.*, 18: 785: (2002).
17. Hara H, Kobayashi H, Maeda M, Ueno A, Kobayashi Y, *Anal. Chem.*, 73: 5590 (2001).
18. Bantan T, Milacic R, Mitrovic B, *Anal. Bioanal. Chem.*, 365: 545 (1999).
19. Mitrovic B, Milacic R, *Sci. Total Environ.*, 258: 183 (2000).
20. Martin I, Lopez-Gonzalvez MA, Gomez M, Camara C, Palacios MA, *J. Chromatogr. B, Biomed. Appl.*, 666: 101 (1995).
21. Tsalev DL, Sperling M, Welz B, *Analyst*, 123: 1703 (1998).
22. Mattusch J, Wennrich R, *Anal. Chem.*, 70: 3649 (1998).
23. Zheng J, Kosmus W, Pichler-Semmelrock F, Kock M, *J. Trace Elem. Med. Biol.*, 13: 150 (1999).
24. Slejkovec Z, Salma I, van Elteren JT, *Fresenius' J. Anal. Chem.*, 366: 830 (2000).
25. Miller GP, Norman DI, Frisch PL, *Water Res.*, 34: 1397 (2000).
26. Gettar RT, Garavaglia RN, Gautier EA, Batistoni DA, *J. Chromatogr. A*, 884: 211 (2000).
27. Bissen M, Frimmel FH, *Fresenius' J. Anal. Chem.*, 367: 51 (2000).
28. Lindemann T, Prange A, Dannecker W, Neidhart B, *Fresenius' J. Anal. Chem.*, 368: 214 (2000).
29. He B, Jiang GB, Xu X, *Fresenius' J. Anal. Chem.*, 368: 803 (2000).
30. Schmidt AC, Reisser W, Mattusch J, Popp P, Wennrich R, *J. Chromatogr. A*, 889: 83 (2000).
31. Roig-Navarro AF, Martinez-Bravo Y, Lopez FJ, Hernandez F, *J. Chromatogr. A*, 912: 319 (2001).
32. Martinez-Bravo Y, Roig-Navarro AF, Lopez FJ, Hernandez F, *J. Chromatogr. A*, 926: 265 (2001).
33. Day JA, Montes-Bayon M, Vonderheide AP, Caruso JA, *Anal. Bioanal. Chem.*, 373: 664 (2002).
34. Schlegel D, Mattusch J, Dittrich K, *J. Chromatogr. A*, 683: 261 (2002).
35. Van Hulle M, Zhang C, Zhang X, Cornelis R, *Analyst*, 127: 634 (2002).
36. Zhang X, Cornelis R, Kimpe JD, Mees L, *Anal. Chim. Acta*, 319: 177 (1996).
37. Larsen EH, Pritzl G, Hansen SH, *J. Anal. At. Spectrom.*, 8: 1075 (1993).
38. Johnson RL, Aldstad JH III, *Analyst*, 127: 1305 (2002).

39. Cornelis R, Zhang X, Mees L, Christensen JM, Byrialsen K, Dyrschel C, *Analyst*, 123: 2883 (1998).
40. Sheppard BS, Caruso JA, Heitkemper DT, Wolnik KA, *Analyst*, 117: 971 (1992).
41. van Elteren JT, Slejkovec Z, *J. Chromatogr. A*, 789: 339 (1997).
42. Pantsar-Kallio M, Manninen PKG, *J. Chromatogr. A*, 779: 139 (1997).
43. Chausseau M, Roussel C, Gilon N, Mermet JM, *Fresenius' J. Anal. Chem.*, 366: 476 (2000).
44. Vela NP, Heitkemper DT, Stewart KR, *Analyst*, 126: 1011 (2001).
45. Jackson BP, Shaw Allen PL, Hopkins WA, Bertsch PM, *Anal. Bioanal. Chem.*, 374: 203 (2002).
46. Bohari Y, Lobos G, Pinochet H, Pannier F, Astruc A, Potin-Gautier M, *J. Environ. Monit.*, 4: 596 (2002).
47. Kohlmeyer U, Kuballa J, Jantzen E, *Rapid Commun. Mass Spectrom.*, 16: 965 (2002).
48. Scancar J, Milacic R, *Analyst*, 127: 629 (2002).
49. Byrdy FA, Olson LK, Vela NP, Caruso JA, *J. Chromatogr. A*, 712: 311 (1995).
50. Saverwyns S, Van Hecke K, Vanhaecke F, Moens L, Dams R, *Fresenius' J. Anal. Chem.*, 363: 490 (1999).
51. Vanhaecke F, Saverwyns S, Wannemacker GD, Moens L, Dams R, *Anal. Chim. Acta*, 419: 55 (2000).
52. Milacic R, Scancar J, Tusek J, *Anal. Bioanal. Chem.*, 372: 549 (2002).
53. Inoue Y, Sakai T, Kumagai H, *J. Chromatogr. A*, 706: 127 (1995).
54. Pobozy E, Wojasinska E, Trojanowicz M, *J. Chromatogr. A*, 736: 141 (1996).
55. Collins CH, Pezzin SH, Rivera JFL, Bonato PS, Windmöller CC, Archundia C, Collins KE, *J. Chromatogr. A*, 789: 469 (1997).
56. Collins KE, Bonato PS, Archundia C, de Queiroz MELR, Collins CH, *Chromatographia*, 26: 160 (1988).
57. Pantsar-Kallio M, Manninen PKG, *J. Chromatogr. A*, 750: 89 (1997).
58. Sikovec M, Novic M, Hudnik V, Franko M, *J. Chromatogr. A*, 706: 121 (1995).
59. Sikovec M, Franko M, Novic M, Veber M, *J. Chromatogr. A*, 920: 119 (2001).
60. Shum SCK, Houck RS, *Anal. Chem.*, 65: 2972 (1993).
61. Laborda F, Chakraborti D, Mir JM, Castillo JR, *J. Anal. At. Spectrom.*, 8: 643 (1993).
62. Pitts L, Fisher A, Worsfold P, Hill SJ, *J. Anal. At. Spectrom.*, 10: 519 (1995).
63. Kannamkumarath SS, Wrobel K, Wrobel K, Vonderheide A, Caruso JA, *Anal. Bioanal. Chem.*, 373: 454 (2002).
64. Bird SM, Ge H, Uden PC, Tyson JF, Block E, Denoyer E, *J. Chromatogr. A*, 789: 349 (1997).
65. Pedersen GA, Larsen EH, *Fresenius' J. Anal. Chem.*, 358: 591 (1997).
66. Vilano M, Padro A, Rubio R, Rauret G, *J. Chromatogr. A*, 819: 211 (1998).
67. Zhang Y, Frankenberger WT Jr., *Sci. Total Environ.*, 26: 39 (2001).
68. Gomez-Ariza JL, Sanchez-Rodas D, Caro de la Torre MA, Giraldez I, Morales E, *J. Chromatogr. A*, 889: 33 (2000).
69. Michalke B, *J. Chromatogr. A*, 1058: 203 (2004).
70. Cardellicchio N, Avalli S, Ragone P, Riviuello P, 22nd International Symposium on Chromatography, September, pp. 13, Rome (1998).
71. Weber G, Neumann G, Romheld V, *Anal. Bioanal. Chem.*, 373: 767 (2002).
72. Ulrich N, Shaked P, Zilberstein D, *Fresenius' J. Anal. Chem.*, 368: 62 (2000).
73. Miravet R, López-Sánchez JF, Rubio R, *J. Chromatogr. A*, 1052: 121 (2004).

74. Takaya M, Sawatari K, *Ind. Health.*, 32: 165 (1994).
75. Nachtigall D, Artelt S, Wünsch G, *J. Chromatogr. A*, 775: 197 (1997).
76. Al-Rashdan A, Heitkemper D, Caruso JA, *J. Chromatogr. Sci.*, 29: 98 (1991).
77. Pobozy E, Glod B, Kaniewska J, Trojanowicz M, *J. Chromatogr. A*, 718: 329 (1995).
78. Svete P, Milacic R, Mitrovic B, Pihlar B, *Analyst*, 126: 1346 (2001).
79. Wang Q, Huang B, Guan Z, Yang L, Li B, *Fresenius' J. Anal. Chem.*, 370: 1041 (2001).
80. Rubio A, Padro A, Rauret G, *Fresenius' J. Anal. Chem.*, 351: 331 (1995).
81. Velez D, Ybanez N, Montoro R, *J. Anal. At. Spectrom.*, 11: 271 (1996).
82. Beauchemin D, Siu KWM, McLaren JW, Berman SS, *J. Anal. At. Spectrom.*, 4: 285 (1089).
83. Rubio R, Padro A, Alberti J, Rauret G, *Mikrochim. Acta*, 109: 39 (1992).
84. Rauret G, Rubio R, Padro A, *Fresenius' J. Anal. Chem.*, 340: 157 (1991).
85. Heitkemper D, Creed J, Caruso JA, Fricke FL, *J. At. Anal. Spectrom.*, 4: 279 (1989).
86. Jimenez de Blas O, Vicente Gonzalez S, Seisdedos Rodriguez R, Hernandez Mendez J, *J. AOAC Int.*, 77: 441 (1994).
87. Heng-Bin H, Yang-Bin L, Shi-Fen M, Zhe-Ming N, *J. Anal. At. Spectrom.*, 8: 1085 (1993).
88. Hansen SH, Larsen EH, Pritzl G, Cornett C, *J. Anal. At. Spectrom.*, 7: 629 (1992).
89. Larsen EH, Pritzl G, Hansen SH, *J. Anal. At. Spectrom.*, 8: 557 (1993).
90. Hakala E, Pyy L, *J. Anal. At. Spectrom.*, 7: 191 (1992).
91. Lamble KJ, Hill SJ, *Anal. Chim. Acta*, 334: 261 (1996).
92. Falk K, Emons H, *J. Anal. At. Spectrom.*, 15: 643 (2000).
93. Bin H, Gui-bin J, Xiao-bai X, *Fresenius' J. Anal. Chem.*, 368: 803 (2000).
94. Roychowdhry SB, Koropchank JA, *Anal. Chem.*, 62: 484 (1990).
95. Morita M, Uehiro T, Fuwa K, *Anal. Chem.*, 53: 1806 (1981).
96. Alberti J, Rubio R, Rauret G, *Fresenius' J. Anal. Chem.*, 351: 415 (1995).
97. Teräsahde P, Pantsar-Kallio M, Manninen PKG, *J. Chromatogr. A*, 750: 83 (1996).
98. Thomas P, Sniatecki K, *Fresenius' J. Anal. Chem.*, 351: 410 (1995).
99. Thomas P, Sniatecki K, *J. At. Anal. Spectrom.*, 10: 615 (1995).
100. Bavazzano P, Perico A, Rosendahl K, Apostoli P, *J. At. Anal. Spectrom.*, 11: 521 (1996).
101. Shibata Y, Morita M, *Anal. Sci.*, 5: 107 (1989).
102. Demesmay C, Olle M, Porthault M, *Fresenius' J. Anal. Chem.*, 348: 205 (1994).
103. Goessler W, Kuehnelt D, Schlagenhaufen C, Kalcher K, Abegaz M, Irgolic KJ, *J. Chromatogr. A*, 640: 73 (1993).
104. Lopez-Gonzalvez MM, Gomez MM, Palacios MA, Camara C, *Chromatographia*, 43: 507 (1996).
105. Blais JS, Momplaisir GM, Marshall WD, *Anal. Chem.*, 62: 1161 (1990).
106. Corr JJ, Larsen EH, *J. Anal. At. Spectrom.*, 11: 1215 (1996).
107. Sule PA, Ingle JD Jr., *Anal. Chim. Acta*, 326: 85 (1996).
108. Gjerde DT, Wiederin DR, Smith FG, *J. Chromatogr. A*, 640: 73 (1993).
109. Pantsar-Kallio MP, Manninen PKG, *J. Chromatogr. A*, 750: 89 (1996).
110. Derbyshrine M, Lamberty A, *Anal. Chem.*, 71: 4203 (1999).
111. Beere HG, Jones P, *Anal. Chim. Acta*, 293: 237 (1994).
112. Barnowski C, Jakubowski N, Stuewer D, Broekaert JAC, *J. Anal. At. Spectrom.*, 12: 1155 (1997).

113. Powell MJ, Boomer DW, Wiederin DR, *Anal. Chem.*, 67: 2474 (1995).
114. Collins CH, Lancas FM, *Radiochem. Radioanal. Lett.*, 56: 117 (1983).
115. Fukakusa Y, Yoshimura K, Waki H, Ishiguro SI, *J. Chromatogr. A*, 719: 365 (1996).
116. Collins CH, Collins RE, Ackerhalt RE, *J. Radioanal. Chem.*, 8: 263 (1971).
117. Gütlich P, Harbottle G, *Radiochim. Acta*, 5: 70 (1996).
118. Thompson ME, Connick RE, *Inorg. Chem.*, 20: 2279 (1981).
119. Blais JS, Huyghues-Despointes A, Momplaisir GM, Marshall WD, *J. Anal. At. Spectrom.*, 6: 225 (1991).
120. Gibson JAE, Willett IR, *Commun. Soil Sci. Plant Anal.*, 22: 1303 (1991).
121. Kanai Y, *Analyst*, 115: 809 (1990).
122. Urasa IT, Mavura WJ, Lewis VD, Nam SH, *J. Chromatogr. A*, 547: 211 (1991).
123. Smichowski P, Madrid Y, Guntinas MBD, Camara C, *J. Anal. At. Spectrom.*, 10: 815 (1995).
124. Beer HD, Coetzee PP, *Fresenius' J. Anal. Chem.*, 348: 806 (1994).
125. Tomlinson MJ, Wang J, Caruso JA, *J. Anal. At. Spectrom.*, 9: 957 (1994).
126. Small H, Stevens TS, Bauman WC, *Anal. Chem.*, 47: 1801 (1975).
127. Smith RE, *Ion Chromatography Applications*, CRC Press, Boca Raton, FL (1988).
128. Kolla P, Köhler J, Schomburg G, *Chromatographia*, 23: 465 (1987).
129. Röllin S, Kopatjtic Z, Wernli B, Magyar B, *J. Chromatogr. A*, 739: 139 (1996).
130. Nesternko PN, Zukova OS, Spigun OA, Jones P, *J. Chromatogr. A*, 813: 47 (1998).
131. Shaw MJ, Jones SJP, Nesternko PN, *J. Chromatogr. A*, 876: 127 (2000).
132. Rey M, Pohl C, International ion chromatography symposium, Nice (2000).
133. Hajos P, Revesz G, Horvath O, Peear J, Sarzanini C, *J. Chromatogr. Sci.*, 34: 291 (1996).
134. Bruzzoniti MC, Mentasti E, Sarzanini C, *Anal. Chim. Acta*, 353: 239 (1997).
135. Otha K, Tanaka K, *Anal. Chim. Acta*, 373: 189 (1998).
136. Strelow FEW, Victor AH, *Talanta*, 37: 1155 (1990).
137. Sun Q, Wang H, Mou S, *J. Chromatogr. A*, 708: 99 (1995).
138. Williams T, Barnett NW, *Anal. Chim. Acta*, 264: 297 (1992).
139. Al-Shawi AW, Dhal R, *Anal. Chim. Acta*, 333: 23 (1996).
140. Bruzzoniti MC, Mentasti E, Sarzanini C, Braglia M, Cocito G, Kraus J, *Anal. Chim. Acta*, 332: 49 (1996).
141. Tanaka K, Otha K, Haddad PR, Fritz JS, *J. Chromatogr. A*, 804: 179 (1998).
142. Thomas P, Finnie JK, Williams JG, *J. At. Anal. Spectrom.*, 12: 1367 (1997).
143. Lintschinger J, Schramel P, Hatalak-Rauscher A, Wendler I, Michalke B, *Fresenius' J. Anal. Chem.*, 362: 313 (1998).
144. Pantsar-Kallio M, Manninen PKG, *Anal. Chim. Acta*, 318: 335 (1996).
145. Zoorob GK, Caruso JA, *J. Chromatogr. A*, 773: 157 (1997).
146. Michalke B, Witte H, Schramel P, *J. Anal. At. Spectrom.*, 16: 593 (2001).
147. Ammann AA, *Anal. Bioanal. Chem.*, 372: 448 (2002).
148. Gammelgaard B, Jons O, Bendahl L, *J. Anal. At. Spectrom.*, 16: 339 (2001).
149. Kerl W, Becker JS, Dannecker W, Dietze HJ, *Fresenius' J. Anal. Chem.*, 362: 433 (1998).
150. Mateja ikovec M, Franko M, Novi M, Veber M, *J. Chromatogr. A*, 920: 119 (2001).
151. Vela NP, Olson LK, Caruso JA, *Anal. Chem.*, 65: 585A (1993).
152. Busch M, Seubert A, *Fresenius' J. Anal. Chem.*, 366: 351 (2000).
153. Roehl R, Alforque MM, *At. Spectrosc.*, 11: 210 (1990).
154. Suyani H, Creed J, Davidson T, Caruso JA, *J. Chromatogr. Sci.*, 27: 139 (1989).

155. Kawabata K, Kishi Y, Kawaguchi O, Watanabe Y, Inoue Y, *Anal. Chem.*, 63: 2137 (1991).
156. Hajos P, Horvath O, Revesz G, Peear J, Sarzanini C, in Dyer A, Hudson M, Williams P (Eds.), *Progress in Ion Exchange*, RSC, Cambridge, UK, p. 144 (1991).

8 Speciation of Metal Ions by Ion-Pair, Micellar Electrokinetic, Size Exclusion, Chiral, Capillary Electro-, and Supercritical Fluid Chromatographic Methods

Of course, gas chromatography (GC), reversed-phase high-performance liquid chromatography (RP-HPLC), and ion chromatography (IC) are the well-known and reputed techniques for metal ion speciation in environmental and biological samples, but due to the increase in our knowledge and the developments in science, other forms of chromatography have emerged and are now used for this purpose. These newly developed modalities of chromatography used for metal ion speciation are ion-pair (IPC), micellar electrokinetic (MEKC), size exclusion (SEC), chiral (CC), capillary electro- (CEC), and supercritical fluid (SFC) chromatography. Basically, these are modified versions of liquid chromatography, and their basic principles are designed for better metal ion speciation. Moreover, sometimes these techniques are useful for metal ion speciation that is not possible by RP-HPLC, IC, and GC. These chromatographic methods have certain advantages over GC, HPLC, and IC and are gaining importance day by day. High speeds, inexpensiveness, sensitivity, selectivity, and good reproducibility are some of their salient features.[1–3] Although their use in metal speciation is still limited,[4] improvements are under way, and they will definitely replace other chromatographic techniques in metal ion separation in environmental and biological samples for specific applications. Due to the growing interest in these newly developed chromatographic methods, attempts were made to describe metal ion speciation using IPC, MEKC, SEC, CC, CEC, and SFC in this chapter.

8.1 ION-PAIR CHROMATOGRAPHY

Ion-pair chromatography (IPC) is similar to RP-HPLC, as reversed-phase stationary phases are used in both cases. It can be used effectively for the speciation of cations, anions, and neutral species simultaneously. A counterion is added in the mobile phase in RP-HPLC, and the technique is termed IPC. IPC is also known by the names of soap chromatography, ion interaction chromatography, and dynamic ion-exchange chromatography. The counterion is typically referred to as the ion-pair reagent, consisting of a polar head and a nonpolar tail. This type of chromatography was used to speciate charged and uncharged metallic species. Generally, alkylammonium salts, alkyl sulfates or alkylsulfonates, dihydroxybenzene and analogs, and ethylenediaminetetraacetic acid (EDTA) are used as the ion-pairing reagents. Generally, the concentrations of these ion-pair reagents ranged from 0.001 to 0.005 *M*. Another approach to IPC is the use of common reversed phases coated with suitable hydrophobic agents, such as alkylsulfonates or alkylsulfates with sufficiently long alkyl groups. Octadecyl-bonded silica permanently coated with sodium dodecyl sulfate (SDS) in the presence of ion-pairing agents was considered for the separation of transition metal ions. The ion-pairing reagents used should be soluble in the mobile phase, univalent, aprotic, and nondestructive to the column. The ion-pairing reagents are suitable for the coupling of an inductively coupled plasma–mass spectrometry (ICP-MS) detector because of their good volatility. An important advantage of ion-pair reversed chromatography over reversed-phase HPLC is that it facilitates the separation of ionic as well as uncharged species; ionic species are not retained on a conventional reversed-phase column. Moreover, higher efficiencies and greater versatility are seen with ion-pair reversed chromatography than with fixed-site ion-exchange columns.[5]

Boucher et al.[6] reported the separation of As(III), As(V), monomethylarsonate (MMA), dimethylarsinate (DMA), arsenobetaine (AsB), and arsenocholine (AsC) using a C_{18} column. The mobile phase used was tetrabutylammonium hydroxide, with detection by ultraviolet (UV) and amperometric devices. Le and Ma[7] reported the speciation of arsenite, arsenate, MMA, DMA, AsB, AsC, and $Me_4As(I)$ species using μBondclone C_{18} and μBondapak C_{18} columns. The authors used a mobile phase containing 10 m*M* propanesulfonate, 4 m*M* malonic acid, and 0.1% methonol (pH 3). The detection was achieved by hydride generation atomic absorption spectrometry (HG-AAS) and ICP-MS detectors. Hwang and Jiang[8] speciated As(III), As(V), MMA, and DMA metal ion species using reversed-phase ion-pair chromatography (RP-IPC). The eluate was delivered to the hydride generation system after a prereduction with L-cysteine at 95°C in dilute nitric acid. Martin et al.[9] developed an online ion-pair chromatographic–microwave-assisted oxidation–hydride generation–atomic absorption spectrometric (HG-AAS) system (using columns of different kinds) for the determination of arsenite, arsenate, DMA, MMA, AsB, and AsC in environmental samples. Do et al.[10] proposed an ion-pair reversed-phase liquid chromatography for the speciation of arsenic and selenium metal ions in the environment and in mammals. The studied metal species were arsenite, arsenate, MMA, DMA, selenite, selenate, selenocystamine, selenocystine, selenomethionine, and selenoethionine. In order to study the retention behavior of these compounds and to estimate optimal conditions for the chromatographic separation, central composite designs were used to evaluate the influence of the eluent parameters, such as pH, tetrabutylammonium phosphate (TBA) concentration, and sodium hydrogenphosphate amounts. The retention factors of each species and the selectivity were established as response criteria. Response surfaces and isoresponse curves were drawn from the mathematical models and enabled one to determine the optimal conditions and to visualize the method robustness. The predicted optimal zone was situated at pH 5.5 to 6.5, 4 m*M* Na_2HPO_4, and 3 to 4 m*M* TBA. Regression models suggested linearity for the studied compounds in the range of 25 to 200 μg selenium and arsenic per liter investigated. Kohlmeyer et al.[11] developed a method using IPC-ICP-MS for the speciation of inorganic arsenic (arsenite, arsenate) along with organic arsenic compounds (monomethylarsonic acid, dimethylarsinic acid, arsenobetaine, arsenocholine, trimethylarsine oxide, tetramethylarsonium ion, and several arsenosugars) in fish, mussel, oyster, and marine algae samples.

Ho and Uden[12] used tetra-*n*-alkylammonium bromide ion-pairing reagent in methanol–water for the speciation of methyl-, ethyl-, benzyl-, and phenylmercury species. The authors studied the effect of tetramethylammonium (TMA), tetraethylammonium (TEA), and tetrabutylammonium (TBA) ions on the retention of mercury species. Al-Rashdan et al.[13] speciated TTML, TTEL, and triphenyl lead (TPhL) by ion polar chromatography coupled with ICP-MS techniques. The authors used methanol–water containing 4 m*M* sodium pentanesulfate as the mobile phase. Furthermore, the same group[14] reported Pb(II), triethyllead chloride, triphenyllead chloride, and tetraethyllead chloride using sodium pentane sulfonate as the pairing reagent (2 m*M*). The detection was carried out by ICP-MS. Houck et al.[15] analyzed Se(IV) and Se(VI) on a C_{18} column using methanol–water–tetrabutylammonium ion as the mobile phase. The detection was achieved by ICP-MS with a detection limit of 10 to 20 ng/mL. Yiang and Jiang[16] reported the separation of TMSe, Se(IV), and Se(VI) using IPC. The authors reported 0.17 to 0.76 ng/mL as the detection limits for these species using an ICP-MS detector. Jen et al.[17] used ion-pair liquid chromatography with UV detection to determine selenium(IV) and selenium(VI) ions simultaneously. The chromatographic behavior of the selenium species was examined in detail. Factors affecting chromatographic separation and quantitative determination, as well as potential interference, were systematically optimized. B'Hymer et al.[18,19] reported the speciation of arsenic species using ion-pair RP-HPLC. The authors used a C_{18} reversed-phase column with a 100% aqueous mobile phase containing citric acid and pentane sulfonic acid sodium salt. The detection was achieved using an ICP-MS detector with 1.6 to 3.6 μg/L detection limits. Furthermore, the same group[20] analyzed different selenium species in yeast solution on Phenomenex C_8 column. The authors used 10% methanol and 1% trifluoroacetic acid as the mobile phase. Zheng et al.[21] speciated eight selenium compounds, namely, selenite [Se(IV)], selenate [Se(VI)], selenocystine (SeCys), selenourea (SeUr), selenomethionine (SeMet), selenoethionine (SeEt), selenocystamine (SeCM), and trimethylselenonium ion ($TMSe^+$) by using mixed ion-pair reagents containing 2.5 m*M* sodium 1-butanesulfonate and 8 m*M* tetramethylammonium hydroxide as a mobile phase. The separation of these anionic, cationic, and neutral organic selenium compounds on a LiChrosorb RP-18 reversed-phase column took only 18 min at a flow rate of 1 mL/min with isocratic elution, and baseline separation among the six organic Se compounds was achieved. ICP-MS was employed as the element-specific detection method. A comparison of ICP-MS signal intensity obtained with a Barbington-type nebulizer and with an ultrasonic nebulizer (USN) was made. Different signal enhancement factors were observed for the various selenium compounds when an USN was used. Kotrebai et al.[22] reported the speciation of 13 seleno species with three different ion-pairing reagents. Gayon et al.[23] speciated selenium species in human urine on a Nucleosil 120 C_{18} column using 30 m*M* ammoniumformate (pH 3)–methanol (95:5, v/v) as the mobile phase containing tetrabutylalkylammonium acetate (TBAA). Kannamkumarath et al.[24] speciated selenium metal ion in lipid extract, low molecular weight, and protein fractions using IPC. The detection was achieved by ICP-MS. The speciation of different metal ion species by IPC is summarized in Table 8.1. To show the nature of the chromatograms of the separated metal ion species by IPC, a typical example of the chromatograms of arsenic speciation is shown in Figure 8.1.[25] Separated metallic species are quantitatively determined by retention times (t), capacity (k), separation (α), and resolution factors (R). The quantitative estimation is carried out by comparing the area of identified peak with the area of the peak obtained by the standard metallic species of the known concentration. In the case of ion-pair HPLC, the detection limits of metallic species vary from one ion-pairing reagent/derivatizing reagent to another. Shum et al.[30] studied the effect of ion-pairing reagent on the shape of MeHg(I) species. The authors used 5 m*M* concentrations of sodium pentane sulfonate (S5), sodium heptanesulfonate (S7), and sodium dodecanesulfonate (S12). These results are shown in Figure 8.2, and it is obvious from this figure that the peak shape varied from one ion-pairing reagent to another.

8.2 MICELLAR ELECTROKINETIC CHROMATOGRAPHY

In IPC, if the counterion is a surfactant molecule, a micellar is formed. This type of liquid chromatography is referred to as micellar electrokinetic chromatography (MEKC). This technique was introduced by Terabe et al.[31] in 1984. Basically, a surfactant is a molecule possessing two zones of different polarities, which shows special characteristics in solution. The surfactant in micellar chromatography has a long-chain hydrocarbon tail and charged head. The formation of micelles occurs

TABLE 8.1
Speciation of Metal Ions by Different Liquid Chromatographic Modalities

Metal Ions	Sample Matrix	Columns	Detection	Ref.
		Ion-Pair Chromatography		
As species	—	C_{18} column	ICP-MS	18, 19
As(III), As(V), MMA, DMA, arsenobetaine, and arsenocholine	—	C_{18} column	UV and amperometric	6
Arsenite, arsenate, MMA, DMA, AsB, AsC, and $Me_4As(I)$	—	μBondclone C_{18} and μBondapak C_{18}	HG-AAS and ICP-MS	7
As(III), As(V), MMA, and DMA	—	C_{18} column	HG-AAS	8
As(III), AS(V), MMA, and DMA	—	C_{18} column	UV-HG-QF-AAS	26, 27
Arsenic species	Biological samples	C_{18} column	ICP-MS	28
TTML, TTEL, and triphenyllead (TPhL)	—	C_{18} column	ICP-MS	13
Methyl-, ethyl-, benzyl-, and phenylmercury	—	C_{18} column	—	12
Selenium species	Urine	Nucleosil 120 C_{18}	—	23
Selenium species	Yeast	Phenomenex C_8	—	20
Selenium species	Urine	C_{18} column	ICP-MS	15
TMSe, Se(IV), and Se(VI)	—	C_{18} column	ICP-MS	16
Tin species	Sediment	C_{18} column	ICP-MS	29
		Size Exclusion Chromatography (SEC)		
Aluminum species	Tea	Superose 12 HR	UV/visible	38
Aluminum species	Water	Superdex-75-HR 10/30 and Superdex-Peptide-HR 10/30	—	39
Aluminum species	Soil	Superdex HR75 10/30	ICP-AES	40
As(III), As(V), MMA, and DMA	Soil and sediment	Hamilton PRP-X-100	ICP-MS	45
As(III), As(V), MMA, DMA, and arsenobenite	Urine	IonPac AS14	ICP-MS	46
As(III), As(V), MMA, DMA, and arsenobenite	Vegetables	IC-PAK	ICP-MS	47
As(III), As(V), MMA, DMA, and arsenobenite	Apple	PRP-X-100	ICP-MS	48
Arsenic species	Biological samples	—	ES-QTOF-MS	41

TABLE 8.1 (Continued)
Speciation of Metal Ions by Different Liquid Chromatographic Modalities

Metal Ions	Sample Matrix	Columns	Detection	Ref.
Cadmium species	—	RP size exclusion column	ICP-MS	49
Cr(III) and Cr(VI)	Water	IC-Pak	ICP-MS	50
Cr(III) and Cr(VI)	Water	IonPac AG-5	ICP-MS	45
Pertechnetate (TcO_4), Tc-diethylenetriamine-pentaacetate (Tc-DTPA), and $[Tc(V)O_2(1,4,8,11\text{-tetraaza-cyclotetradecane}]^+$	—	Sephadex G-25, HEMA-SEC Bio 1000, and Zorbax GF-250	UV/visible	42
Selenium, arsenic, copper, cadmium, and zinc metal ions species	Biological samples	AS7	ICP-MS	43
Se-Met, Se-Cys, selenite, and selenate	Plants	Polyspher IC-AN-2	ICP-MS	47
Se-Urea, Se-Met, Se-Eth, and Se-Cys	Bacteria	AS11	ICP-MS	51
Selenite, Se-Met, and trimethylselenonium	Urine	IonPac CS5	ICP-MS	52
		Chelation Chromatography		
Aluminum species	—	—	ICP-MS	53
Cr(III) and Cr(VI)	—	—	Photometry	54
Cr(III) and Cr(VI)	—	—	ET-AAS	55

in aqueous mobile phase when the concentration of the counterions exceeds a critical micelle concentration. Approximately 40 to 100 ions aggregate to form roughly spherical particles with the hydrophobic tail oriented toward the center, and the hydrophobic head oriented toward the outside of the micellar particle. In this way, a second phase is formed, and uncharged species may be solubilized into the micelle. The metallic species has different polarities partitioned between aqueous phase, stationary phase, micellar hydrophobic phase, and the micellar hydrophillic phase. Therefore, both ionic and uncharged species may be separated easily. Surfactants are divided into three categories — ionic (cationic and anionic), nonionic, and zwitterionic. The most important surfactants are sodium dodecyl sulfate, sodium tetradecyl sulfate, sodium decanesulfonate, sodium *N*-lauryl-*N*-methyllaurate, sodiumpolyoxyethylene dodecyl ether sulfate, sodium *N*-dodecanoyl-L-valinate, sodium cholate, sodium deoxycholate, sodium taurocholate, sodium taurodeoxycholate, potassium perfluoroheptanoate, dodecyltrimethylammonium chloride, dodecyltrimethylammonium bromide, tetradecyltrimethylammonium bromide, and cetyltrimethylammonium bromide. Surfactants molecules

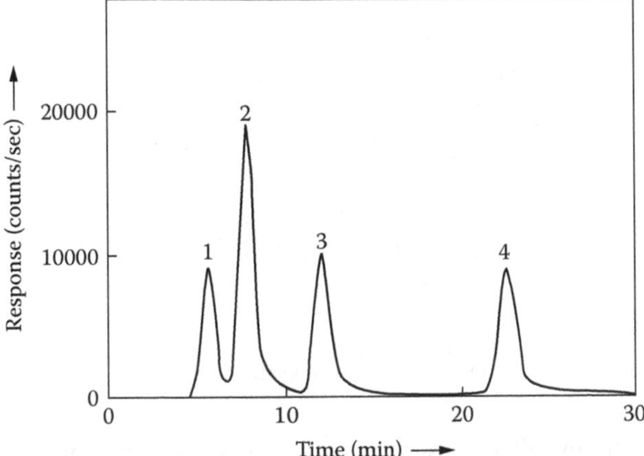

FIGURE 8.1 Chromatograms of arsenic by ion-pair chromatography: (1) monomethylarsonate (MMA), (2) dimethylarsinate (DMA), (3) arsenobenite, and (4) arsenocholine. (From Kannamkumarath, S.S., Wrobel, K., Wrobel, K., Vonderheide, A., and Caruso, J.A., *Anal. Bioanal. Chem.*, 373: 454 (2002). With permission.)

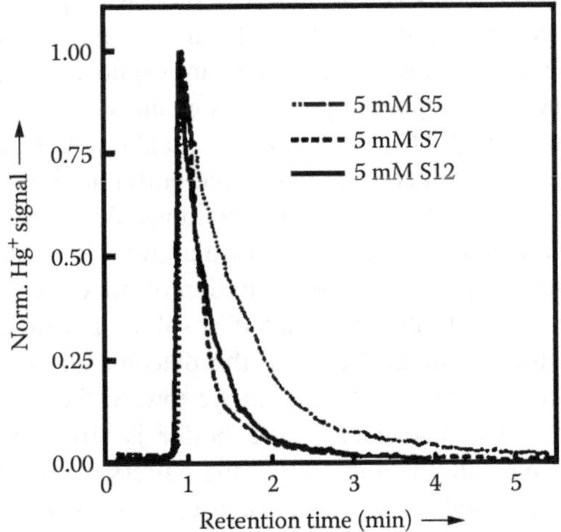

FIGURE 8.2 Effect of the ion-pairing reagents on the peak shape of MeHg(I) organo metal ion details. (From Shum, S.C.K., Pang, H., and Houk, R.S., *Anal. Chem.*, 64: 2444 (1992). With permission.)

containing long alkyl chains (hydrophobic group) and charged or neutral heads (polar group) aggregate in aqueous solutions above their critical micellar concentration (CMC) and form micelles, as shown in Figure 8.3. These types of aggregates are spherical in shape, with polar and hydrophobic groups at outer and core regions, respectively. This micelle works as a pseudo-stationary phase that possesses a self-mobility different than that of the surrounding aqueous phase.

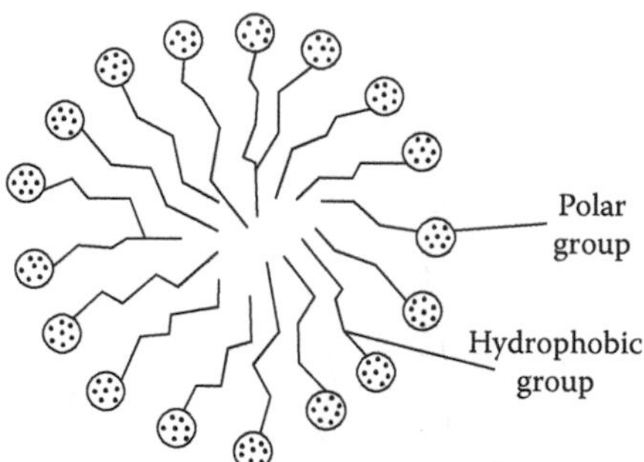

FIGURE 8.3 Schematic representation of a micelle.

Accordingly, the micellar phase acts as a stationary phase, while the aqueous phase acts as the mobile phase in MEKC. The distribution of metal ion species occurs between the micellar and aqueous phases. The solute and micellar interactions are of three types: the solute is adsorbed on the surface of the micelle by electrostatic or dipole interactions, the solute behaves as a cosurfactant by participating in the formation of the micelle, and the solute is incorporated into the core of the micelle.[32] The extent of these interactions depends on the metal ion species and the micelle. Highly polar metal ions will mainly be adsorbed on the surface of the micelle, and low polar solutes are supposed to interact at the core of the micelle. The most important properties of the surfactants are their critical micellar concentration (CMC), aggregation number, and Kraft point. CMC depends on temperature, salt concentration, and buffer additives. The aggregation number indicates the number of surfactant molecules taking part in micelle formation. The Kraft point is the temperature above which the solubility of the surfactant increases steeply due to the formation of micelles. The best surfactant for MEKC possesses good solubility in a buffer solution, forming a homogeneous micellar solution that is compatible with the detector and has a low viscosity. Basically, the micelle and buffer tend to move toward the positive and negative ends, respectively. The movement of the buffer is stronger than the micelle movement, and as a result, the micelle and the buffers move toward the negative end. Separation in this mode of liquid chromatography depends on the distribution of analytes between the micelle and the aqueous phases.

Due to these salient features of MEKC, it has been utilized for metal ion speciation. Only a few reports are available in the literature, as it is a newly developed modality of liquid chromatography. Ding et al.[33] reported the speciation of arsenic metal ions in human urine by using micellar chromatography with an aqueous mobile phase containing 10% propanol and 0.02 *M* borate buffer. The authors detected arsenic species by using an ICP-MS assembly. The micellar chromatography was found to be suitable for urine samples, as proteins found in urine were dissolved in the micelle, avoiding any interference in speciation and detection and, hence, no need for the deproteinization of biological samples. Recently, Hong et al.[34] speciated dimethylarsenic acid (DMA), monomethylar-

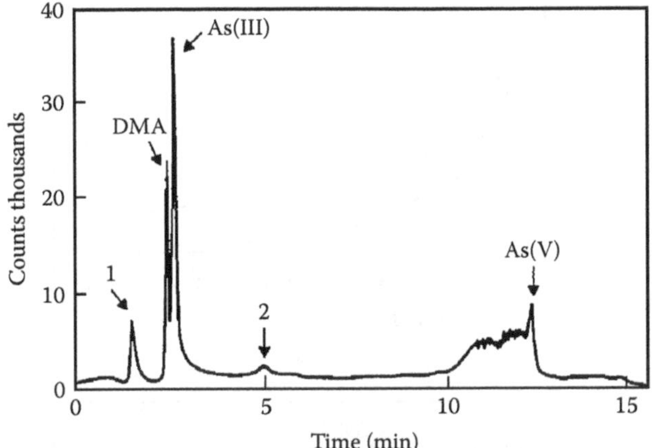

FIGURE 8.4 Chromatograms of arsenic speciation in human urine using micellar electrokinetic chromatography (MEKC). Mobile phase: 0.05 *M* cetyltrimethylammonium bromide (CTAB) and 10% methanol (pH 10.2). Column: Hamilton RP-18. (From Ding, H., Wang, J., Dorsey, L.G., and Caruso, J., *J. Chromatogr. A*, 694: 425 (1995). With permission.)

sonic acid (MMA), As(III), and As(V) species by micellar liquid chromatography. Linear dynamic ranges for the four species were three orders of magnitude, and detection limits were in the picogram range with ICP-MS detection. The micellar mobile phase, which consisted of 0.05 *M* cetyltrimethylammonium bromide, 10% propanol, and 0.02 *M* borate buffer, showed good compatibility with ICP-MS. This method allowed for the direct injection of urine samples onto the chromatographic system without extensive pretreatment and presented no interference from chlorine in the matrix. Suyani et al.[35] reported the application of micellar electrokinetic chromatography for the speciation of alkyltin compounds using 0.1 *M* sodium dodecyl sulfate mobile phase on a C_{18} column. The authors used other surfactants, such as dodecyltrimethylammonium bromide and polyoxyethylene(23)dodecanol, but sodium dodecyl sulfate resulted in the best speciation. LaFuente et al.[36] analyzed selenium species in urine, while Infante et al.[37] speciated cadmium metallic species using micellar HPLC. The specific chromatograms of arsenic speciation using MEKC are shown in Figure 8.4.[33]

The optimization of speciation in MEKC can be achieved as in the case of other modalities of liquid chromatography. Ding et al.[33] attempted to optimize the speciation of arsenite, arsenate, MMA, and DMA arsenic species in urine samples, which was achieved using different concentrations of cetyltrimethylammonium bromide (CTAB) surfactant. Various concentrations used were in the range of 0.02 to 0.10 *M*. The variation of the retention factors with varied concentrations of CTBA is shown in Figure 8.5a for all four arsenic species. It is clear from this figure that retention factors are decreasing with higher concentrations of CTAB, indicating poor separation. The same authors also tried to optimize speciation by using different amounts of organic modifier (i.e., *n*-propanol), and the results are shown in Figure 8.5b, which again indicates poor

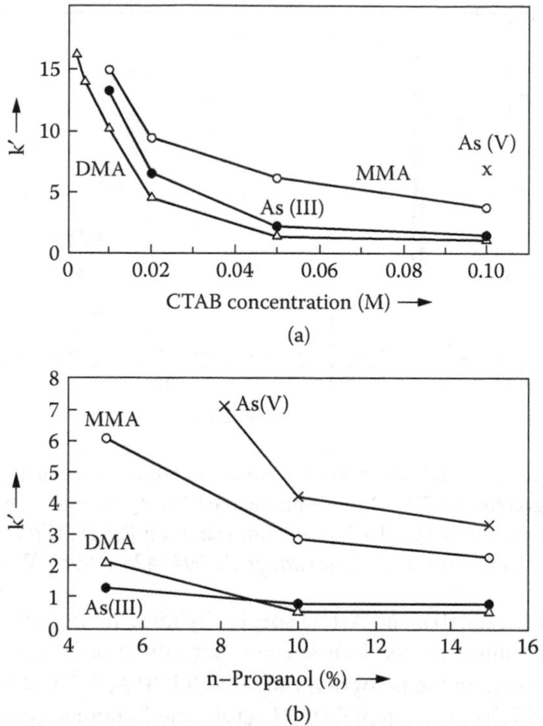

FIGURE 8.5 Effect of (a) cetyltrimethylammonium bromide (CTAB) concentrations (with 5% *n*-propanol) and (b) percentage of *n*-propanol (with 0% *M* CTAB and 10 pH) on the retention factors of As(III), As(BV), dimethylarsinate (DMA), and monomethylarsonate (MMA) arsenic species. (From Ding, H., Wang, J., Dorsey, L.G., and Caruso, J., *J. Chromatogr. A*, 694: 425 (1995). With permission.)

speciation of arsenic species at higher concentrations of *n*-propanol. The effect of pH on the retention factor was also studied, and these findings are plotted in Figure 8.6a, which clearly shows higher values of retention factors for all four species at higher pH. Furthermore, the effect of *n*-propanol on the sensitivity of the detector (ICP-MS) was also studied by these authors (Figure 8.6b). It can be observed from Figure 8.6b that the sensitivity of detection for all arsenic species increased at higher concentrations of *n*-propanol. The sensitivity of all four species was in the order DMA > As(III) > As(V) > MMA. Similarly, the optimization of other metal ion species in MKEC can be achieved by varying the experimental conditions.

8.3 SIZE EXCLUSION CHROMATOGRAPHY

Some polymeric-based columns act as sieves to separate ionic species based on their sizes. Typically, the smaller species are held back, and the larger species

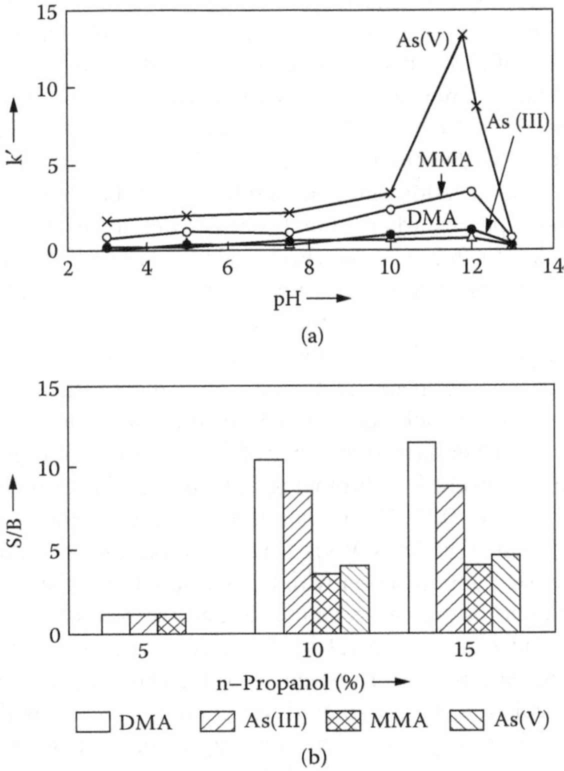

FIGURE 8.6 Effect of (a) pH (with 0.05 *M* cetyltrimethylammonium bromide [CTAB]) on the retention factors of As(III), As(BV), dimethylarsinate (DMA), and monomethylarsonate (MMA) arsenic species and (b) the different percentages of *n*-propanol on the detector sensitivity for As(III), As(BV), DMA, and MMA arsenic species. (From Ding, H., Wang, J., Dorsey, L.G., and Caruso, J., *J. Chromatogr. A*, 694: 425 (1995). With permission.)

pass through. This type of chromatography is called size exclusion chromatography (SEC), and the separation is useful for larger molecules, such as metallic proteins, peptides polymers, and so forth. However, some other inorganic and organometallic species were separated using this modality of chromatography. There are two types of SEC chromatography — gel permeation chromatography (GPC) and gel filtration chromatography (GFC). GPC uses an organic mobile phase, while GFC uses an aqueous mobile phase. In SEC, the stationary phase consists of a network of pores where ionic species are separated depending on their sizes. Because the retention time is not controlled by chemical interactions, the mobile phase does not play a crucial role, while it retains the conformational structure of the species that resemble its native form. The typical mobile phases in SEC are buffers with neutral pH. First, starch was used as the starting stationary phase in SEC, which was not stable in aqueous mobile phases. At a later date, starch was replaced by new and more durable materials. Sephadex is the most popularly used stationary phase in SEC. The speciation of metal ions in seawater or brine becomes difficult when using RP-HPLC or IC, as the high ionic strength of seawater or brine samples swamped the ion-exchange columns. For these kinds of applications, the selectivity of the separations can be enhanced by using chelating reagents in the mobile phase, and the modality is called chelation chromatography. In chelation chromatography, either the chelating reagents are chemically bonded phases, where complexation reactions in the stationary phase (ion exchange due to free or protonated chelating groups act as ion-exchange resin sites) or chelating agents in the mobile phase are responsible for the separation. Two main approaches can be followed to obtain a proper stationary phase: chemical bonding of the chelating group to the substrate and coating of a substrate with a ligand that is permanently trapped onto the substrate. This mode of chromatography is not popular due to its limited application.

Flaten and Lund[38] described the speciation of aluminum organic species in tea infusions by size exclusion chromatography using a Superose 12 HR column with a mobile phase of 0.12 *M* l-1 tris(hydroxymethyl)aminomethane adjusted to pH 5.5 and detection by UV-visible spectrophotometry. Aluminum was detected after postcolumn reaction with pyrocatechol violet at 585 nm. The tea samples of different origins were analyzed, and the results indicated that aluminum was bound to the same relatively narrow size-range of large organic molecules in the tea infusions, irrespective of the origin of the tea. Hils et al.[39] described the speciation of trace amounts of aluminum present in the percolating water of forest soil through the online coupling of size exclusion chromatography. Size fractionations of the organic aluminum species were obtained by using Superdex-75-HR 10/30 and Superdex-Peptide-HR 10/30 columns. Mitrovic and Milacic[40] speciated aluminum species in forest soil extracts by SEC with UV and inductively coupled plasma–atomic emission spectrometry (ICP-AES) detection. Size exclusion chromatography was performed on a Superdex HR75 10/30 column. Isocratic elution with 0.15 *M* NaCl in Tris-HCl buffer (pH, 5.5) was applied over 100 min at a flow rate of 0.35 mL/min. The separated peaks were detected by ICP-AES detector. McSheehy et al.[41] used ion chromatography–electrospray quadrupole/time-of-flight mass spectrometry (IC-ES-QTOF-MS) for the speciation of arsenic metal in a kidney of the Tridacna clam. The species were isolated by three-dimensional size exclusion anion-exchange, cation-exchange chromatography. A total of 15 organoarsenic species were separated and identified, and 13 of these possessed the dimethylarsinoyl group. Arsenobetaine and dimethylarsinic acid were also detected with the major species (accounting for up to 50% water-soluble arsenic) of 5-dimethylarsinoyl-2,3,4-trihydroxypentanoic acid. Harms et al.[42] presented an ion-exclusion chromatographic method for the speciation of technetium metal ion — pertechnetate (TcO_4^-), Tc-diethylenetriaminepentaacetate (Tc-DTPA), and $[Tc(V)O_2(1,4,8,11\text{-tetraazacyclotetradecane}]^+$. The columns used were Sephadex G-25, HEMA-SEC Bio 1000, and Zorbax GF-250 with the mobile phases of 5.8 and 13 m*M* NaCl separately with the detection by UV/visible spectrometer. Jackson et al.[43] used SEC coupled to ICP-MS to investigate the speciation of Se, As, Cu, Cd, and Zn in tissue extracts from a largemouth bass (*Micropterus salmoides*) collected from a coal fly ash basin using an AS7 column. Ion chromatography with ICP-MS using an AS7 column and HNO_3 gradient elution indicated that the selenium and arsenic species in the liver and gill extracts had similar retention times, but these retention times did not correspond to retention times for As(III), As(V), dimethylarsenate, arsenobetaine, Se(IV), Se(VI), selenomethionine, or selenocystine.

Recently, Lindermann and Hintelman[44] reported the speciation of organoselenium compounds by coupling an SEC column with a reversed-phase column. The authors used Sephadex G-50 as the size exclusion column and Hypersil Hypercarb as the reversed-phase column, simultaneously, in combination. The mobile phase used was 300 m*M* formic acid in methanol with ICP-MS detection. In a column switching technique, the direction of the flow of the mobile phase is changed by valves so that the effluent (or a portion of it) from the primary column is passed to a secondary column for a definite period of time. The stability of metal ion complexes with chelating reagents in chelation chromatography is also controlled by the pH of the mobile phase. A schematic representation of metal ion speciation by SEC is shown in Figure 8.7. The speciation of different metal ion species by size exclusion chromatography is summarized in Table 8.1.

8.4 CHIRAL CHROMATOGRAPHY

In chiral chromatography, the mobile or stationary phase contains chiral molecules, which are used for the resolution of enantiomers. Various chiral molecules (chiral selectors) used in liquid chromatography are polysaccharides, cyclodextrins, antibiotics, Pirkle type, ligand exchangers, crown ethers, imprinted polymers, and several other types of chiral compounds. There are several organic racemates containing metal ions, and hence, the enantiomers of such types of molecules may be considered metal ion species. The speciation of such a type of enantiomer is very difficult due to their similar physicochemical properties. However, the pharmacological activities of these organometallic species may be different, and, therefore, their speciation is very important. Such types of organic metallic compounds are amino acids, proteins, and enzymes. For detailed information on chirality and chiral resolution by liquid chromatography, readers should consult the books of Ali and Aboul-Enein.[1,56]

Mendez et al.[57] used chiral chromatography (CC) for the speciation of derivatized selenomethionine enantiomeric species in a β-cyclodextrin stationary phase. The authors used the ICP-MS technique for the detection purpose. A chiral crown ether chiral selector has also been used for the speciation of selenoamino acids.[58] Macrocyclic glycopeptide antibiotics stationary phases were also used for organometallic species. Ponce de Leon et al.[59] used a teicoplanin-based chiral column to speciate selenomethionine enantiomers in selenized yeast. The chromatograms of the speciation of selenoamino acid enantiomeric species on chiral crown ether column are shown in Figure 8.8.[58]

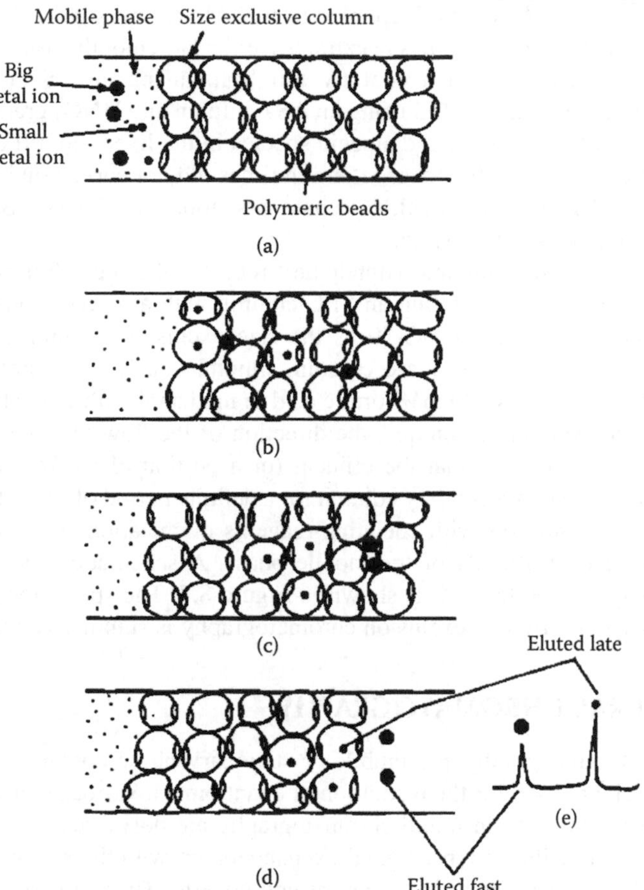

FIGURE 8.7 Metal ion speciation by size exclusion chromatography (SEC) and chelation chromatography.

8.5 CAPILLARY ELECTROCHROMATOGRAPHY

A hybrid technique between HPLC and CE was developed in 1990 and is called capillary electrochromatography (CEC).[60–73] CEC is expected to combine the high peak efficiency characteristic of electrically driven separations with high separation selectivity. CEC experiments can be carried out on wall-coated open tubular capillaries or capillaries packed with particulate or monolithic silica or other inorganic materials as well as organic polymers. The chromatographic band-broadening mechanisms are different in the individual modes. The chromatographic and electrophoretic mechanisms work simultaneously in CEC, and several combinations are possible. An illustrative quantitative treatment on the effect of capillary diameter and migration mode on the separation results in open tubular CEC was given by Vindevogel and Sandra.[74] The separation occurs on the mobile

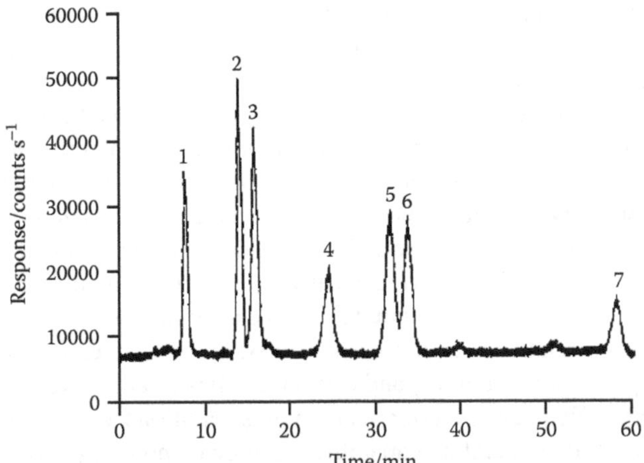

FIGURE 8.8 Chromatograms of selenoamino acids using chiral crown ether column with 0.1 *M* $HClO_4$ (pH 1) as the mobile phase. Peaks: (1) selenocystine, (2) selenomethionine, (3) *meso*-selenocystine, (4) D-selenocystine, (5) D-selenomethionine, (6) L-selenothionine, and (7) D-selenothionine. (From Suton, K.L., Ponce de Leon, C.A., Ackley, K.L., Suton, R.M., Stalcup, A.M., and Caruso, J., *Analyst*, 125: 281 (2000). With permission.

phase/stationary phase interface, and the exchange kinetics between the mobile and stationary phases are important. Again, in this mode of liquid chromatography, the separation principle is a mixture of liquid chromatographic and capillary electrophoresis; therefore, a variety of metallic species can be speciated with great accuracy, selectivity, and sensitivity. No report was published on metal ion speciation by using this modality of liquid chromatography. Metallic species may be distributed between the mobile and stationary phases through adsorption and the partition phenomenon. Moreover, the electrophoretic nature of this technique may be highly useful for metallic species with different charges. The possibilities of different interactions of metallic species with the capillary wall may be an asset of this technique in the metal ion speciation area. Therefore, in the future, CEC may be an excellent technique for metal ion speciation.

8.6 SUPERCRITICAL FLUID CHROMATOGRAPHY

In the past three decades, supercritical fluids have been used as mobile phases in liquid chromatography, and the modality is known as supercritical fluid chromatography (SFC). When both temperature and pressure of a system exceed critical values, that is, values of critical temperature (Tc) and critical pressure (Pc), the fluid is known to be critical in nature. These fluids have the mixed properties of both liquid and gas. Supercritical fluids are highly compressible as gases, and density and viscosity can be maintained by changing the pressure and temperature conditions, as in the case of liquids. In chromatographic systems, the solute diffusion coefficients are often an order of higher magnitude in supercritical fluids than in traditional liquids. However, the viscosities are lower than those of liquids.[75] At temperatures below Tc and pressures above Pc, the fluid becomes a liquid. On the other hand, at temperatures above Tc and pressures below Pc, the fluid behaves as a gas. Therefore, supercritical fluids can be used as part of mixtures of liquid and gas. The most commonly used supercritical fluids (SFs) are carbon dioxide, nitrous oxide, and trifluoromethane.[75] Its compatibility with most detectors, low critical temperature and pressure, low toxicity and environmental burden, and low cost, make carbon dioxide the supercritical fluid of choice. The main drawback of supercritical carbon dioxide as a mobile phase is its inability to elute more polar compounds. This can be improved by adding organic modifiers to the relatively apolar carbon dioxide. SFC is carried out either in packed or open tubular columns and capillaries. Easier and faster method development, high efficiency, superior and rapid separations of a wide variety of analytes, extended temperature capability, analytical and preparative scale equipment improvements, and a selection of detection options are the main features of this technique.[76] Organometallic species that cannot be vaporized for GC analysis, yet have no functional groups for sensitive detection with the usual HPLC detectors, can be speciated and detected using SFC. Compared with GC, the strongly enhanced solvation strength of, for example, supercritical carbon dioxide, allows the oven temperature to be reduced significantly. There is no report on metallic speciation using SFC; however, due to its salient features, this technique may be useful for metal ion speciation in the future.

8.7 MECHANISMS

In ion-pair HPLC, a secondary chemical equilibrium of ion-pair formation is responsible for controlling retention and selectivity. The mechanisms involve push–pull effects, with mobile phases containing strong driving cation and a small amount of complexing agent, and pure pull effect with a mobile phase containing very weak driving cation and higher concentrations of complexing agent. In micellar chromatography, metallic species with different polarities partitioned among aqueous, stationary, micellar hydrophobic, and micellar hydrophilic phases, and as a result, were eluted at different retention times. Suyani et al.[35] suggested that the electrostatic interactions are responsible for the speciation of alkyltin compounds using sodium dodecyl sulfate as a surfactant molecule. The working principle of SEC depends on the sizes of metallic species. The column packing material acts as a sieve to separate ionic species based on their sizes. Typically, the smaller species are held back, and the larger species pass through. The speciation of metallic species occurs through permeation and filtration processes in GPC and GFC, respectively.

8.8 LC VERSUS GC

GC has been used widely for metal ion speciation in environmental and biological samples. The use of many element-specific detectors in GC makes it a good technique for this purpose. In spite of this, GC suffers from some drawbacks, as it cannot be used for the speciation of inorganic metallic species or requires the derivatization of inorganic metallic ions before analysis, which is a tedious, time-consuming, and costly process. In addition, some organometallic species decompose at the high working temperature of a GC machine, leading to poor sensitivity. On the contrary, liquid chromatography does not suffer from such types of limitations. The different modalities of liquid chromatography are useful for metal ion speciation under different situations. A wide range of mobile and stationary phases makes these forms of liquid chromatography the ideal choice. In addition, the successful combination of liquid chromatography with element-specific detectors is an asset for metal ion speciation. The use of capillary columns in CEC is another advancement in liquid chromatography. The greater efficiency of CEC is due to higher theoretical plate numbers. Therefore, LC is growing rapidly in popularity in the field of metal ion speciation in environmental and biological samples and is slowly replacing GC.

8.9 CONCLUSION

Metal ion speciation is achieved satisfactorily by using GC, RP-HPLC, and IC modalities, but certain metal ion species in some matrices could not be speciated by these techniques, which compelled scientists to modify these forms of chromatography. The newly developed liquid chromatographic modalities are IPC, MEKC, SEC, CC, CEC, and SFC. These liquid chromatographic approaches are useful and successful for metal ion speciation in special cases. For example, chiral organometallic species cannot be speciated by any modality except chiral chromatography. These chromatographic techniques have many of the advantages of RP-HPLC and IC and are useful for metal ion speciation with more sensitivity, selectivity, and reproducibility, particularly in unknown matrices. These techniques are in their development stages and have not yet been used in large-scale application. But in the future, it is expected that they may be assets to the area of metal ion speciation.

REFERENCES

1. Ali I, Aboul-Enein HY, *Chiral Pollutants: Distribution, Toxicity and Analysis by Chromatography and Capillary Electrophoresis*, John Wiley & Sons, Chichester (2004).
2. Wu CS (Ed.), *Handbook of Size Exclusion Chromatography and Related Techniques*, 2nd ed., Vol. 91, Marcel Dekker, New York (2003).
3. Krull IS, Stevenson RL, Mistry K, Swartz ME, *Capillary Electro-Chromatography and Pressurized Flow Capillary Electrochromatography: An Introduction*, HBN Publishing, New York (2000).
4. Cornelis R, Caruso J, Crew H, Heumann K (Eds.), *Handbook of Elemental Speciation: Techniques and Methodology*, John Wiley & Sons, Chichester (2003).

5. Poole CF, Poole SK, *Chromatography Today.* Elsevier Science, Amsterdam (1995).
6. Boucher P, Accominotti M, Vallon JJ, *J. Chromatogr. Sci.*, 34: 226 (1996).
7. Le XC, Ma M, *J. Chromatogr. A*, 764: 55 (1997).
8. Hwang CJ, Jiang SJ, *Anal. Chim. Acta*, 289: 205 (1994).
9. Martin I, Lopez-Gonzalvez MA, Gomez M, Camara C, Palacios MA, *J. Chromatogr. B, Biomed. Appl.*, 666: 101 (1995).
10. Do B, Robinet S, Pradeau D, Guyon F, *J. Chromatogr. A*, 918: 87 (2001).
11. Kohlmeyer U, Kuballa J, Jantzen E, *Rapid Commun. Mass Spectrom.*, 16: 965 (2002).
12. Ho YS, Uden PC, *J. Chromatogr. A*, 688: 107 (1994).
13. Al-Rashdan A, Heitkemper DT, Caruso JA, *J. Chromatogr. Sci.*, 29: 98 (1991).
14. Al-Rashdan A, Vela NP, Caruso JA, Heitkemper DT, *J. Anal. At. Spectrom.*, 7: 551 (1992).
15. Houck RS, Shum SCK, Wiederin DR, *Anal. Chim. Acta*, 250: 61 (1991).
16. Yiang KL, Jiang SJ, *Anal. Chim. Acta*, 307: 109 (1995).
17. Jen JF, Yang YJ, Cheng CH, *J. Chromatogr. A*, 791: 357 (1997).
18. B'Hymer C, Caruso JA, Winter conference on plasma spectrochemistry, January 5, paper M6, Scottland (1988).
19. B'Hymer C, Sutton KL, Caruso JA, *J. Anal. At. Spectrom.*, 13: 855 (1998).
20. B'Hymer C, Caruso JA, *J. Anal. At. Spectrom.*, 15: 1531 (2000).
21. Zheng J, Ohata M, Furuta N, Kosmus W, *J. Chromatogr. A*, 874: 55 (2000).
22. Kotrebai M, Tyson JF, Block E, Uden PC, *J. Chromatogr. A*, 866: 51 (2000).
23. Gayon JMM, Thomas C, Feldmann I, Jakuboswski N, *J. Anal. At. Spectrom.*, 16: 457 (2001).
24. Kannamkumarath SS, Wrobel K, Wrobel K, Vonderheide A, Caruso JA, *Anal. Bioanal. Chem.*, 373: 454 (2002).
25. B'Hymer C, Brisbin JA, Sutton KL, Caruso JA, *Am. Lab.*, Feb. issue, 17 (2000).
26. Zhang X, Cornelis R, DeKimpe J, Mees L, Vanderbiesen V, DeCubber A, Vanholder R, *Clin. Chem.*, 42: 1231 (1996).
27. Zhang X, Cornelis R, DeKimpe J, Mees L, *Anal. Chim. Acta*, 319: 177 (1996).
28. Beauchemin D, Siu KWM, McLaren JW, Berman SS, *J. Anal. At. Spectrom.*, 4: 285 (1989).
29. Yang HJ, Jiang SJ, Yang YJ, Hwang CJ, *Anal. Chim. Acta*, 312: 141 (1995).
30. Shum SCK, Pang H, Houk RS, *Anal. Chem.*, 64: 2444 (1992).
31. Terabe S, Otsuka K, Ichikawa A, Ando T, *Anal. Chem.*, 56: 111 (1984).
32. Terabe S, *J. Pharm. Biomed.* Anal., 10: 705 (1992).
33. Ding H, Wang J, Dorsey LG, Caruso J, *J. Chromatogr. A*, 694: 425 (1995).
34. Hong D, Jiansheng W, Dorsey JG, Caruso JA, *J. Chromatogr. A*, 694: 425 (2002).
35. Suyani H, Heitkemper D, Creed J, Caruso J, *Appl. Spectros.*, 43: 962 (1989).
36. LaFuente JMG, Dlaska M, Sanchez MLF, Sanz-Medel A, *J. Anal. At. Spectrom.*, 13: 423 (1998).
37. Infante HG, Fernandez-Sanchez ML, Sanz-Medel A, *J. Anal. At. Spectrom.*, 15: 519 (2000).
38. Flaten AK, Lund W, *Sci. Total Environ.*, 14: 21 (1997).
39. Hils A, Grote M, Janeen E, Eichhorn J, *Fresenius' J. Anal. Chem.*, 364: 457 (1999).
40. Mitrovic B, Milacic R, *Sci. Total Environ.*, 258: 183 (2000).
41. McSheehy S, Szpunar J, Lobinski R, Haldys V, Tortajada J, Edmonds JS, *Anal. Chem.*, 74: 2370 (2002).

42. Harms A, van Elteren JT, Claessens HA, *J. Chromatogr. A*, 755: 219 (1996).
43. Jackson BP, Shaw Allen PL, Hopkins WA, Bertsch PM, *Anal. Bioanal. Chem.*, 374: 203 (2002).
44. LindermannT, Hintelman H, *Anal. Chem.*, 74: 4602 (2002).
45. Zheng J, Kosmus W, *J. Liq. Chromatogr.*, 21: 491 (1998).
46. Lintschinger J, Schramel P, Hatalak-Rauscher A, Wendler I, Michalke B, *Fresenius' J. Anal. Chem.*, 362: 313 (1998).
47. Broeck KVD, Vandecasteele C, Geuns JMC, *Anal. Chim. Acta*, 361: 101 (1998).
48. Caruso JA, B'Hymer C, Heitkemper DT, *Analyst*, 126: 136 (2001).
49. Crews HM, Dean JR, Ebdon L, Massey RC, *Analyst*, 114: 895 (1989).
50. Jackson BP, Miller WP, *J. Anal. At. Spectrom.*, 13: 1107 (1998).
51. Michalke B, Witte H, Schramel P, *J. Anal. At. Spectrom.*, 16: 593 (2001).
52. Gammelgaard B, Jons O, Bendahl L, *J. Anal. At. Spectrom.*, 16: 339 (2001).
53. Kozuh N, Milacic R, Gorenc B, *Ann. Chim.*, 86: 99 (1996).
54. Milacic R, Stupar J, Kozuh N, Korosin J, Glazer I, *J. Am. Leather Chem. Assoc.*, 87: 221 (1992).
55. Kozuh T, Stupar J, Milacic R, Gorenc B, *Int. J. Environ. Anal. Chem.*, 56: 207 (1994).
56. Aboul-Enein HY, Ali I, *Chiral Separations by Liquid Chromatography and Related Technologies*, Marcel Dekker, New York (2003).
57. Mendez SP, Gonzalez EB, Fernandez-Sanchez ML, Sanz Medel A, *J. Anal. At. Spectrom.*, 13: 893 (1998).
58. Suton KL, Ponce de Leon CA, Ackley KL, Suton RM, Stalcup AM, Caruso J, *Analyst*, 125: 281 (2000).
59. Ponce de Leon CA, Suton KL, Caruso JA, Uden P, *J. Anal. At. Spectrom.*, 15: 1103 (2000).
60. Mendez SP, Gonzalez EB, Sanz Medel A, *J. Anal. At. Spectrom.*, 15: 1109 (2000).
61. Baeyens WRG, Van der Weken G, Aboul-Enein HY, Reygaerts S, Smet E, *Biomed. Chromatogr.*, 14: 58 (2000).
62. Pirkle WH, Murray PG, *J. High Resolut. Chromatogr.*, 16: 285 (1993).
63. Pfeiffer CC, *Science*, 124: 29 (1956).
64. Pirkle WH, Pochapsky TC, *Chem. Rev.*, 89: 347 (1989).
65. Groombridge JJ, Jones CG, Bruford MW, Nichols RA, *Nature*, 403: 615 (2000).
66. Rogozhin SV, Davankov VA, German Patent 1932190 (1969).
67. Rogozhin SV, Davankov VA, *Chem. Lett.*, 490 (1971).
68. Davankov VA, *J. Chromatogr. A*, 666: 55 (1994).
69. Mayer S, Schurig V, *J. High Resolut. Chromatogr.*, 15: 129 (1992).
70. Li S, Lloyd D, *Anal. Chem.*, 65: 3684 (1993).
71. Mayer S, Schurig V, *J. Liq. Chromatogr.*, 16: 915 (1993).
72. Schurig V, Jung M, Mayer S, Fluck M, Negura S, Jakubetz H, *J. Chromatogr. A*, 694: 119 (1994).
73. Schurig V, Jung M, Mayer S, Negura S, Fluck M, Jakubetz H, *Angew Chem.*, 106: 2265 (1994).
74. Vindevogl J, Sandra P, *Electrophoresis*, 15: 842 (1994).
75. Weast RC (Ed.), *Handbook of Chemistry and Physics*, 54th ed., CRC Press, Boca Raton, FL (1973).
76. Terfloth G, *J. Chromatogr. A*, 906: 301 (2001).

9 Speciation of Metal Ions by Capillary Electrophoresis

Various chromatographic techniques, such as reversed-phase high-performance liquid chromatography (RP-HPLC), ion chromatography (IC), ion-pair chromatography (IPC), chelation chromatography (CC), and gas chromatography (GC), have been used for the speciation of metal ions, but there is still a great need for speciation techniques that are fast and inexpensive. This demand is fulfilled, to some extent, by a newly developed technique called capillary electrophoresis (CE). Since the early 1990s, interest in using CE for speciation analysis increased rapidly.[1–5] Ultraviolet (UV) detection is mainly used for the speciation of metal ions, but CE has been combined with element-specific detectors, such as atomic absorption spectrometry (AAS), inductively coupled plasma spectrometry (ICP), mass spectrometry (MS), and so forth.[6–11] Therefore, in the last decade, a newly developed technique called capillary electrophoresis (CE) has been used as one of many tools for the speciation of metal ions. More recently, reviews on CE selectivity, sensitivity, and application for metal ion speciation were published.[12–16] CE has certain advantages in comparison to other chromatographic techniques. The advantages of CE include simplicity, low cost of running, high speed of analysis, unique selectivity, and high degree of matrix independence (due to high theoretical plate numbers, i.e., normally ranging from 200,000 to 70,000), all of which make it a suitable technique for the speciation of metal ions. Due to the importance of metal ion speciation and the ease of CE operation, scientists are considering CE to be one of the foremost methods for metal ion speciation. The mechanisms of the speciation of metal ions in CE are based on the differences in electrophoretic mobility of the metal ions. Under CE conditions, the migration of the metal ions is controlled by the sum of intrinsic electrophoretic mobility (μ_{ep}) and electroosmotic mobility (μ_{eo}), due to the action of electroosmotic flow (EOF). Details of the basic principle of the separation process by CE is discussed in Chapter 1. In this chapter, the art of metal ion speciation by CE is discussed, including the applications, optimization, detection, metal ions complexation, separation efficiency, and mechanisms of the speciation.

9.1 APPLICATION OF CAPILLARY ELECTROPHORESIS FOR THE SPECIATION OF METAL IONS

Opposite to the large molecules, small cations have higher charge/density (q_i/r_i) ratios and larger ionic mobilities. As a principally striking electrophoretic property, it should lead to highly efficient rapid separation. But in practice, this seldom occurs unless special precautions are taken. The mobilities of metal ion species are almost equal due to their similar sizes, but these species in different oxidation states have different sizes; hence, this may be considered the advantage of speciation in CE. Moreover, the different charges on these species may again be an advantage of their speciation in CE, as the migration of cations in CE is controlled by the charges on the cations. The speciation of metal ion species can be carried out by several modes of CE. The various modes of CE include capillary zone electrophoresis (CZE), micellar electrokinetic capillary electrophoresis (MECE) or micellar electrokinetic capillary chromatography (MECC), capillary isotachphoresis (CI), capillary gel electrophoresis (CGE), ion-exchange electrokinetic chromatography (IEEC), capillary isoelectric focusing (CIEF), affinity capillary electrophoresis (ACE), capillary electrochromatography (CEC), separation on microchips (MC), and nonaqueous capillary electrophoresis (NACE). However, most metal ion speciation has been carried out by CZE mode.

9.1.1 Arsenic (As)

Due to the acute toxicity of arsenic metal ion, the speciation of arsenic was carried out extensively.[17–31] The speciation of arsenic species has been studied using EOF as the modifier to separate inorganic and organometallic species simultaneously.[24,32–35] Lin et al.[24] analyzed arsenate, arsenite, dimethylarsinate (DMA), and monomethylarsonate (MMA) in coal fly ash using sodium chromate as the background electrolyte (BGE). The authors also discussed the advantages of CE as an efficient and sensitive separation method. In the same year, Olesik et al.[21] coupled CE with inductively coupled plasma (ICP) spectrometry for the speciation of As(III) and As(V) species using 0.06 *M* calcium chloride as the background electrolyte. Michalke and Schramel[36] reported the speciation of DMA and MMA species. The authors used phosphate buffer as the background electrolyte.

9.1.2 Chromium (Cr)

Timerbaev et al.[7,37] developed a CE methodology for the speciation of Cr(III) and Cr(VI), adding 1-2-cyclohexanediaminetetraacetic acid (CDTA) in the sample solution and using phosphate buffer as the background electrolyte. Jung et al.[38] used cetyltrimethylammonium bromide (CTBA) as the liganding agent for chromium speciation and obtained sharp peaks with short analysis time. Padarauskas and Schwedt[39] described chromium speciation using diethylenediaminepentaacetic acid (DTPA) as the chelating agent. Olesik et al.[21] used 0.06 *M* calcium chloride as the background electrolyte for the speciation of Cr(III) and Cr(VI) species by CE. Baraj et al.[40] presented a comparative method for the speciation of chromium metal ion. Recently, Chen et al.[41] reported a simple method for the speciation of chromium species where Cr(III) was chelated with ligands to form anionic complexes. Nitrilotriacetic acid, *N*-2-hydroxyethylenediaminetriacetic acid, ethylenediaminetetraacetic acid, diethylenetriaminepentaacetic acid, and 2,6-pyridinedicarboxylic acid (PDCA) were investigated as Cr(III) complexing ligands. Of all the ligands studied, 2,6-PDCA with Cr(III) gave good UV response and high selectivity for Cr(III). In addition, the conditions for precolumn derivatization, including pH, concentration ratio [Cr(III)/2,6-PDCA], and the stability of Cr(III) complexes, were also examined. The separation of anionic forms of Cr(III) and Cr(VI) was achieved. The electrolyte contained 30 m*M* phosphate, 0.5 m*M* tetradecyltrimethylammonium bromide, 0.1 m*M* 2,6-PDCA, and 15% (v/v) acetonitrile at pH 6.4. Conradi et al.[8] reported the speciation of chromium using CE. Other authors also reported the analysis of Cr(III) and Cr(VI) species.[38,39] Recently, King et al.[42] speciated chromium metal ion [Cr(III) and Cr(VI) species] in a water sample. These species were analyzed as chromate and chromium pyridinedicarboxylate utilizing simple precapillary complexation with 2,6-pyrinedicarboxylic acid (PDCA). Two species were separated using 30 m*M* phosphate buffer (pH 6.4) containing 10 m*M* *N*-carboxymethylated polyethyleneimine to suppress the EOF. The detection was achieved using an UV photodiode array detector.

9.1.3 Iron (Fe)

Olesik et al.[21] reported the speciation of iron using CE coupled with plasma spectrometry. The authors used sodium acetate as the background electrolyte. Schäffer et al.[43] investigated speciation methodology of the analysis of iron(II) and iron(III) after selective complexation of these species by *o*-phenanthroline and ethylenediaminetetraacetic acid (EDTA) added in excess to the sample solution. The cationic and anionic natures of the ferrous and ferric complexes, respectively, allow for their easy electrophoretic analysis in a single run in the presence of an electroosmotic flow. It was ascertained that the addition of the mixed complexing solution does not alter the iron(II)/iron(III) ratio of the sample, and no significant evolution of this ratio was observed over a period of 10 days. When only EDTA is added as the complexing agent, CE allows for the rapid direct determination of total iron content, which provides a means for checking total iron recovery. Both methods were successfully applied to monitor a wet chemistry process converting iron(II) into iron(III). Kuban et al.[44] also described the speciation of iron metal ion using a direct mode of UV detection. Magnuson et al.[29] and Timerbaev et al.[45] also reported the speciation of iron metal ion using CE.

9.1.4 Mercury (Hg)

Medina et al.[46] speciated inorganic mercury and organomercury species. Furthermore, an improvement in mercury speciation was carried out by Carro-Diaz et al.[47] using sample stacking as the injection technique. The method was tested using different reference materials with certified methylmercury content. Dithiozone sulfonate (DzS) complexes of organic and inorganic mercury were prepared and speciated by Jones and Hardy.[48] Liu and Hian[9,49] described the simultaneous method of speciation of mercury using nitriloacetic acid (NTA) and triethylenetetraaminehexaacetic acid (TTHA) as the ligands. Furthermore, the selectivity of speciation was increased using SDS. Silva da Rocha et al.[50] developed a method for the speciation of inorganic and organic mercury species. The background electrolyte used was sodium borate. Diaz et al.[51] used cysteine as the complexing agent for the speciation of mercury using CE. According to the authors, in a weakly basic buffer, 1:2 organomercurial–cysteine complexes were formed, and the separation was accomplished, but baseline separation of methyl- and ethylmercury could not be achieved.

9.1.5 Platinum (Pt)

Michalke and Schramel[32] developed an online combination of CE with inductively coupled plasma–mass spectrometry (ICP-MS) with a homemade nebulizer as the interface. The high resolution power of CE was used for the separation of metal species including platinum. After optimization, electropherograms with high resolution were obtained with typical analysis times below 22 min. Quality control aspects concerning species stability during the analytical procedure and stability of electrical current during nebulization were studied. A possible interfering suction flow was estimated and found to be negligible. The same group[52] used a powerful online coupling of CE and ICP-MS for platinum speciation. The authors reported that aqueous extracts of a clay-like humic soil treated with platinum-contaminated tunnel dust and platinum model compounds were examined to elucidate transformation processes of platinum species in soil. Furthermore, this group[36,53] reported an improvement in the coupling of CE with MS for the speciation of platinum species. Hamáek and Havel speciated thiocyanate complexes of Pt(II) and Pt(IV) species by CE.[54] When the method optimized for the determination of Pt(II) was applied to the drugs Cykloplatin, Ribocarbo (containing carboplatin), and Platidiam (containing cisplatin), agreement between the platinum content and the declared value was obtained. Samples of vehicle exhaust particulates (National Institute for Environmental Studies, Japan) were also analyzed.

9.1.6 Selenium (Se)

Liu and Hian[9,49] described the simultaneous speciation for selenium metal ion using NTA and TTHA as the ligands. Liu et al.[25] coupled CE with ICP-MS for the speciation of selenium species. The interface was built using a direct-injection nebulizer (DIN) system. In this interface, the CE capillary was placed concentrically inside the fused silica DIN sample introduction capillary. A makeup liquid was pumped at a flow rate of 15 μL/min into the fused-silica DIN sample transfer capillary through a connector. The makeup liquid flowing outside the CE capillary served continuous and stable electrical contact at the exit terminus of the CE capillary, allowing the DIN system to be optimized independently of the CE system for efficient sample introduction to the ICP-MS system. The makeup liquid along with the liquid from the electroosmotic flow generated inside the CE capillary was nebulized by the DIN system directly in the ICP torch. The developed CE-DIN-ICP-MS system was found to be suitable for the speciation of selenium metal ion. Furthermore, Schramel et al.[55] combined CE with MS via an electrospray ionization (ESI) interface. The commercially available interface was hardly able to produce stable electrospray conditions over an extended period of time, mainly caused by an insufficient positioning of the CE capillary inside the ESI stainless steel tip. A device was developed that allowed for an infinitely variable adjustment of the capillary to be made. The optimum position for stable electrospray conditions was set to 0.4 to 0.7 mm outside the ESI tip. Off-line ESI-MS investigations of free metal ions selenocystamine and selenomethionine were carried out in order to assess the suitability of the technique for metal speciation. The ionization process did not alter their structure and would mostly be detected as singly charged molecular ions. The application of CE-ESI-MS for selenium speciation using an alkaline buffer system (Na_2CO_3-NaOH) gave unsatisfactory results. However, the nonvolatile electrolyte affects the ESI process dramatically. Finally, the CE method was used, with an acidic background electrolyte (2% acetic acid), for the separation of three organometallic selenium species — selenomethionine (SeM), selenocystamine (SeCM), and selenocystine (SeC). Selenium species were sufficiently separated and appeared at 6.49 min (SeCM), 19.47 min (SeM), and 20.60 min (SeC) migration times, respectively. Furthermore, the same group[56] reported the coupling of CE and ESI-MS for the speciation of selenium metal ion. The authors presented online combinations of the separation and detection units for the identification and characterization of metal species. The high resolution power of CE was used for the separation of three selenium species, whereas either ESI-MS or inductively coupled plasma mass spectrometry (ICP-MS) were taken for molecule- or element-specific detection. The authors claimed that this work gave an overview of the possibilities and limitations of using two combined systems for the speciation investigations. In order to show the power of two complementary techniques, a CE method using 5% acetic acid as the background electrolyte was applied to the separation of selenomethionine (SeM), selenocystine (SeC), and selenocystamine (SeCM) species. Michalke and Schramel[32] described an online CE and ICP-MS for the speciation of selenium species. The authors used phosphate buffer as the background electrolyte. Furthermore, the same author[36] reported the speciation of selenium species by coupling CE with ICP-MS. Recently, Gilon and Potin-Gautier[57] reported the speciation of seleno compounds. The background electrolyte used was chromate with trimethyltetradecylammonium hydroxide as the electroosmotic flow modifier. Recently, Dzierzgowska et al.[58] reported the speciation of selenium metal ion using pyromellitic electrolyte (pH 8.8) and hexamethonium hydroxide as the electroosmotic flow modifier. The detection was achieved by UV with 0.17 and 0.29 μg/mL as the detection limits for Se(VI) and Se(IV) species, separately, and respectively.

9.1.7 Miscellaneous Metal Ions

Hamáek and Havel[54] reported the speciation of palladium metal ion species in the form of their thiocyanate complexes with higher sensitivity than in the form of their chloro complexes. The possibility of simultaneous determination of Pd(II) in the form of thiocyanate complexes was also demonstrated. Aleksenko et al.[59] speciated rhodium(III) species occurring in different acidic environments. The separation was optimum under acidic electrolyte conditions in which the complexed rhodium species were at their most stable state, and the electroosmotic flow approached zero, thereby aiding resolution. The distribution of metal complexes was highly dependent on the nature and concentration of the acid and the age of the rhodium stock solutions. On dilution, Rh(III) tends to be readily hydrolyzed, giving rise to a wider variety (and a varied distribution) of complexed forms. In 0.1 *M* HCl, four differently charged chloro complexes — $RhCl_4(H_2O)_2^-$, $RhCl_3(OH)(H_2O)_2^-$, $RhCl_3(H_2O)_3$, and $RhCl_2(H_2O)_4^+$ — were separated and identified. When a stock solution in 11 *M* HCl was run, Rh produced a major peak ascribed to $RhCl_6^{3-}$ and two slowly migrating peaks from ions in which one or two of the chloride ligands were probably replaced by water and hydroxyl ion, as a result of hydrolysis. Aquatic cationic species were found to be predominant in $HClO_4$ and HNO_3 solutions, whereas only negatively charged forms of Rh(III) occurred in sulfuric acid. This speciation information opens new possibilities for assessing the catalytic activity of Rh in kinetic reactions. Sanchez et al.[60] reported a fast and effective CE method for the speciation of rhodium species. At least five species, some of which seem to be oligomeric, were formed in solution during the aquation process at pH > 1. Fast hydration of $RhCl_6^{3-}$ makes this species impossible to detect. The first species detected in optimized conditions was $RhCl_5(H_2O)_2$, although $RhCl_4(H_2O)_2$ was the main species during the first stage of the aquation process. When equilibrium was reached, either $RhCl_3(H_2O)_3$ or a cationic complex, $RhCl_2(H_2O)_4^+$, was formed as the main species. Matrix-assisted laser desorption ionization time-of-flight mass spectrometry (MALDI-TOF-MS) was used as a novel technique to elucidate the structure of the rhodium aqua/chloro complexes formed in solution. Results obtained by CE are confirmed by spectrophotometry. Recently, Kitazumi et al.[61] developed a CE method for the sensitive determination of V(V) and V(IV) by using Mo(VI)-P(V) reagent, which reacted with a mixture of trace amounts of V(V) and V(IV) to form the $[P(V^{V}Mo_{11})O_{40}]^4$ and $[P(V^{IV}Mo_{11})O_{40}]^5$ complexes in 0.1 *M* monochloroacetate buffer (pH 2.2). Because V-substituted Keggin anions possessed high molar absorptivities in the UV region, and the peaks due to their migrations were well separated in the electropherogram, the precolumn complex formation reaction was applied to the simultaneous CE determination of V(V) and V(IV) with direct UV detection at 220 nm. Timerbaev et al.[62] also reported the speciation of vanadium metal ion using CE. CE was also used for the speciation of Sn,[21,63]

Sr,[21] Pd,[54] Cu,[28] and Al[48] metal ions. Recently, Yeh and Jiang[64] speciated iron(III/II), vanadium(V/IV), and chromium(III/VI) metal ion species using capillary electrophoresis–dynamic reaction coupled with ICP-MS (CE-DRC-ICP-MS). The authors used precomplexation with EDTA using counterelectroosmotic and coelectroosmotic modes. The detection limits were in the range of 0.1 to 0.5, 0.4 to 1.3, and 1.2 to 1.7 ng/mL for vanadium, chromium, and iron, respectively. The developed method was used to speciate these metal ions into wastewater.

Obviously, CE is a very useful technique for the speciation of charge species, however, its application has been increased for the speciation of neutral metallic species by the modification of the background electrolyte (BGE). Certain surfactants were used as the BGE which form the micelle, and the technique is called micellar electrokinetic capillary electrophoresis (MECE), as the separation mechanisms slightly shifted from CE to the chromatographic process, and it was introduced in 1984 by Terabe et al.[65] The most commonly used surfactants are sodium dodecyl sulfate (SDS), cetyltrimethylammonium bromide (CTAB), and tetradecyltrimethylammonium bromide.[18,28,34] Ng et al.[34] reported the separation of organolead and organoselenium compounds using this mode of CE. The optimization of the separation of trimethyl- and triethyllead and phenylselenium and diphenylselenium species was achieved by changing the concentration of SDS. Liu and Lee[9] reported an enhancement in the detection by increasing the concentration of SDS for the speciation of lead, mercury, and selenium metal ions. Similarly, Li et al.[66] used SDS above its critical micellar concentration and reported improved detection for organotin and lead compounds. The separation of metallic species was also improved by using a complexation of these species in MECE.[14] The speciation of metal ions by CE is summarized in Table 9.1. Typical electropherograms of the speciation of arsenic species are shown in Figure 9.1.[23] The electrophoretic speciation is characterized by the migration times (t), electrophoretic mobility (μ_{ep}), separation (α), and resolution factors (Rs).

9.2 OPTIMIZATION OF THE SPECIATION

To achieve maximum speciation of metal ions by CE, the optimization of CE conditions is a critical issue for the analytical chemist. Several parameters are required to control speciation. The optimization factors may be categorized into two classes. The independent parameters are under the direct control of the operator. These parameters include the choice of buffer, pH of the buffer, ionic strength of the buffer, voltage applied, temperature of the capillary, dimension of the capillary, BGE additives, and some other parameters. Dependent parameters are directly affected by the independent parameters and are not under the direct control of the operator. These types of parameters are field strength (V/m), EOF, Joule heating, BGE viscosity, sample diffusion, sample mobility, sample charge, sample size and shape, sample interaction with capillary and BGE, and molar absorptivity. Therefore, the optimization of metal ion speciation can be controlled by varying all these parameters, as discussed below.

TABLE 9.1
Speciation of Metal Ions by Capillary Electrophoresis and Micellar Electrokinetic Capillary Electrophoresis

Metal Ions	Sample Matrix	Electrolytes	Detection	Detection Limit	Ref.
		Capillary Electrophoresis (CE)			
Arsenic (As)					
Arsenic species	Drinking water	0.025 m*M*, phosphate buffer, pH 6.8	Direct UV, 190 nm	<2 mg/L	17
	Water	50 m*M* CHES, 20 m*M* LiOH	Conductivity	0.4 mg/L	18
	Tin mining process water	15 m*M* phosphate buffer, 1 m*M* CTAB 50 m*M* CHES, 0.03% triton X-100 20 m*M* LiOH, pH 9.4	Conductivity	0.4 mg/L	18
As(III) and As(V)	—	Chromate-NICE-Pak OFM anion-BT, pH 10	Indirect UV, 254 nm	—	19
	—	5 m*M* chromate, 0.25 m*M* cetyltrimethyl ammonium bromide, pH 10	Indirect UV, 254 nm	10^4 *M*	20
	—	5 m*M* $K_2Cr_2O_4$, 0.25 m*M* CTAB, pH 10	Indirect UV, 254 nm	10^4 *M*	20
As(III), As(V), and dimethylarsenic acid	—	60 m*M* calcium chloride (pH 6.7), cetyltrimethyl ammonium bromide, pH 10	ICP-MS	1 ppb	21
As(III), As(V), monomethyl-, and dimethylarsenic acid	—	75 m*M* Na_2HPO_4–25 m*M* $Na_2B_4O_7$, pH 7.8	Direct UV, 195 nm	0.8 ppm	22
As(III), As(V), and dimethylarsenic acid	—	75 m*M* Na_2HPO_4–25 m*M* $Na_2B_4O_7$, pH 7.8	Direct UV, 190 nm	0.8–3.7 ppm	23
As(III), As(V), monomethyl-, and dimethylarsenic acid	—	6 m*M* chromate-NICE-Pak OFM anion-BT, pH 8	Indirect UV, 274 nm	0.4 ppm	24
As(III), As(V), monomethyl-, and dimethylarsenic acid	—	2.3 m*M* pyromellitic acid-6.5 NaOH–1.6 m*M* triethanolimine-75 m*M* hexamemethonium bromide, pH 7.7	ICP-MS	20–100 ppt	25

TABLE 9.1 (Continued)
Speciation of Metal Ions by Capillary Electrophoresis and Micellar Electrokinetic Capillary Electrophoresis

Metal Ions	Sample Matrix	Electrolytes	Detection	Detection Limit	Ref.
As(III), As(V), monomethyl-arsonic acid, dimethylarsenic acid and monophenyl-arsonic acid	—	7 m*M* Na_2HPO_4-NaH_2PO_4, pH 5.8	Direct UV, 190 nm	—	27
As(III), As(V), monomethyl-arsonic acid, dimethylarsenic acid, and monophenyl-arsonic acid	—	25 m*M* $Na_2HPO_4NaH_2PO_4$, pH 6.8	Direct UV, 190 nm	—	18
As(III), As(V), monomethyl-arsonic acid, dimethylarsenic acid, and monophenyl-arsonic acid	—	15 m*M* $Na_2HPO_4NaH_2PO_4$, pH 5.8	Direct UV, 190 nm	—	28
As(III), As(V), hexafluoroarsenate, dimethylarsinic acid, *p*-aminobenzene-arsonic acid, and monophenyl-arsonic acid	—	15 m*M* Na_2HPO_4, pH 6.5	Direct UV, 200 nm	60–90 ppb	18
As(III), As(V), hexafluoroarsenate, dimethylarsinic acid, *p*-aminobenzene-arsonic acid, and monophenyl-arsonic acid	—	50 m*M* 2-(Cyclohexylamino)-ethanesulfonic acid-0.03% triton X-100–20 m*M* LiOH, pH 9.4	Conductivity, 200 nm	40–76 ppb	18
As(III), As(V), monomethylarsonic acid, dimethylarsinic acid, arsenobetaine, and arsenocholine	Urine and sewage sludge	—	ICP-MS	15 µg/L	31
Arsenic species	Drinking water	20 m*M* KHP, 20 m*M* boric acid, pH 9.03	Hydride generation, ICP-MS	6 ng/L	29, 30

TABLE 9.1 (Continued)
Speciation of Metal Ions by Capillary Electrophoresis and Micellar Electrokinetic Capillary Electrophoresis

Metal Ions	Sample Matrix	Electrolytes	Detection	Detection Limit	Ref.
	Tap and drinking waters	75 m*M* dihydrogen phosphate–25 m*M*, tetraborate, pH 7.65	Direct UV, 195 nm	12 μg/L	22
Arsenic species	—	—	Proton-induced x-ray emission		67
As(III), As(V), MMA, DMA, AsBet, AsCh, and pAS	—	Borate buffer, pH 9.3	Direct UV, 192 nm		68
Chromium (Cr)					
Cr(IV) and Cr(VI)	Rinse water from chromium platings	1 m*M* CDTA, 10 m*M* formate buffer, pH 3.8	Direct UV, 214 and 254 nm	10 μg/L	7, 37
	Chromium plating water	10 m*M* formate buffer, 1 m*M* CDTA, pH 3	Direct UV, 214 nm	10 ppb	7, 37
	—	10 m*M* phosphate buffer, 0.5 m*M* HTMAB, 30 m*M* NaAc, 5 m*M* Na_2SO_4, pH 7	Indirect UV, 254 nm	15 ppm	48
	Electroplating water	10 m*M* formate buffer, pH 3	Indirect UV, 214 nm	50 ppb	7
	Wastewater	20 m*M* Na_2HPO_4, 0.05 m*M* TTAOH	Direct UV, 214 and 254 nm	—	39
Cr(IV) and Cr(VI)	—	EDTA with buffer	Direct UV	—	38
Iron (Fe)					
Fe(II) and Fe(III)	Electroplating waters	20 m*M* phosphate buffer, pH 7	Direct UV, 214 nm	10^5 *M*	3
Fe(II) and Fe(III)	—	NaAc, 0.12 m*M* 5-Br-PAPS, pH 4.9	Indirect UV, 550 nm	10^8 *M*	69
	—	100 m*M* borate buffer, pH 9.2	Indirect UV, 214 nm	10^6 *M*	62
	—	20 m*M* borate buffer, 1 m*M* CDTA, 0.5% EtOH, pH 9	Indirect UV, 214 nm	10^5 *M*	70
	—	100 m*M* borate buffer, pH 9	Indirect UV, 185–214 nm	ppb Level	12
	—	100 m*M* borate buffer, 1 m*M* CDTA, pH 9	Indirect UV, 254 nm	0.6 ppm	71

TABLE 9.1 (Continued)
Speciation of Metal Ions by Capillary Electrophoresis and Micellar Electrokinetic Capillary Electrophoresis

Metal Ions	Sample Matrix	Electrolytes	Detection	Detection Limit	Ref.
	Environmental samples	20 m*M* phosphate buffer, 2 m*M* NaCN, pH 9.4	Indirect UV, 214 nm	25 ppb	71
	—	7.5 m*M* salicylic acid, 0.5 m*M* EDTA, 0.2 m*M* HTMAB, pH 4	Indirect UV, 200 nm	10^4 *M*	72
	—	20 m*M* Na_2HPO_4, KH_2PO_4, 2 m*M* NaCN, pH 9.4	Indirect UV, 214 nm	10^7 *M*	8
	—	5 m*M* Na_2HPO_4, 5 m*M* triethanol-imine, 0.8 m*M* hexamethonium bromide	Indirect UV, 214 nm	10^7 *M*	73
Selenium (Se)					
Se(IV) and Se(V)	Thermal water	Chromate, 0.5 m*M* TTAOH, pH 10.5	Indirect UV, 254 nm	10 µg/L	57
	—	5 m*M* chromate–0.25 m*M* cetyltri-methylammonium bromide, pH 10	Indirect UV, 254 nm	10^4 *M*	20
	—	75 m*M* Na_2HPO_4–25 m*M* $Na_2B_4O_7$, pH 7.8	Direct UV, 195 nm	1.9–202 ppm	22
	—	2.3 m*M* pyromellitic acid–6.5 m*M* NaOH–1.6 m*M* triethanolamine–75 m*M* hexameth-onium bromide, pH 7.7	ICP-MS	100–300 ppt	25
Se(IV) and Se(V)	—	50 m*M* 2-(cyclo-hexylamino)-ethane-sulfonic acid–0.03% triton X-100–20 m*M* LiOH, pH 9.4	Conductivity	65–85 ppb	18
	—	200 m*M* HPO_4^{2}, H_2PO_4, pH 6	ICP-MS	10^7 *M*	32
Se(IV) and Se(V)	—	10 m*M* Na_2CO_3, pH 11.5, adjusted with KOH	ICP-MS	10^7 *M*	32

TABLE 9.1 (Continued)
Speciation of Metal Ions by Capillary Electrophoresis and Micellar Electrokinetic Capillary Electrophoresis

Metal Ions	Sample Matrix	Electrolytes	Detection	Detection Limit	Ref.
Se(IV), Se(VI), selenocystine, and selelenomethionine	—	80 m*M* Na_2HPO_4 2 m*M* tetradecyltri-methylammonium bromide	Direct UV, 200 nm	—	18, 74
Selenium species	Drinking water	20 m*M* KHP, 20 m*M* boric acid (pH 9.03) hydrodynamically modified EOF	Hydride generation ICP-MS	6 ng/L	29, 30
Miscellaneous Metal Ions					
Co(III) and Co(IV)	—	NaAc, 0.12 m*M*, 5-Br-PAPS, pH 4.9	Indirect UV, 550 nm	10^8 *M*	69
	—	20 m*M* Na_2HPO_4, 5 m*M* DTPA, pH 8 or 8.5	Indirect UV, 214 nm	10^6 *M*	39
Hg(II), CH_3Hg^+, and CH_3CH_2Hg+	—	25 m*M* $Na_2B_4O_7$.10 H_2O, pH 9.3	ICP-MS	81–275 ppb	50
MeHg,EtHg,PhHg, and Hg(II)	—	Dithiozone sulfonate	UV	1 µg/L	47
PhHg, Hg(I), and Hg(II)	—	Tap water	UV	2.48 µg/L	9
PhHg and Hg(II)	Seawater	—	UV	0.3 µg/L	49
Ir(II) and Ir(III)	—	4 m*M* H^+, 23 m*M* Cl, pH 2.4	Indirect UV, 214 nm	—	52
Lead species	—	—	Proton-induced x-ray emission	—	67
TML, TEL, DPhL, and Pb(II)	—	Tap water	UV	2.48 µg/L	9
TML, TEL, DPhL, and Pb(II)	Seawater	—	UV	0.3 µg/L	49
Pd(II)	—	0.1 *M* KSCN, pH 3	UV 305	ppb Level	54
Pt(II) and Pt(IV)	—	4 m*M* H^+, 23 m*M* Cl, pH 2.4	Indirect UV, 214 nm	—	52
	Soil	50 m*M* phosphate buffer (pH 6)	ICP-MS	1 µg/L	75
Pt(II) and Pt(IV)	—	0.1 *M* KSCN, pH 3	UV 305	ppb Level	54
Rhodium(III)	—	10 m*M* KCl-HCl	UV 200 nm	0.18 mg/mL	59
Rhodium(III)	—	40 m*M* KCl and 1 m*M* HCl	MS, UV	10.0 mg/L	60
DMT, DVT, and TBT	—	—	UV	ppm Level	63
TMT, TET, TPT, TBT, and TPhT	—	α-Cyclodextrin	Indirect UV	2 µ*M*	76

TABLE 9.1 (Continued)
Speciation of Metal Ions by Capillary Electrophoresis and Micellar Electrokinetic Capillary Electrophoresis

Metal Ions	Sample Matrix	Electrolytes	Detection	Detection Limit	Ref.
V(IV) and V(V)	Electroplating bath	20 m*M* Na_2HPO_4, 5 m*M* DTPA, pH 8 or 8.5	Indirect UV, 214 nm	10^6 *M*	39
V(IV) and V(V)	—	0.1 *M* monochloro-acetate, pH 2.2	UV, 220 nm	10^7 *M*	61
		Micellar Electrokinetic Chromatography (MECC)			
As(III), As(V), monomethyl-, and dimethylarsenic acid	—	10 m*M* dodecyltrimethyl ammonium phosphate, pH 8	Direct UV, 190 nm	15–90 ppb	9
As(III), As(V), MMA, and DMA	—	Dodecyltrimethylam monium phosphate	UV	0.045 µg/L	43
Hg(II) and phenyl-mercury (II)	—	Na_2HPO_4-$Na_2B_4O_7$, 2.5 m*M* TTHA, 2 m*M* SDS, pH 7.5	Indirect UV, 220 nm	ppb Level	49
Fe(II) and (Fe(III)	—	1 m*M* ammonium phosphate buffer, 75 m*M* SDS, 0.1 m*M* MPAR, pH 8	Indirect UV, 254 nm	10^7 *M*	45
Pb(II), triethyllead(IV), trimethyllead(IV), and diphenyl-lead(IV)	—	Na_2HPO_4-$Na_2B_4O_7$, 2.5 m*M* TTHA, 2 m*M* SDS, pH 7.5	Indirect UV, 220 nm	ppb Level	49
Organotin	—	SDS	—	—	66
TML and TEL, PhSe, and DPhSe	—	SDS	—	—	34
Se(IV), phenyl-selenium(II), and diphenylselenium-(II)	—	Na_2HPO_4-$Na_2B_4O_7$, 2.5 m*M* TTHA, 2 m*M* SDS, pH 7.5	Indirect UV, 220 nm	ppb Level	49

9.2.1 Composition of the Background Electrolytes

As reported earlier, a high voltage is maintained in CE using buffers as the BGE. It is also important to know that the conductivity of the BGE should be higher than the conductivity of the sample, which can be obtained by using buffers as the BGE. Buffers are also useful to control the pH of the BGE throughout the experiments. Therefore, buffers are used as BGEs in most CE applications. The most commonly used buffers are phosphate, acetate, borate, ammonium citrate, tris, CHES, MES, PIPES, and HEPES, which are used at different concentrations and pHs. The electrolyte identity and concentration must be chosen carefully to

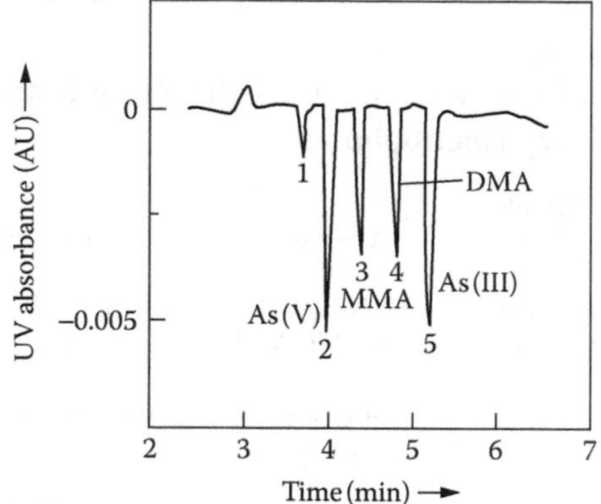

FIGURE 9.1 Electropherograms of the arsenic species by CE, BGE: 6 m*M* chromate-NICE-Pak Anion-BT (1:40, v/v) and detection by UV at 274 nm. (From Lin, L., Wang, J., and Caruso, J., *J. Chromatogr. Sci.*, 33: 177 (1995). With permission.)

achieve optimum metal ion speciation. The selection of the BGEs depends on their conductivity and the type of metal ions to be studied. The relative conductivities of different electrolytes can be estimated from their condosities (defined as the concentration of sodium chloride, which has the same electrical conductance as the substance under study).[77] A wide variety of electrolytes can be used to prepare the buffers for CE. Low-UV-absorbing components are required for the preparation of the buffers, if detection is to be achieved by a UV detector. The volatile components are required in MS or ICP detection methods. These conditions substantially limit the choice to a moderate number of electrolytes. The pH of the BGE is another factor that determines the choice of the buffers. For low pH buffers, phosphate and citrate are commonly used, although the latter absorbs strongly at wavelengths <260 nm. Basic buffers, such as borate, tris, and CAPS, are used as suitable BGEs. A list of useful buffers along with pHs and working wavelengths is given in Table 9.2.[78]

Lin et al.[24] used 6 m*M* chromate solution (pH 8) with NICE-Pak OFM Anion-BT (1:40, v/v) as the BGE for the speciation of arsenic. On the other hand, Liu and workers[25] used a complex BGE for the speciation of arsenic and selenium metal ions. This BGE was comprised of 2.3 m*M* pyromellitic acid, 6.5 m*M* sodium hydroxide, 1.6 m*M* triethanolamine, and 0.75 m*M* hexamethonium hydroxide solutions. Olesik et al.[21] used 0.04 *M* sodium acetate (pH 8.2) as the BGE for the speciation of arsenic, iron, chromium, strontium, and tin metal ions. Michalke et al.[32,36,52,53] used phosphate buffers for the speciation of platinum, arsenic, and selenium metal ions. Michalke et al.[56] also used dilute acetic acid as the BGE for the speciation of some metal ions. Hamaek and Havel[54] used potassium thiocyanate adjusted to pH 3 by $HClO_4$ as the BGE for the speciation of platinum and palladium metal ions. An acetate buffer of pH 3.8 was used as the BGE for the speciation of chromium.[40] Silva da Rocha et al.[50] used 25 m*M* sodium borate (pH 9.3) as the BGE for the speciation of mercury. A monochloroacetate buffer

TABLE 9.2
Commonly Used Buffers with Suitable pHs and Wavelengths for Metal Ion Speciation in Capillary Electrophoresis

Buffers	pH	Wavelength (nm)
Phosphate	1.14–3.14	195
Citrate	3.06–5.40	260
Acetate	3.76–5.76	220
MES	5.15–7.15	230
PIPES	5.80–7.80	215
Phosphate	6.20–8.20	195
HEPES	6.55–8.55	230
Tricine	7.15–9.15	230
Tris	7.30–9.30	220
Borate	8.14–10.14	180
CHES	9.50–10.00	<190

Note: CHES: 2-(*N*-cyclohexylamino)ethanesulfonic acid, HEPES: *N*-2-hydroxyethylpiperazine-*N*-2-ethanesulfonic acid, MES: morpholinoethanesulfonic acid, PIPES: piperazine-*N,N*-bis(2-ethanesulfonic acid).

Source: From Oda, R.P., and Landers, J.P., in *Hand Book of Capillary Electrophoresis*, Landers, J.P., Ed., CRC Press, London, 1994. With permission.

of pH 2.2 was used for vanadium speciation by Kitazumi et al.[61] Chen et al.[41] used 30 m*M* phosphate, 0.5 m*M* tetradecyltrimethylammonium bromide and 0.1 m*M* 2,6-PDCA + 15% acetonitrile (v/v), pH 6.4, as the background electrolyte for the speciation of vanadium metal ion. Recently, Sanchez et al.[60] used the different mixtures of 40 m*M* KCl, 1 m*M* HCl, and water as the BGEs for the speciation of rhodium metal ion.

9.2.2 pH of the Background Electrolytes

In metal ion speciation, pH plays a crucial role, as the oxidation states of metal ions are pH dependent. Moreover, the complexes' formations of metal ion species with suitable ligands are also controlled by pH. In addition, the increase in the buffer pH from 4 to 9 may result in an increase in EOF, and, therefore, by increasing the pH, the analysis time may be reduced. It is also important to note that the buffer pH may be altered in a secondary manner by other parameters such as temperature, ion depletion, and so forth. The literature reported herein indicates that a wide range of pHs was used for the speciation of metal ions. Some reports indicate the speciation at acidic levels, while other reports indicate speciation at basic pHs, which clearly reveals that the pH need depends upon the

TABLE 9.3
Effect of pH on Migration Times of Arsenic Species

Arsenic Species	Migration Times (min)		
	pH 6.5	pH 8	pH 10
As(V)	4.97	3.95	3.46
As(III)	7.01	5.21	3.89
MMA	5.69	4.33	3.68
DMA	6.11	4.78	3.68

Source: From Lin, L., Wang, J., and Caruso, J., *J. Chromatogr. Sci.*, 33, 177, 1995. With permission.

type of buffer used, metal ions to be speciated, and other CE parameters. For example, Liu et al.[25] reported the speciation of selenium at pH 7.7 (buffer containing pyromellitic acid, sodium hydroxide, triethanolamine, and hexamethonium hydroxide), while the same metal ion was speciated at acidic pH (dilute acetic acid) by Aleksenko et al.[59] Similarly, arsenic was speciated at pH 6.7 and 8 by Olesik et al.[21] and Lin et al.,[24] respectively.

The effect of pH on the migration times of arsenic species was studied by Lin et al.,[24] and the results are given in Table 9.3. It may be concluded from Table 9.3 that the migration times decreased by increasing pH values. Furthermore, it may be observed from this table that the separation and resolution factors decreased by increasing pH. The effect of pH on the peak areas of vanadium metal ion species was studied by Kitazumi et al.[61] The results are presented in Figure 9.2, and we may conclude that the peak area for $PV^{V}Mo_{11}$ is greater around pH 2.2, and the corresponding value for $PV^{IV}Mo_{11}$ is practically constant in the pH range of 2 to 3.0. As a result, a pH of 2.2 was chosen by the authors for the simultaneous CE determination of V(V) and V(IV) species. Aleksenko et al.[59] studied the effect of pH from 1.8 to 9 on the speciation of rhodium metal ion. The effect of pH on the speciation of rhodium metal ion is shown in Figure 9.3. It may be seen that the best speciation was achieved at pH 1. The authors reported a drastic change in the migration time by changing the pH (above 4.5) of the BGE. Carrier electrolytes with pH <2.5 were found to be unsuitable, because their use was accompanied by a significant increase in the current. Furthermore, the authors reported that at intermediate pHs, between 2.5 and 4, the migration times of the complexes were largely unaffected. For example, the observed mobility of a positively charged form ascribed to $RhCl_2(H_2O)_4^+$ in KCl-HCl electrolyte systems of pH 2.5, 2.8, and 4 was almost constant (3.5×10^{-4}, 3.4×10^{-4}, and 3.4×10^{-4} $cm^2 V^{-1} s$, respectively). It is also important to mention that the number and relative content of rhodium species in the equilibrium mixture did not change noticeably with pH over the specified range. Recently, Sanchez et al.[60] reported a different pattern of rhodium metal ion speciation at different

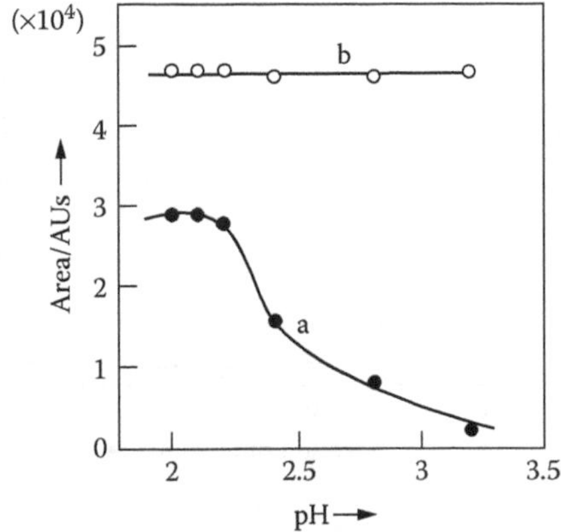

FIGURE 9.2 Effect of pH on the peak areas of (a) $PV^{V}Mo^{11}$ and (b) $PV^{IV}Mo^{11}$ species. The test solution was 5 m*M* Mo(VI) to 0.50 m*M* P(V) to 0.1 *M* monochloroacetate buffer containing 1.10^4 *M* V(V) and 1.10^4 *M* V(IV), BGE: 0.1 *M* monochloroacetate buffer. (From Kitazumi, I., Nakashima, Y., and Himeno, S., *J. Chromatogr. A*, 939: 123 (2001). With permission.)

pH levels. The existence of metal ion species was pH dependent; hence, different types of species appeared at different pHs. For instance, Baraj et al.[40] reported the presence of different chromium species at different pHs (Figure 9.4a and Figure 9.4b). The authors also reported the presence of only one species of chromium metal ion at basic medium (pH 10), while in acidic medium and at a higher concentration of chromium, $HCrO_4^-$ and $Cr_2O_7^{-2}$ species existed. Contrarily, at acidic conditions and lower concentrations of chromium, mainly the $HCrO_4^-$ form was reported.

9.2.3 Ionic Strength of the Background Electrolytes

Generally, EOF decreases by increasing the ionic strength, and, consequently, an increase in the separation time occurs. The increasing ionic strength also increases the current at a constant voltage to the point where adequate thermostating of the capillary becomes a concern. The increasing ionic strength decreases metal ions and capillary wall interactions. Briefly, the increasing ionic strength results in short analysis times with sharp peaks, but sometimes, the higher ionic strength results in poor speciation. Therefore, the selection of the ionic strength depends on several parameters, such as capillary length and diameter, applied voltage, and efficiency of the capillary thermostating condition. The use of moderately high ionic strength buffers is useful for the suppression of ionic interactions between metal ion species and ionized silanol groups on the capillary wall. It is very important to mention here that at high concentrations of buffers, the excessive Joule heating occurs, which is responsible for the nonreproducibility of speciation. Buffers problematic in this concern are those with electrolytes, such as chloride, citrate, and sulfate. However, the heating problem can be solved by

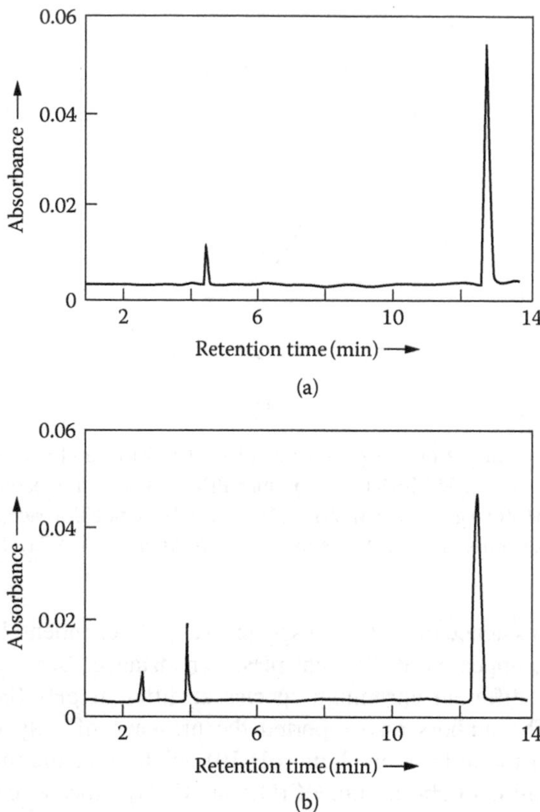

FIGURE 9.3 Effect of pH of hydrochloric acid solution on the speciation of Rh(III) on (a) pH 1, (b) pH 3.5. Rh concentration 0.18 mg/mL; BGE: 10 m*M* KCl-HCl, pH 4. (from Aleksenko, S.S., Gumenyuk, A.P., Mushtakova, S.P., and Timerbaev, A.R., *Fresenius' J. Anal. Chem.*, 370: 865 (2001). With permission.)

decreasing the applied voltage, increasing the length, and decreasing the internal diameter of the capillary. Therefore, the optimization of metal ion speciation may be achieved by varying the ionic strength of BGE. Much work remains to be done on the optimization of metal ion speciation by ionic strength variation. However, Aleksenko et al.[59] studied the effect of ionic strength on the speciation of rhodium metal ion. The authors reported that by increasing the electrolyte concentration from 10 to 70 m*M*, the mobility of charged rhodium forms underwent only a minor alteration with no peak area change.

9.2.4 Voltage Applied

Voltage may be varied as per the requirement of metal ion speciation. Generally, an increase in voltage results in an increase of EOF, short migration time, sharp peaks, and improved speciation. However, sometimes, high voltage may result

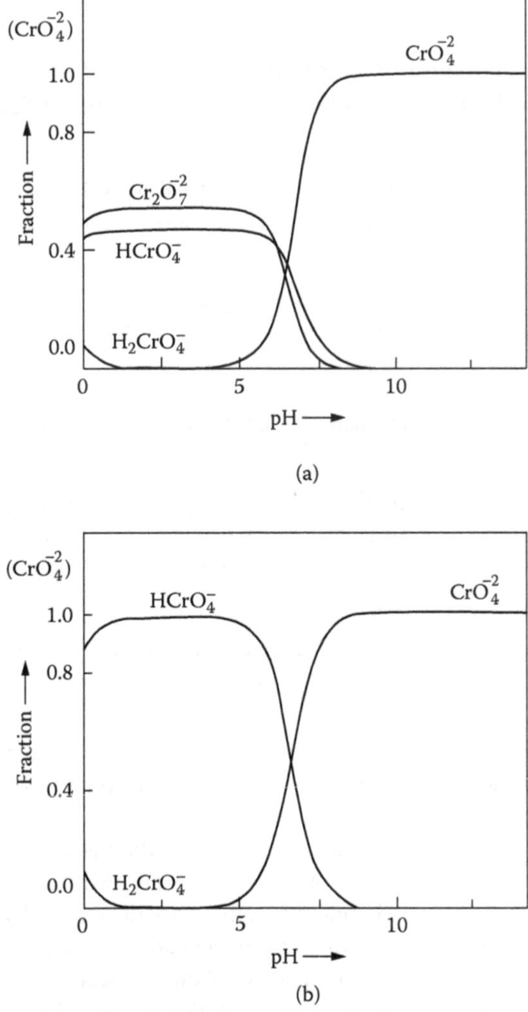

FIGURE 9.4 The species–pH distribution diagrams of (a) 2000 µg/mL Cr(VI) and (b) 20 µg/mL Cr(VI). (From Baraj, B., Niencheski, L.F.H., Soares, J.A., Martinez, M., and Merkoci, A., *Fresenius' J. Anal. Chem.*, 367: 12 (2000). With permission.)

in partial speciation. Therefore, it is advisable to start with a moderate value of voltage, that is, between 10 to 15 kV. It is also important to keep in mind that sometimes, if the sample matrix ionic strength is much greater than the EOF, the increased production of Joule heat cannot be efficiently dissipated. The heating of the capillary results in a decreased solution viscosity, and the reproducibility is lost. The nonreproducibility is observed due to an increase in the ionic mobility and metal ions diffusion. The magnitude of the applied voltage also depends on the type of buffer used. Nelson et al.[79] reported no heating of

TABLE 9.4
Effects of Voltage, Type of Electrolytes, and Concentrations of the Metal Ion Species on the Peak Areas Due to Trapping of These Species into the Stern Layer

Electrolyte	Voltage (kV)	Peak Area (Voltage Applied)/Peak Area (No Voltage)			
		Co^{+2} 0.02 ppm	Co^{+2} 0.2 ppm	Co^{+2} 2 ppm	Co^{+2} 20 ppm
NaCl	4	0.27	0.50	0.60	0.66
$CaCl_2$	4	0.92	1.06	0.98	0.99
NaCl	10	0.22	0.37	0.32	0.47
$CaCl_2$	10	0.40	0.87	0.72	0.82
		Li^{+} 0.02 ppm	Li^{+} 0.2 ppm	Li^{+} 2 ppm	Li^{+} 20 ppm
NaCl	4	0.77	0.81	0.90	0.91
$CaCl_2$	4	0.98	0.96	1.02	0.98
NaCl	10	0.64	0.74	0.88	0.89
$CaCl_2$	10	0.51	0.69	0.61	0.86

Note: NaCl: 0.04 *M*, $CaCl_2$: 0.023 *M*.

Source: From Olesik, J.W., Kinzer, J.A., and Olesik, S.V., *Anal. Chem.*, 67, 1, 1995. With permission.

the capillary up to 30 kV as the applied voltage when borate buffer was used, while heating of the capillary was observed even at 10 and 12 kV using CAPS and phosphate buffers, respectively.

Olesik et al.[21] studied the effect of voltage on the behavior of cobalt(II) and lithium(I) metal ions. The results of this finding are given in Table 9.4. It may be concluded from Table 9.4 that the sensitivity decreases by increasing the voltage. This may be due to the trapping of the cations in the Stern layer formed at higher voltage.[21] Therefore, care must be taken to select the voltage, especially for the speciation of the metal ions with smaller charges. However, little effect is observed on the sensitivity of metal ion species with higher charges, as these types of metal ions successfully compete with the negatively charged sites of the Stern layer.

9.2.5 Temperature

The speciation of metal ions is also controlled by temperature, as at high temperatures, the viscosity of BGE decreases, which results in the short analysis time. In the hydrostatic mode of sample introduction, the sample volume sometimes increases at a higher temperature, which results in poor speciation. Concurrent changes in buffer pH and peak broadening also occur at high temperatures. Contrarily, some of the advantages were also observed at elevated experimental temperatures. Therefore, the temperature affects the speciation of metal ions, but temperature was not used as a routine optimization parameter, as the control of temperature is difficult in the present model of CE.

9.2.6 Stationary Phases (Capillaries)

Mainly, the speciation of metal ions depends on the stationary phase, that is, length and diameter of the capillary. The fused silica capillaries were used for speciation purposes.[21,40] Silica gel derivatives, such as dimethylpolysiloxane, phenylpolysiloxane, and so forth, were used for to prepare fused silica gel capillaries. These capillaries are available in 20 cm × 50 μm to 107 × 50 μm sizes. It is recommended that the analysis should be started with 20 to 30 cm length, and the optimization should be achieved by varying the length of the capillary. Lower and higher values of the internal diameter and length of the capillary, respectively, were suggested for complete metal ion species separations. The literature cited here indicates the use of fused silica capillaries of different sizes and diameters, but no report is available on the effect of the length and the diameter of the capillaries on metal ions speciation. However, the optimization can be achieved to some extent by changing the dimensions of the fused silica capillaries.

9.2.7 Background Electrolyte Additives

Different types of buffers may be used as BGEs in metal ions speciation by CE, but the use of some organic solvents known as BGE additives or organic modifiers may be useful for optimum speciation. The addition of the organic modifiers may change the physicochemical properties of BGEs, which results in the interaction of the metal ion species with the capillary wall, decrease of conductivity, decrease of thermal diffusion, stability of the complexes formed between the metal ions and the ligands, and so forth. Therefore, speciation may be optimized by varying the concentration and the type of organic modifier. The improved separation of metal ion species may be due to the combined effect of the above changes (due to the addition of organic solvent) in CE conditions. The most important organic modifiers are acetonitrile, methanol, ethanol, and 1,4-diaminobutane. Care should be taken to adjust the concentrations of these organic modifiers, as higher concentrations may precipitate the buffer constituents, which may block the capillary. To the best of our knowledge, no reports were published on the optimization of metal ion speciation by organic modifiers, but the use of organic solvents as BGE modifiers may be useful for speciation.

9.2.8 Chelating Reagents

The speciation of metal ions by CE can be facilitated by using complexing agents as BGEs. The chelation in CE enhances speciation in two ways — it changes the physicochemical properties of metal ions and makes the speciation easy. Second, some chelating agents are UV sensitive; hence, UV transparent metal ions, after complexation, can be detected effectively by UV mode. Sometimes, the derivatization of metal ion species is also carried out with suitable derivatizing reagents, which results in improved speciation and detection sensitivity. The energies of specific interactions involved in derivatization are usually much higher than those in complexation, and, reactions taking place in derivatization are irreversible.[80] Covalent bonds are formed in derivatization,[81] and, therefore, derivatization is a more permanent modification than complexation. The most commonly used derivatizing reagents are diethyldithiocarbamate (DDTC), nitriloacetic acid (NTA), ethylenediaminetetraacetic acid (EDTA), 1,2-cyclohexanediaminetetraacetic acid (CDTA), diethylenediaminepentaacetic acid (DTPA), tartrate, oxalate, citrate, amino acids, and cyclodextrins.[14] In MEKC, no transfer of electron is involved in the micelle formation, and nonspecific van der Waals forces are involved in this. The anionic surfactants, for example, SDS and other complexing agents such as amino acids and polyaminocarboxylic acids, receive much attention in MEKC.

CE conditions are adjusted in such a way so as to support the complex formation of sample metal ions with the chelating agents. A good complexing agent must be highly pure; soluble in BGE; should form stable, sensitive, and single-state complexes; and should have good UV absorbance. The complexing agent is either added to the electrolyte (*in situ*, online complexation) or in the sample before introduction to the capillary column (off-line complexation).[22] With an online complexation method, the mobility of cations toward the negative electrode can be selectively modulated due to partial complexation within the capillary; hence, metal complexes of different charges and sizes are formed and good separation results. In off-line complexation, a water sample containing metal ions is treated with a ligand before being injected into CE. This method is only applicable if a metal-ion–ligand complex is more stable and does not feel any effects of the experimental conditions of CE.

9.2.9 Detection

Detection is an essential issue for the efficiency of metal ion speciation by CE. Despite the use of many detectors in CE for metal ion speciation, the sensitivity of detection is still a challenging problem. The narrow diameter capillaries used in CE speciations limit detection sensitivity due to mass loading limitations. Therefore, attempts were made to make the detection devices more sensitive. The most important detectors in CE speciations are nonelement-specific, namely, UV, electrochemical, fluorescence, and element-specific detectors, viz., AAS, AES, ICP, MS, and so forth.[14] In the last few years, some reviews[82–85] appeared on the detection of metal ions in CE. However, many experimental reports deal with detection devices for the speciation of different metal ions (Table 9.1). The different detection methods available for metal ion speciation are described herein.

The concentration sensitivity of common UV detection is restricted insofar as the capillary diameter limits the optical path length. Obviously, direct detection is hardly a universal detection technique in metal ion speciation, because most metal ions are non-UV absorbing. However some reports were published on the direct detection of metal ion species (Table 9.1). Jandik and Bonn[86] and Foret et

TABLE 9.5
Visualizing/Organic Compounds (with Chromophores) Used for Detection of Metal Ions in CE

Metal Ions	Visualizing Agents	Detection Wavelength	Ref.
Fe(II), Cu(II), Co(II), Zn(II), Pb(II), V(IV), and Ni(II)	4,7-Dimethyl-1,10-phenanthroline, 1,10-phenanthroline, α,β,γ,δ-tetrakis(4-carboxyphenyl)porphyrine, 2-nitroso-1-naphthol-4-sulfonic acid, 2-(5′-bromo-2′-pyridylazo)-5-diethylaminophenol, 2-(5-bromo-2-pyridylazo)-5-(*N*-propyl-*N*-sulfopropyl-amino) phenol	220–587 nm	69, 89, 90
Al(III), Ba(II), Sr(II), U(IV), La(III), and lanthanoides	1-(2,4-Dihydroxy-1-phenylazo)-8-hydroxy-3,6-naphthalenesulfonic acid, 3,6-bis-(2-sulfophenylazo-4,5-dihydroxy-2,7-naphthalenesulfonic acid, arsenazo III	450–654 nm	91, 92
Fe(II), Co(II), and Ni(II)	2-(5-Nitro-2-pyridylazo)-5-(*N*-propyl-*N*-sulfopropylamino) phenol	560 nm	93
Fe(II, III), Cu(II), Co(II, III), Zn(II), Ni(II), and Cr(III)	4-(2-Pyridylazo) resorcinol	500 nm	45, 94, 95

al.[87] discussed the detection of metal ion species by direct UV detectors. The selection of UV wavelength depends on the type of buffer and pH. Due to these drawbacks, indirect UV detection was developed and used for metal ions speciation by CE. The main advantage of indirect UV detection is its universal applications, and, thus, it is most commonly used for the speciation of metal ions. In this mode of detection, metal ion species are derivatized by some suitable agents called visualizing agents. The most commonly used visualizing agents are Cu(II) salt, chromate, aromatic amines, and heterocyclic compounds (e.g., benzylamine, 4-methylbenzylamine, dimethylbenzylamine, imidazole, *p*-toludine, pyridine, creatinine, ephedrine, and anionic chromophores [benzoate and anisates]).[22,85,88] The derivatization of metal ion species can be achieved by pre- or postcolumn processes. In precolumn derivatization, metal ions are derivatized before loading onto the CE machine or the derivatizing reagent is added into BGE. Contrarily, in postcolumn derivatization, metal ion species are allowed to derivatize after their separation but before entering the detector. Various visualizing reagents for the detection of metal ions in CE are summarized in Table 9.5.

In order to achieve a high sensitivity in indirect detection, the peak width of the cations should be minimized. The interaction of cations and visualization agents with the capillary wall should be suppressed. The visualization agents should exhibit mobility close to that of the cations, its UV absorbance should be as high as possible, and the detector noise should be as small as possible. Furthermore, the sensitivity of UV detection was increased using a double-beam laser as the light source.[96] A detailed discussion on UV detection of metal ions in CE was presented in various reviews.[97–106] Of course, the indirect UV method of detection is almost universal but suffers from some problems. The limit of detection of this mode is up to the parts per billion (ppb) level only, and hence, other modes of detection were explored.

Indirect fluorescence detection was also used in the analysis of metal ions,[4,107,108], but this mode of detection was not used frequently for metal ion speciation in CE. Desbene and Morin[108] used quinine sulfate as the fluorescence agent for the detection of metal ions using a laser-induced fluorescence technique, but the detection limit was not good. Direct fluorescence detection is possible if the ligands used for the analyte complexation reaction contain a chromophore or a fluorophore.[109] 8-Hydroxyquinoline-5-sulfonic acid and lumogallion possess fluorescent properties and were used for metal ion detection in CE by using fluorescent detectors.[84] Dithiozone sulfonate forms complexes with inorganic mercury.[48] Overall, fluorescence detectors do not yet receive wide acceptance in CE for metal ions analysis, although their gains in sensitivity and selectivity over photometric detectors (UV-visible) are significant, and they are commercially available.

Another approach for detection is based on electrochemical properties. This mode of detection includes conductometry, amperometry, and potentiometry as alternative methods. But this mode has not been adequately explored. The electrochemical mode of detection, involving an electrical signal, is simpler than the UV or fluorescence methods. The conductivity of BGE is the universal property, and it can be used as the mode of metal ion detection in CE. Inorganic and organic buffers of low conductivities (e.g., borate, MES-histidine, etc.) are suitable for conductivity detection. However, other buffers were used with the conductivity detector. A higher ionic strength is favorable for the operation of conductivity detectors. Huang et al.,[110] in 1987, first reported the use of a conductivity detector. Jones et al.[111] reported the sensitivity of a conductivity detector to be ten times higher than that of UV detection. The limit of detection of this mode is up to 10^{-5} to 10^{-7} *M*. On the other hand, amperometric detection relies on redox processes of the metal ion species, so the method is not universal, as only electroactive species can be detected. The limit of detection is about 10^{-5} *M*, which is supposed as a low limit of detection.[73] Amperometric detection, in CE, was introduced by Wallingford and Ewing in 1987.[112] Metal ions such as lead, cadmium, copper, tellurium, mercury, and other heavy metal ions were detected using this mode.[113] In potentiometric detection, the potential due to metal ions is measured by the ion-selective electrode, and this method was introduced in 1991 by Haber et al.[114] The applications for this detector include detection of sodium, potassium, calcium, lithium, barium, strontium, cesium, and other alkali and alkaline earth metals.[85]

Despite the development of various detection methods in CE, the problem of detection, especially the limit of detection, is associated with them. Therefore, a new approach to detection was used through the element-specific detectors.[82,84,85] The coupling of AAS with CE was reported for the speciation of several metal ions. Baraj et al.[40] reported the interface of AAS with CE, and the developed system was used for the speciation of the chromium metal ion species. The coupling of CE with ICP-MS allowed us to overcome the problem of detection to an extent. Several groups explored this arrangement for the speciation of metal ions.[21,25,114–117] These studies indicated that the major problems, such as the low flow rate of CE separations that limit the choice of a nebulizer, a postcapillary detector design with accompanying problems in applying a high voltage across the capillary and keeping the electrode grounded, and separation buffer incompatibilities with the ionization process, have been overcome. Olesik et al.[21] developed an interface (Figure 9.5a) to generate a fine aerosol that can be delivered from the end of the capillary to ICP with a good transport efficiency and minimum dead volume with a limit of detection up to 10^{-8} *M*. However, a laminar flow induced in the capillary by a pneumatic nebulizer interface, which generated a parabolic-shaped velocity profile, was found to be a factor limiting detectability. Liu et al.[25] demonstrated the feasibility of using a commercially available direct-injection nebulizer system (Figure 9.5b) ensuring 100% sample introduction to ICP-MS. With this interface, no suction and no significant band broadening were observed. The detection limit was of the order of picograms/milliliter (pg/mL). Michalke and Schramel[32] and Silva da Rocha[50] also presented the modules of the nebulizer interfaces of ICP-MS with CE (Figure 9.5c and Figure 9.5d). Various reports are available on metal ion speciation using the CE-ICP-MS detection mode.[29,32,33,53,118–121] Despite the suitable interfaces available in the ICP-MS detector with CE, the problem of nebulizer adjustment persists. Therefore, a hydride generation system was interfaced with a CE machine. Magnuson et al.[29] incorporated a hydride generation system as the interface for the speciation of arsenic, and the nebulization problem was removed. Electrospray ionization mass spectrometry (ESI-MS) detection (including ion-spray–MS detection, which can be regarded as a pneumatically assisted version of electrospray) can lead to the attainment of structure-selective information including metal ion species. ESI-MS also provides information on the direct detection and quantification of the metal ion species.[56] Therefore, the interface of CE with ESI was reported by several workers for metal ion speciation studies.[122–124] Schramel et al.[55] used a CE-ESI-MS setup for the speciation of selenium metal ion. The interface CE-ESI-MS setup is shown in Figure 9.6. To present the pattern of separation and detection by the CE-ESI-MS setup, the electropherograms of the selenium species are shown in Figure 9.7. Schramel et al.[56] also reported the online coupling of CE with an ESI-MS detection unit. The authors reported the successful speciation of selenium and copper metal ions.

In addition to these, some other devices, such as chemiluminescence, atomic emission spectrometry (AES), refractive index, radioactivity, and x-ray diffraction, were used as the detectors in CE for metal ions analysis,[84] but their uses are still limited due to certain drawbacks associated with them. The coupling of CE with chemiluminescence (CL)-based postcapillary reactions catalyzed by metal ions, such as the luminol/hydrogen peroxide reaction, is promising when extremely high sensitivity is required.[125] For instance, a detection limit of 5×10^{-13} *M* for cobalt(II) metal ion was achieved by this method.[126]

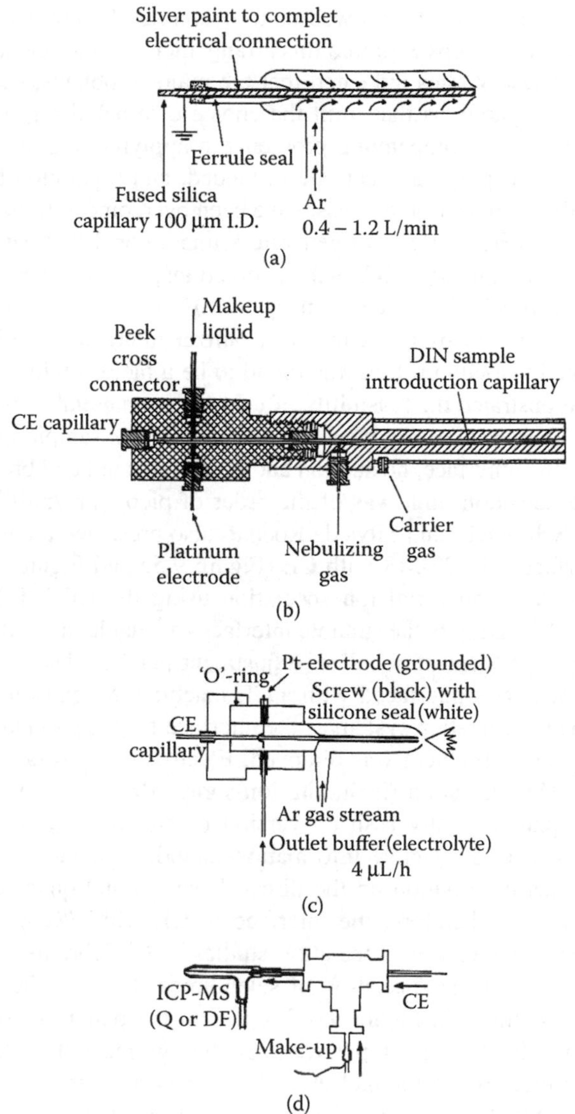

FIGURE 9.5 Interface units with capillary electrophoresis (CE): (a) pneumatic nebulizer, (b) CE–inductively coupled plasma (ICP)–mass spectrometry (MS) interface, (c) homemade nebulizer, and (d) Meinhard nebulizer CE-ICP-MS interface. (Part a from Olesik, J.W., Kinzer, J.A., and Olesik, S.V., *Anal. Chem.*, 67: 1 (1995); Part b from Liu, Y., Lopez-Avila, V., Zhu, J.J., Wiederin, D.R., and Beckert, W.F., *Anal. Chem.*, 67: 2020 (1995); Part c from Michalke, B., and Schramel, P., *Fresenius' J. Anal. Chem.*, 357: 594 (1997); Part d from Silva da Rocha, M., Soldado, A.B., Blanco-Gonzalez, E.B., and Sanz-Medel, A., *Biomed. Chromatogr.*, 14: 6 (2000). With permission.)

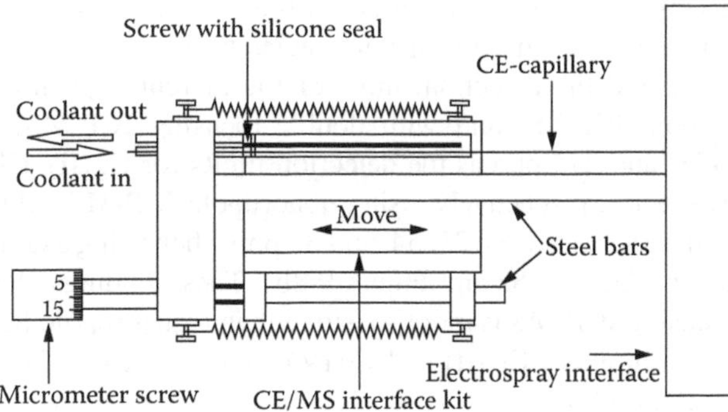

FIGURE 9.6 The modified capillary electrophoresis (CE)–mass spectrometry (MS) interface kit. (From Schramel, O., Michalke, B., and Kettrup, A., *J. Chromatogr. A*, 819: 231 (1998). With permission.)

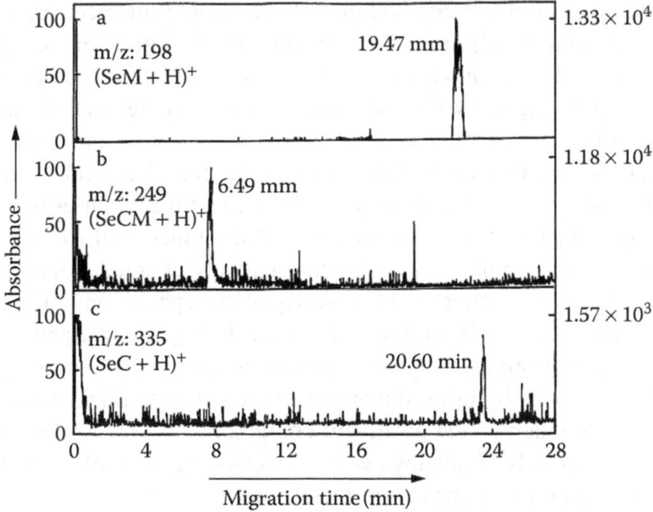

FIGURE 9.7 Electropherograms of SeM, SeC, and SeCM species using 2% acetic acid as the background electrolyte (BGE). (From Schramel, O., Michalke, B., and Kettrup, A., *J. Chromatogr. A*, 819: 231 (1998). With permission.)

The limits of detection for metal ions are different for different detectors and, normally, vary from milligram (mg) to pico level. The limits for different detectors are summarized in Table 9.1. The limits of the detection by the direct and indirect UV detectors (at ppb levels) are also reviewed by Timerbaev.[75] Baraj et al.[40] reported a detection limit of 30 to 8 μg/L for Cr(IV) and Cr(III) species. Recently, Kitazumi et al.[61] reported 10^{-7} *M* as the detection limit for vanadium metal ion species. Olesik et al.[21] reported a detection limit of the magnitude of 0.06 ppb when using a CE-ICP-MS setup. Michalke and Schramel[32] reported a detection limit in the range of 1 to 20 μg/L for platinum and selenium using an ICP-MS detector. The same detector was reported with a detection limit up to picogram level for arsenic and selenium metal ion species by Liu et al.[25] Recently, Silva da Rocha[50] compared the detection limits of the mercury species by detection with a quadrupole ICP-MS and with double-focusing ICP-MS. The authors reported 81, 128, and 275 ppb as the detection limits for Hg(II), $CH_3Hg(I)$, and $CH_3CH_2Hg(I)$ species, respectively, using quadrupole ICP-MS, while the detection limits for these species were 25, 54, and 84 ppb when using a double-focusing ICP-MS detector (Figure 9.8a and Figure 9.8b). These findings clearly indicate that double-focusing ICP-MS is more sensitive than a quadrupole ICP-MS detector. The detection limits of Cr(III) and Cr(IV) were reported to be 30 to 8 μg/L using AAS by Baraj et al.[40]

9.2.10 Other Parameters

In addition to the parameters discussed above, other factors can be controlled to optimize the speciation of metal ions by CE. Among these most important parameters are the reversal of polarity, volume of the sample injected, use of the EOF modifiers, and prederivatization of metal ion species with suitable ligands. In a normal CE machine, the anode (+) is always at the inlet and the cathode (–) at the outlet. In this format, EOF is toward the cathode (detector). If set in reverse polarity, the direction of EOF is away from the detector, and only negatively charged metal ion complexes with electrophoretic mobility greater than the EOF will pass the detector. This format is typically used with capillaries that are coated with substances that reverse the net charge of the inner wall (reverse EOF), or when the metal ion complexes are all net negatively charged. Aleksenko et al.[59] attempted to observe the effect of the polarity on the speciation of rhodium metal ion, and the results of their findings are shown in Figure 9.9a and Figure 9.9b. It may be observed from these figures that no detection was achieved by reversing the polarity electrode (by using the anode at the detector side). Again, this may be due to the attraction of the positively charged metal ions toward the cathode (no detector). Similarly, Semenova et al.[7] reversed polarity of the electrodes to optimize the speciation of chromium metal ion.

Sometimes, partial separations of metal ion species are observed due to sample overload. Under such circumstances, maximum separation can be achieved by reducing the concentration of the sample. Care should be taken to minimize the concentration or the volume of the loading sample so that detection can be achieved. Olesik et al.[21] studied the effect of the amount loading (time of injection) of the chromium metal ion on the peak width. The authors reported that the peak widths were directly proportional to the amount loading. In this area, field-amplified stacking injections (FASIs) are most promising. This takes advantage of electrophoretic migration and electroosmosis and has been demonstrated to achieve over 100-fold enhancement in detection for charged species.[127,128] FASI has been used for the enhancement detection of several metal ion species, such as mercury, lead, and selenium.[14]

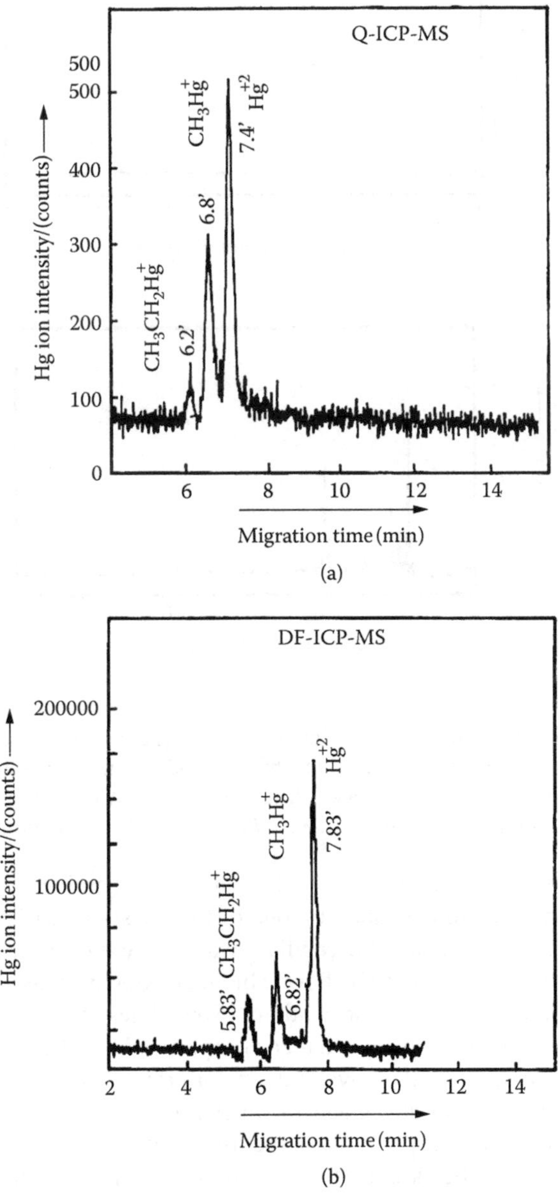

FIGURE 9.8 Electropherograms of the species of organic and inorganic mercury with (a) quadrupole inductively couple plasma–mass spectrometry (ICP-MS) and (b) double-focusing ICP-MS modes of detection. (From Silva da Rocha, M., Soldado, A.B., Blanco-Gonzalez, E.B., and Sanz-Medel, A., *Biomed. Chromatogr.*, 14: 6 (2000). With permission.)

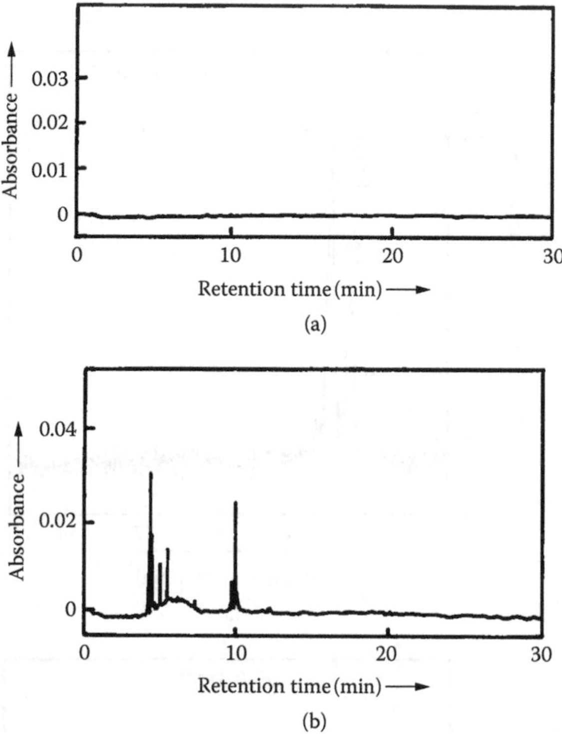

FIGURE 9.9 Electropherograms of rhodium(III) complexes in 1×10^3 *M* H_2SO_4 recorded with (a) positive and (b) negative polarities; background electrolyte (BGE): 50 m*M* phosphate buffer, pH 2.5. (From Aleksenko, S.S., Gumenyuk, A.P., Mushtakova, S.P., and Timerbaev, A.R., *Fresenius' J. Anal. Chem.*, 370: 865 (2001). With permission.)

Some chemicals were also reported to alter the chemistry of the capillary wall. For instance, diaminoalkanes are one of the classes of buffer additives that appear to alter the properties of the capillary wall. The use of these EOF modifiers was reported to decrease the EOF flow. This methodology may be used for the optimization of the metal ion speciations. Sometimes, metal ion species are derivatized with a suitable ligand and then are speciated by CE. The derivatization of metal ion species alters the physicochemical properties of metal ion species, which may be used as an advantage for their speciation. The literature reveals many reports on metal ion speciation in the form of their complexes with suitable ligands.[21,41,43,52,54,55,59] Olesik et al.[21] speciated chromium in the form of $Cr(OH)_3$ and $Cr_2O_7^{-2}$, while Chen et al.[41] used EDTA, DTPA, NTA, and pyridine-2,6-dicarboxylic acid (2,6-PDCA) complexes for the speciation of chromium metal ion. The electropherograms of these complexes are given in Figure 9.10a through Figure 9.10e, respectively. A comparison of the speciation of these complexes can be carried out from these figures, which clearly indicate that EDTA is the best ligand for this study. Schramel et al.[55] used Cu-EDTA complexes for copper

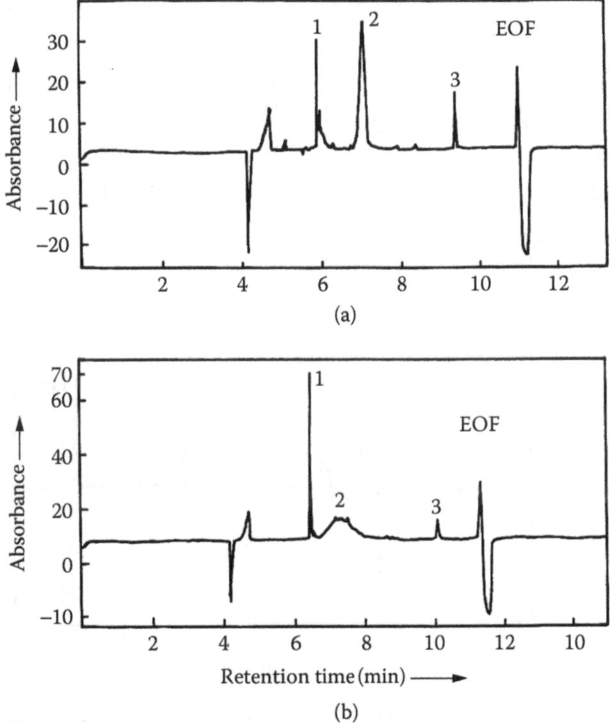

FIGURE 9.10 Electropherograms of Cr(III) complexed with (a) EDTA, (b) HEDTA, (c) DTPA, (d) NTA, and (e) 2,6-DPCA ligand showing 1 = free ligand, 2 = Cr(III) complex (I), and 3 = Cr(III); background electrolyte (BGE): 25 m*M* sodium phosphate, 0.25 m*M* TTBA, 15% (v/v) acetonitrile at pH 6.4. (From Chen, Z., Naidu, R., and Subramanian, A., *J. Chromatogr. A*, 927: 219 (2001). With permission.) *Continued.*

speciation (Figure 9.11). Similarly, vanadium was allowed to react with Mo(VI)-P(V) reagent, and the corresponding complexes formed were speciated.[61] Recently, Sanchez et al.[60] speciated rhodium metal ion by preparing hydro–chloro complexes of the metal ion. Therefore, the derivatization of the metal ion species is one of the best means for the optimization of the speciation. The complexation of metal ions is also helpful for metal ion speciation in MEKC. Ng et al.[34] used cyclodextrins as the complexing agent for the speciation of organometallic compounds by MEKC.

9.2.11 Optimization by Dependent Variables

The improvements in metal ion speciation may also be achieved by dependent parameters. that is, by field strength (V/m), EOF, Joule heating, BGE viscosity, sample diffusion, sample mobility, sample charge, sample size and shape, sample interaction with capillary and BGE, and molar absorptivity. As discussed previ-

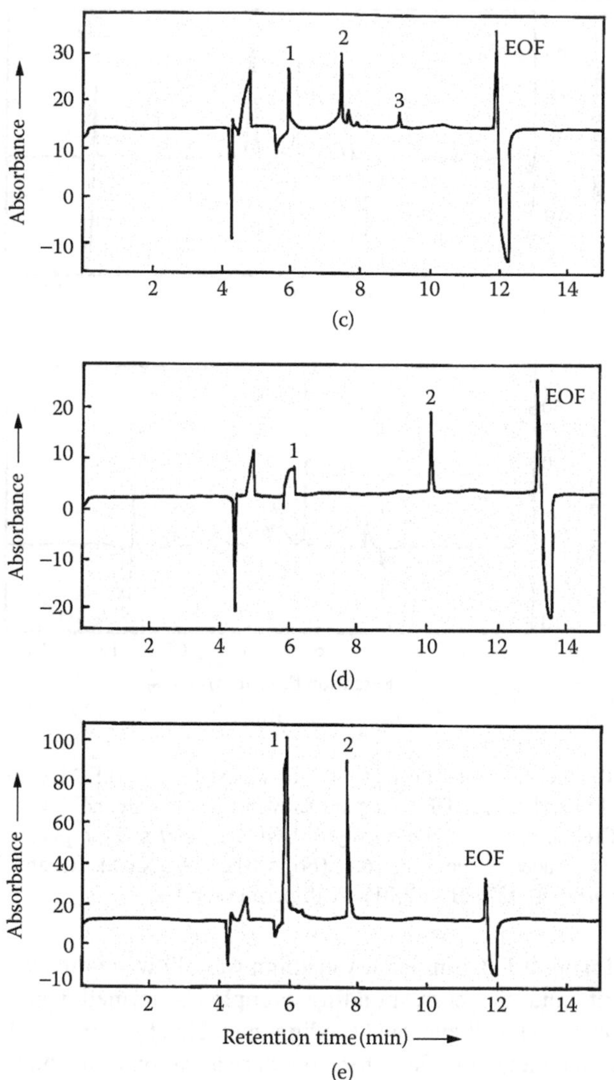

FIGURE 9.10 *Continued.*

ously, these parameters are directly linked to the independent parameters; therefore, the variation in the variable parameters can be achieved through the independent variables. For example, the optimization of metal ion speciation may be achieved by changing the temperature of the capillary, which may result in the change of viscosity, EOF, and so forth. Additionally, the different sizes of metal ion species and physicochemical properties are responsible for the different migration times in CE, which may result in an improved separation process. Briefly, variation in the independent parameters may be useful to control the

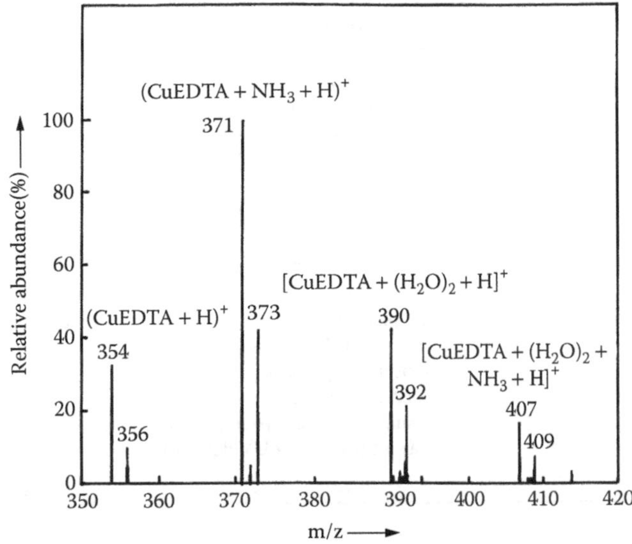

FIGURE 9.11 Electrospray ionization–mass spectrometry (ESI-MS) spectrum of $CuCl_2.H_2O$ in water (500 mg Cu/L). (From Schramel, O., Michalke, B., and Kettrup, A., *J. Chromatogr. A*, 819: 231 (1998). With permission.)

optimization of metal ions via dependent variables. A method development for the speciation of metal ions is shown in Scheme 9.1. As is clear from this scheme, the optimization of metal ions may be achieved by varying CE parameters, starting with the CE machine.

9.3 VALIDATION OF THE METHODS

In spite of the various advantages of CE, it suffers from the drawback of poor reproducibility. Therefore, it is very important to develop speciation methods by CE. Despite this fact, only a few studies deal with the method validation of metal ion speciation by CE. However, some authors demonstrated the application of their developed methods as the best. The accuracy determination of arsenic species was reported by Lin et al.[24] They also reported their method as reproducible with a detection limit of 3.5 pg for As(III). Liu et al.[25] reported the detection of arsenic species at parts per billion (ppb) and parts per trillion (ppt) levels. Aguilar and coworkers[3] described CE as a powerful tool for the analysis of environmental samples. CE as a more attractive method than IC or RP-HPLC for metal ions speciation was described by Silva da Rocha and coworkers.[50] Liu and Hian[49] reported RSD 2.67 to 5.49% for the speciation of lead, mercury, and selenium species. Furthermore, they claimed the applicability of their developed methods for the speciation of these metal ions in real samples. Chen et al.[41] claimed their developed CE method, for the speciation of chromium, as being highly efficient and having a short analysis time. Foret et al.[87] reported their

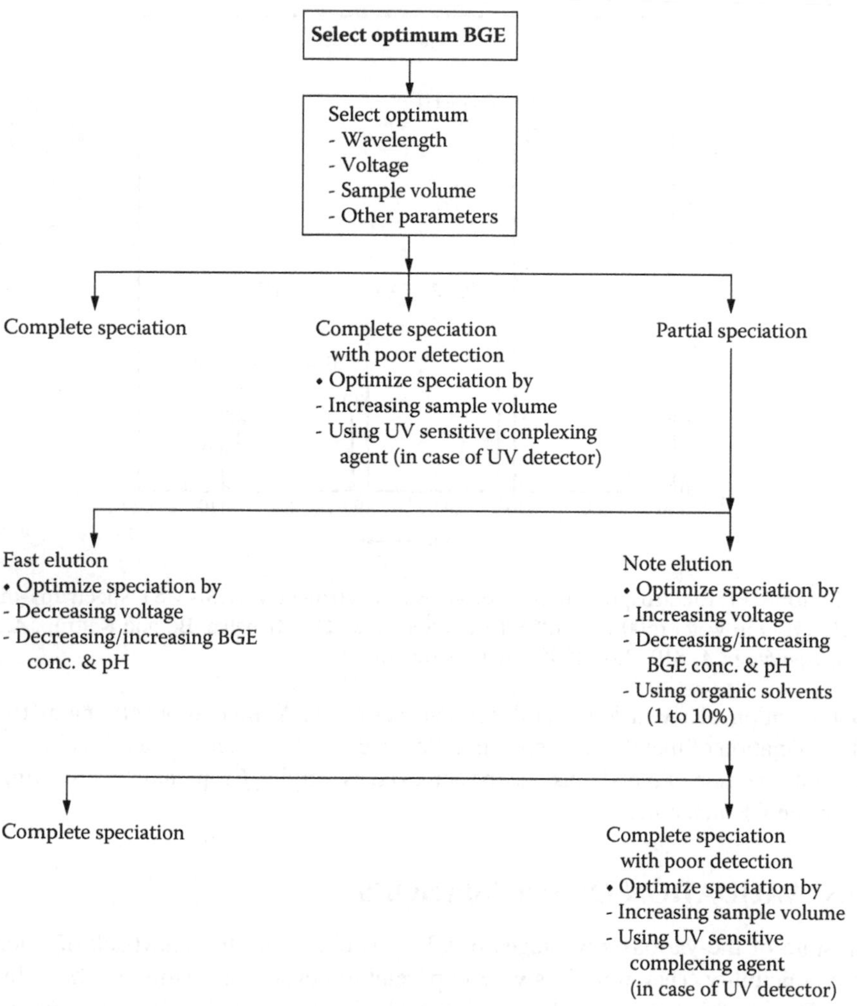

SCHEME 9.1

developed method as being excellent, with RSD < 1%. Other studies also showed CE to have reliable results and recoveries close to 100% or agreement with the results obtained by other methods. Despite this, the precise linearity, sensitivity, and reproducibility of CE methods for metal ions speciation are not superior to those of ion chromatography and other chromatographic methods.

9.4 EFFECT OF INTERFERING IONS

Sometimes, the presence of other metal ions and chemicals interferes with the speciation of metal ions. Kitazumi et al.[61] carried out an extensive study on this issue. They studied the speciation of vanadium metal ions in the presence of a

TABLE 9.6
Effect of Foreign (Interfering) Ions on the Speciation of Vanadium Metal Ion

Ions	Concentration (*M*)	Relative Error (%)	
		V(V)	V(IV)
Cr(III)	1.1×10^{-4}	−1.4	+0.4
Mn(II)	1.1×10^{-4}	−0.9	−0.8
Fe(III)	1.1×10^{-4}	−1.5	−1.5
Co(II)	1.1×10^{-4}	−2.6	+0.6
Ni(II)	1.1×10^{-4}	−0.1	+2.0
Cu(II)	1.1×10^{-4}	+2.3	−1.2
Zn(II)	1.1×10^{-4}	+1.1	−0.4
Al(II)	1.1×10^{-4}	+0.5	+0.3
Ga(II)	1.1×10^{-4}	−0.5	+0.4
Si(IV)	1.1×10^{-4}	−1.5	+2.2
Ge(IV)	1.1×10^{-4}	−1.0	+1.0

Source: Kitazumi, I., Nakashima, Y., and Himeno, S., *J. Chromatogr. A*, 939, 123, 2001. With permission.

variety of other metal ions — Fe(III), Cr(III), Ga(III), Mn(II), Co(II), Ni(II), Cu(II), Zn(II), Al(II), Si(II), and Ge(IV) — and the percentage relative error was calculated (Table 9.6). The authors reported that under the reported conditions, the Mo(VI)-P(V) reagent reacted with Fe(III), and a Fe(III)-substituted molybdophosphate anion $[P(Fe^{III}Mo_{11})O_{40}]^{-6}$ was formed. On the other hand, this reagent reacted with Cr(III) and Ga(III) and the corresponding Anderson-type heteropolyan ions $[XMo_6O_{24}H_6]_{-3}$, where X = Cr(III) and Ga(III) were formed. Their presence caused no serious effect, as the mobilities of the above-mentioned complexes were different than those of the $[P(V^{V}Mo_{11})O_{40}]^{-4}$ and $[P(V^{IV}Mo_{11})O_{40}]^{-5}$ complexes (vanadium species). However, the presence of these metal ions at much greater concentrations may cause negative errors due to the lack of Mo(VI). The remaining metal ions — Mn(II), Co(II), Ni(II), Cu(II), Zn(II), Al(II), Si(II), and Ge(IV) — did not interfere, because no complex formation occurs under the reported conditions.

9.5 MECHANISMS OF METAL ION SPECIATION

In a normal CE machine, the anode and cathode electrodes are situated at the inlet (sample loading) and the detector sides, respectively. Therefore, metal ion species with positive charge move from anode to cathode, where they are detected by a suitable detector. For the separation of anionic organometallic species, the reversal of the CE electrode is required. The movements of metal ion species from anode to cathode are due to the electrophoretic (μ_{ep}) and electroosmotic

(μ_{eo}) mobilities. Therefore, the migrations of metal ion species are controlled by the sum of electrophoretic and electroosmotic mobilities. The electrophoretic mobilites of metal ion species depend on their charges and sizes, while there is no relation among the charges, sizes, and electroosmotic mobility. Metal ion species with different charge/radii ratios migrate by different electrophoretic velocities under the influence of applied voltage. The greater the charge/radius ratio, the greater is the mobility, and the lower is the migration time. Therefore, metal ion species with greater charges are always eluted first, followed by metal ion species with smaller charges. The electroosmotic mobility for all metal ion species remains almost the same, but it helps in their migration toward the cathode. In this way, metal ion species are eluted at different migration times due to the combined effect of different electrophoretic and electroosmotic mobilities. In addition, the interaction of metal ion species with capillary wall, steric effect, van der Waals forces, dispersion interaction, and so forth, play a crucial role for the different mobilities of metal ion species. To make the concept more clear, the separation of arsenic metal ion species [As(III) and As(V)] is shown in Figure 9.12a through Figure 9.12e. Stage 1 corresponds to the loading of an arsenic metal ion species [As(III) and As(V)] sample at the anode end of a CE machine (capillary). The metal ion species start to move toward the cathode at different velocities under the influence of electrophoretic and electroosmotic mobilities and start to separate. The second and third stages indicate the partial separation of As(III) and As(V) species, while the fourth stage (Figure 9.12d) shows the complete separation of both arsenic species. The last stage, Figure 9.12e, indicates no arsenic species in the capillary path, as all metal ion species were eluted at different migration times — first As(V), followed by As(III), detected by the detector, and recorded by the printer.

The neutral organometallic species can be separated by using MEKC, and the theory developed by Terabe et al.[65] is applicable for metal ion separation. In MEKC, a surfactant is a molecule possessing two zones of different polarities, which shows the special characteristics in solution form. The surfactants are divided into three categories — ionic (cationic and anionic), nonionic, and zwitterionic. The surfactant molecules form a micelle, and the partition of metallic species between this micelle and BGE takes place. In this way, the separation mechanisms are shifted slightly toward the chromatographic principle. Briefly, the mechanisms of the separation of metal ion species into CE are simple; hence, optimization of the separation can be achieved easily by knowing the exact mechanisms of separation.

9.6 CAPILLARY ELECTROPHORESIS VERSUS CHROMATOGRAPHY

The chromatographic and electrophoretic methods are considered suitable techniques for metal ion speciation. Some aspects are common in both techniques. For example, the extraction, purification methods, and detectors are concerned.

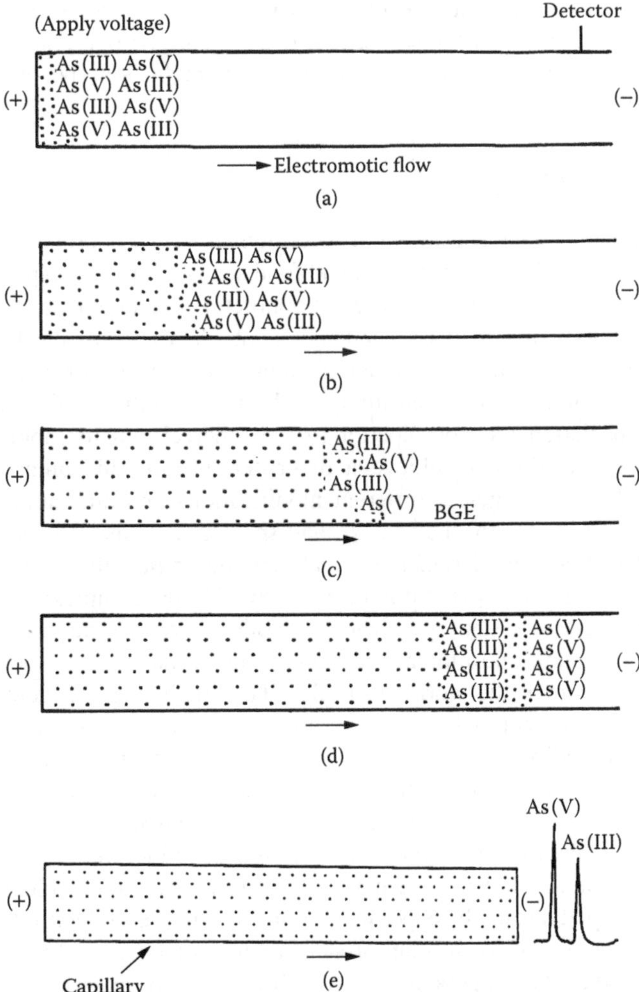

FIGURE 9.12 Mechanisms of the separation of arsenic species — As(III) and As(V) — in capillary electrophoresis with (a) to (e): different stages of the separation of arsenic species.

However, both techniques differ in their applications and working principles. Liquid and gas chromatographic approaches are known as the best for metal ion speciation in comparison to the capillary electrophoretic approach. The reason for this is due to certain drawbacks associated with CE. The chromatographic techniques are universal in use, with very good precision and reproducibility, while the reproducibility of the CE method is still not good. CE is not a useful technique for the speciation of neutral metal ion species. The linking of CE with element-specific detectors is very difficult and requires extra attention due to the low BGE flow rate in CE. The detection limits in chromatographic and electrophoretic methods are comparable. Overall, CE could not yet achieve a respectable place in the field of metal ion speciation. Briefly, chromatographic methods are very popular and have been used worldwide for metal ion speciation in real-world samples.

9.7 CONCLUSION

As discussed above, the speciation of metal ions is very important and essential in the environmental, pharmaceutical, and food sciences areas. Despite the development of several chromatographic, electrophoretic, and atomic absorption spectrometric methods, it remains a challenging job. A review of the literature, as discussed herein, indicates limited applications of CE for the speciation of metal ions. The precision linearity, sensitivity, and reproducibility of CE methods for metal ion speciation are not superior to the other chromatographic methods. Therefore, CE could not achieve a dedicated use in this area. Therefore, many[22,82–85,88,128–130] suggested different modifications and alternations to make CE a method of choice for the analysis and speciation of metal ions in different matrices. To obtain good sensitivity and reproducibility, the selection of the capillary wall chemistry, pH and ionic strength of BGE, complexing and visualizing agents, detectors, and optimization of BGE were also described and suggested.[131–134] In addition to the above points, some other aspects should also be addressed to improve the working of CE. These include the development and wide use of fluorescent and radioactive complexing agents, as the detection by fluorescent and radioactive detectors is more sensitive and reproducible with low limits of detection. Moreover, the use of these detectors is less complex and inexpensive in comparison to the recently developed AAS, ICP-MS, and ESI-MS detectors. To make the CE application more reproducible, the background electrolyte should be developed in such a way that physical and chemical properties remain unchanged during the experimental run. The nonreproducibility of the CE method may be due to the heating of BGE after a long run of the CE machine. Therefore, to keep the temperature constant throughout the experiments, a cooling device should be added to the machine. All the capabilities and possibilities of CE have not been fully explored, but these explorations are under way. Briefly, there is a lot to be developed in order to advance CE for the speciation of metal ions. Definitely, CE will prove itself to be the best technique for the speciation of metal ions and will achieve the status of the best technique for routine metal ion speciation in most laboratories.

REFERENCES

1. Huang X, Gordon MJ, Zare RN, *Anal. Chem.*, 60: 385 (1988).
2. Ebdon L, Steve H, Walton AP, Ward RW, *Analyst*, 113: 1159 (1988).
3. Aguilar M, Huang X, Zare RN, *J. Chromatogr. A*, 480: 427 (1989).
4. Gross L, Yeung ES, *Anal. Chem.*, 62: 427 (1990).

5. Li K, Li FY, *Analyst*, 120: 361 (1995).
6. Pobizy E, Glod B, Kaniewska J, Trojanowicz M, *J. Chromatogr. A*, 718: 329 (1995).
7. Semenova PP, Timerbaev AR, Gagstadter R, Bonn GK, *J. High Resolut. Chromatogr.*, 19: 177 (1996).
8. Conradi S, Vogt C, Wittrisch H, Knobloch G, Werner G, *J. Chromatogr. A*, 745: 103 (1996).
9. Liu WP, Lee HK, *J. Chromatogr. A*, 796: 385 (1998).
10. Ergolic KJ, Stockton RA, Chakarborti D, *Arsenic: Industrial, Biochemical and Environmental Prospectives*, in Lederer WH, Fensterheim RJ (Eds.), Van Nostrand Reinhold, New York (1983).
11. Benramdane L, Bressolle F, Vallon JJ, *J. Chromatogr. Sci.*, 37: 330 (1999).
12. Timerbaev AR, *J. Chromatogr. A*, 792: 495 (1997).
13. Dabek-Zlotorzynskaa E, Lai EPC, Timerbaev AR, *Anal. Chim. Acta*, 359: 1 (1998).
14. Liu W, Lee HK, *J. Chromatogr. A*, 834: 45 (1999).
15. Timerbaev AR, *Electrophoresis*, 23: 3884 (2002).
16. Pyrzynska K, Wierzbicki T, *Talanta*, 64: 823 (2004).
17. Lopez-Sanchez JF, Amran MB, Lakkis MD, Lagarde F, Rauret G, Leroy MJF, *Fresenius' J, Anal, Chem.*, 348: 810 (1994).
18. Schlegel D, Mattusch J, Wennrich R, *Fresenius' J. Anal. Chem.*, 354: 535 (1996).
19. Wildman BJ, Jackson PE, Jones WR, Alden PG, *J. Chromatogr. A*, 546: 459 (1991).
20. Vogt C, Werner GJ, *J. Chromatogr. A*, 686: 325 (1994).
21. Olesik JW, Kinzer JA, Olesik SV, *Anal. Chem.*, 67: 1 (1995).
22. Timerbaev AR, *J. Cap. Elect.*, 1: 14 (1995).
23. Morin P, Amran MB, Favier S, Heimburger R, Leroy MJF, *Fresenius' J. Anal. Chem.*, 342: 357 (1992).
24. Lin L, Wang J, Caruso J, *J. Chromatogr. Sci.*, 33: 177 (1995).
25. Liu Y, Lopez-Avila V, Zhu JJ, Wiederin DR, Beckert WF, *Anal. Chem.*, 67: 2020 (1995).
26. Amran MB, Hegege A, Lagarde F, Leroy MJF, *Anal. Chem. (Warsaw)*, 40: 309 (1995).
27. Albert M, Demesmay C, Porthault M, Rocca JL, *Analyst*, 20: 383 (1992).
28. Albert M, Demesmay C, Rocca JL, *Fresenius' J. Anal. Chem.*, 351: 426 (1995).
29. Magnuson ML, Creed JT, Brockhoff CA, *J. Anal. At. Spectrom.*, 12: 689 (1997).
30. Magnuson ML, Creed JT, Brockhoff CA, *Analyst*, 122: 1057 (1997).
31. Michalke B, Schramel P, *Electrophoresis*, 19: 2220 (1998).
32. Michalke B, Schramel P, *Fresenius' J. Anal. Chem.*, 357: 594 (1997).
33. Michalke B, Schramel P, *Electrophoresis*, 19: 270 (1998).
34. Ng CL, Lee HK, Li SFY, *J. Chromatogr. A*, 652: 547 (1993).
35. Van-den-Broeck K, Vandecasteele C, *Mikrochim. Acta*, 128: 79 (1998).
36. Michalke B, Schramel P, *Analyst*, 26: M51 (1998).
37. Timerbaev AR, Semanova OP, Buchberger W, Bonn GK, *Fresenius' J. Anal. Chem.*, 354: 414 (1996).
38. Jung GY, Kim VS, Lim SB, *Anal. Sci.*, 13: 463 (1997).
39. Padarauskas A, Schwedt G, *J. Chromatogr. A*, 773: 351 (1997).
40. Baraj B, Niencheski LFH, Soares JA, Martinez M, Merkoci A, *Fresenius' J. Anal. Chem.*, 367: 12 (2000).
41. Chen Z, Naidu R, Subramanian A, *J. Chromatogr. A*, 927: 219 (2001).

42. King M, Macka M, Paull B, *Anal. Lett.*, 37: 2771 (2004).
43. Schäffer S, Gareil, P, Dezael C, Richard D, *J. Chromatogr. A*, 740: 151 (2002).
44. Kuban P, Buchberger W, Haddad PR, *J. Chromatogr. A*, 770: 329 (1997).
45. Timerbaev AR, Semenova OP, Jandik P, Bonn GK, *J. Chromatogr. A*, 671: 419 (1994).
46. Medina I, Rubi E, Mejuto MC, Cela R, *Talanta*, 40: 1631 (1993).
47. Carro-Diaz AM, Lorenzo-Ferreira RA, Cela-Torrijos R, *J. Chromatogr. A*, 730: 345 (1996).
48. Jones P, Hardy S, *J. Chromatogr. A*, 765: 345 (1997).
49. Liu W, Hian KL, *Anal. Chem.*, 70: 2666 (1998).
50. Silva da Rocha M, Soldado AB, Blanco-Gonzalez EB, Sanz-Medel A, *Biomed. Chromatogr.*, 14: 6 (2000).
51. Diaz AMC, Lorenzo-Ferreinra RA, Cela-Torrijos R, *Microchim. Acta*, 123: 73 (1996).
52. Lustig S, Michalke B, Beck W, Schramel P, *Fresenius' J. Anal. Chem.*, 360: 18 (1998).
53. Michalke B, Lustig S, Schramel P, *Electrophoresis*, 18: 196 (1997).
54. Hamáek J, Havel J, *J. Chromatogr. A*, 834: 321 (1999).
55. Schramel O, Michalke B, Kettrup A, *J. Chromatogr. A*, 819: 231 (1998).
56. Schramel O, Michalke B, Kettrup A, *Fresenius' J. Anal. Chem.*, 363: 452 (1999).
57. Gilon N, Potin-Gautier M, *J. Chromatogr. A*, 732: 369 (1996).
58. Dzierzgowska M, Pyrzynska K, Pobozy E, *J. Chromatogr. A*, 984: 291 (2003).
59. Aleksenko SS, Gumenyuk AP, Mushtakova SP, Timerbaev AR, *Fresenius' J. Anal. Chem.*, 370: 865 (2001).
60. Sanchez JM, Hidalgo M, Havel J, Salvado V, *Talanta*, 56: 1061 (2002).
61. Kitazumi I, Nakashima Y, Himeno S, *J. Chromatogr. A*, 939: 123 (2001).
62. Timerbaev AR, Semenova OP, Fritz JS, *J. Chromatogr. A*, 756: 300 (1996).
63. Hang F, Fasching JL, Brown PR, *J. Chromatogr. B*, 669: 103 (1995).
64. Yeh CF, Jiang SJ, *J. Chromatogr. A*, 1029: 255 (2004).
65. Terabe S, Otsuka K, Ichikawa A, Ando T, *Anal. Chem.*, 56: 111 (1984).
66. Li K, Li SFY, Lee HK, *J. Liq. Chromatogr.*, 18: 1325 (1995).
67. Vogt C, Vogt J, Wittrisch H, *J. Chromatogr. A*, 727: 301 (1996).
68. Forte G, D'Amato M, Caroli S, *Microchem. J.*, 79: 15 (2005).
69. Motomizu S, Osima M, Kuwabara M, *Analyst*, 119: 1787 (1994).
70. Timerbaev AR, Semenova OP, Fritz JS, *J. Chromatogr. A*, 811: 233 (1998).
71. Buchberger W, Semenova OP, Timerbaev AR, *J. High Resolut. Chromatogr.*, 16: 153 (1993).
72. Seidel BS, Faubel W, *Fresenius' J. Anal. Chem.*, 360: 795 (1998).
73. Buchberger W, Haddad PR, *J. Chromatogr. A*, 687: 343 (1994).
74. Albert M, Demesmay C, Rocca JL, *Analyst*, 21: 403 (1993).
75. Timerbaev AR, *Electrophoresis*, 18: 185 (1997).
76. Whang KS, Whang CW, *Electrophoresis*, 18: 241 (1997).
77. Wolf AV, Morden GB, Phoebe PG, in *CRC Hand Book of Chemistry and Physics*, Weast RC, Astle MJ, Beyer WH (Eds.), 68th ed., CRC Press, Boca Raton, FL p. D219 (1987).
78. Oda RP, Landers JP, in *Hand Book of Capillary Electrophoresis*, Landers JP (Ed.), CRC Press, Boca Raton, FL (1994).
79. Nelson RJ, Paulus A, Cohen AS, Guttmann A, Karger BL, *J. Chromatogr.*, 480: 111 (1989).

80. Nielen MWF, Frei RW, Brinkmann UATh, *Selective Sample Handling and Detection in High Performance Liquid Chromatography*, Elsevier, Amsterdam, chap. 1, (1988).
81. Cagniant D, *Complexation Chromatography*, Marcel Dekker, New York, chap. 1 (1992).
82. Macka M, Haddad PR, *Electrophoresis*, 18: 2482 (1997).
83. Kappes T, Hauser PC, *J. Chromatogr. A*, 834: 89 (1999).
84. Timerbaev AR, Buchberger W, *J. Chromatogr. A*, 834: 117 (1999).
85. Pacakova V, Coufal P, Stulik K, *J. Chromatogr. A*, 834: 257 (1999).
86. Jandik P, Bonn B, *Capillary Electrophoresis of Small Molecules and Ions*, VCH, New York (1993).
87. Foret F, Krivankova L, Bocek P, *Capillary Zone Electrophoresis*, VCH, Weinheim (1993).
88. Valsecchi SM, Polesello S, *J. Chromatogr. A*, 834: 363 (1999).
89. Macka M, Paull B, Anderson P, Hadda PR, *J. Chromatogr. A*, 767: 303 (1995).
90. Yokoyama T, Akamatsu T, Ohji K, Zenki M, *Anal. Chim. Acta*, 364: 75 (1998).
91. Colburn BA, Sepaniak MJ, Hinton ER, *J. Liq. Chromatogr.*, 18: 3699 (1995).
92. Macka M, Nesterenko P, Andersson P, Haddad PR, *J. Chromatogr. A*, 803: 279 (1998).
93. Motmizu S, Mori N, Kuwabara M, Oshima M, *Anal. Sci.*, 10: 101 (1994).
94. Saitoh T, Hoshino H, Yotsuyanagi T, *J. Chromatogr.*, 469: 175 (1989).
95. Regan FB, Meaney MP, Lunte SM, *J. Chromatogr.* B, 657: 409 (1994).
96. Xue Y, Yeung ES, *Anal. Chem.*, 65: 2923 (1993).
97. Hjerten S, Elenbring K, Kilar F, Lio JL, Chen AJC, Siebert CJ, Zhu MD, *J. Chromatogr.*, 403: 47 (1987).
98. Foret J, Fanali S, Oossicini L, Bocek P, *J. Chromatogr.*, 470: 299 (1989).
99. Yeng ES, Kuhr WG, *Anal. Chem.*, 63: 275A (1991).
100. Nielen MWF, *J. Chromatogr. A*, 588: 321 (1992).
101. Wang T, Hartwick RA, *J. Chromatogr. A*, 607: 119 (1992).
102. Ackermans MT, Evetraets FM, Beckers JL, *J. Chromatogr.*, 549: 345 (1991).
103. Poppe H, *Anal. Chem.*, 64: 1908 (1992).
104. Beck W, Engelhardt H, *Chromatographia*, 33: 313 (1992).
105. Beckers JL, *J. Chromatogr. A*, 679: 153 (1994).
106. Kok WT, *Chromatographia*, 51: 52 (2000).
107. Desbene PL, Morin CJ, Desbene AM, Groult RS, *J. Chromatogr. A*, 689: 135 (1995).
108. Desbene PL, Morin CJ, *Spectra Anal.*, 23: 35 (1994).
109. Timerbaev AR, Buchberger W, Semenova OP, Bonn GK, *J. Chromatogr. A*, 630: 379 (1993).
110. Huang X, Pang TKJ, Gordon MJ, Zare RN, *Anal. Chem.*, 59: 2747 (1987).
111. Jones WR, Soglia J, Harber C, Reineck J, Unicam ATI Company, http://www.psu.edu/dept/foodmfg/sources.html
112. Wallingford RA, Ewing AG, *Anal. Chem.*, 59: 1762 (1987).
113. Lu W, Cassidy RM, *Anal. Chem.*, 65: 1649 (1993).
114. Haber C, Silvestri I, Röösli S, Simon W, *Chimia*, 45: 117 (1991).
115. Michalke B, Schramel P, *J. Chromatogr. A*, 750: 51 (1996).
116. Majidi V, Miller-Ihli NJ, *Analyst*, 123: 803 (1998).
117. Tomlison MJ, Lin L, Caruso JA, *Analyst*, 120: 583 (1995).
118. Sutton K, Sutton RMC, Caruso JA, *J. Chromatogr. A*, 789: 85 (1997).

119. Kinzer JA, Olesik JW, Olesik SV, *Anal. Chem.*, 68: 3250 (1996).
120. Michalke B, Schramel P, *J. Chromatogr. A*, 807: 71 (1998).
121. Michalke B, Schramel O, Kettrup A, *Fresenius' J. Anal. Chem.*, 363: 456 (1999).
122. Moseley MA, Jorgenson JW, Shabanowitz J, Hunt DF, Tomer KB, *J. Am. Soc. Mass Spectrom.*, 3: 289 (1992).
123. Garcia F, Henion JD, *J. Chromatogr. A*, 606: 237 (1992).
124. Gale DC, Smith RD, Rapid Commun. *Mass Spectrom.*, 7: 1017 (1993).
125. Gracia-Campana AM, Baeyens WRG, Zhao Y, *Anal. Chem.*, 69: 83A (1997).
126. Huang B, Li JJ, Zhang L, Chen JK, *Anal. Chem.*, 68: 2366 (1966).
127. Chien RL, Burgi DS, *Anal. Chem.*, 64: 489A (1992).
128. Liu BF, Liu LB, Cheng JK, *J. Chromatogr. A*, 834: 277 (1999).
129. Ikuta N, Yamada Y, Yoshiyama T, Hirokawa T, *J. Chromatogr. A*, 894: 11 (2000).
130. Timerbaev AR, Shigun OA, *Electrophoresis*, 21: 4179 (2000).
131. Horvath J, Dolnike V, *Electrophoresis*, 22: 644 (2001).
132. Mayer BX, *J. Chromatogr. A*, 907: 21 (2001).
133. Ga B, Coufal P, Jaro M, Muzika J, Jelink I, *J. Chromatogr. A*, 905: 269 (2001).
134. Fritz JS, Breadmore MC, Hilder EF, Haddad PR, *J. Chromatogr. A*, 942: 11 (2002).

10 Speciation of Metal Ions by Spectroscopic Methods

In spite of the development of advanced chromatographic and electrophoresis methods for metal ion speciation, spectroscopic methods continue to be utilized for metal ion speciation due to certain advantages. The salient features of these techniques are low-cost paraphernalia and inexpensive speciation cost. Metal ion speciation involves the use of various spectroscopic techniques, such as ultraviolet (UV)-visible, flame atomic absorption (FAAS), atomic emission (AES), hydride generation, graphite furnace, inductively coupled plasma (ICP), x-ray, and so forth.[1–4] AES, AAS, and ICP give results that are more precise and exact. X-ray, x-ray fluorescence (XRF), extended x-ray absorption fine structure spectroscopy (EX-AFS), and x-ray absorption near-edge structure (XA-NES) spectroscopic techniques have been used for metal ion speciation. The main goal of spectroscopic techniques is detection, not separation. Therefore, these methods are useful for the detection of metal ions. Therefore, the conversion of one form (by reduction or oxidation) of a particular metal ion into another is required for the speciation purpose. The reduction or oxidation methodology is not universal in nature, and only some metal ions can be speciated by this technique. Due to this drawback, the use of spectroscopic methods is not popular, however, some used spectroscopy for metal ion speciation. In view of these points, we will describe the state-of-the-art of metal ion speciation using the above-mentioned techniques.

10.1 UV-VISIBLE SPECTROSCOPY

Absorption spectroscopy in the ultraviolet and visible region is considered to be one of the oldest physical methods used for the quantitative analysis of inorganic and organic species. The detection of the species is carried out by the absorption of radiation at a fixed value of wavelength (maximum absorption) called λ_{max}. The instrument involves the source of light, filter or monochromator, sample holder, and detector. The sources of light are tungsten and deuterium or hydrogen lamps for visible and UV radiation, respectively. A monochromator is used to select a narrow band from wavelengths of continuous spectra. Previously, single-beam spectrometers were developed that had only one sample holder. In these types of spectrometers, samples and references are used one-by-one, which may

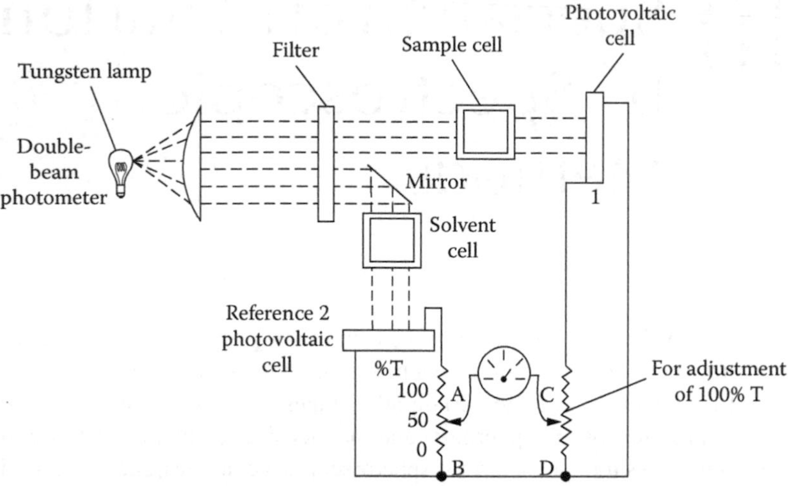

FIGURE 10.1 Double beam UV-visible spectrometer.

result in errors due to variation in the intensity of the source and fluctuation in the detector. This drawback was removed in the double-beam spectrometers that have two sample holders in which the sample and the reference can be placed side by side. This type of spectrometer gives fast and reproducible results. Many of the manufacturers are supplying computerized UV-visible spectrometers that can be handled very easily. The schematic outline of a double-beam UV-visible spectrometer is shown in Figure 10.1. Normally, metal ions are detected by visible spectroscopy after color development with some suitable reagents.

The spectroscopic method of arsenic determination is based on the reaction of ammonium molybdate with arsenate in acidic medium to form an arsenate-containing molybdenum heteropolyacid that can be reduced to molybdenum blue with stannous chloride, hydrazine, or ascorbic acid. The best results can be obtained by using hydrazine sulfate. Molybdenum blue absorbs at 865 nm, which corresponds to arsenate.[5] The concentrations of arsenite can be determined using this method after the oxidation of arsenite to arsenate by oxidizing reagents.[6–10] On the contrary, arsenite reacts quantitatively with potassium iodide in solution with sulfuric acid, liberating iodine, which is extracted in chloroform and detected at a 520 nm wavelength. The detection limit of arsenite was reported to be 2 ppb. The silver diethyldithiocarbamate method is one of the most popular spectroscopic methods for the determination of arsenic. This method is based on the generation of arsine (AsH_3) either with zinc and hydrochloric acid or sodium borohydride in acidic medium. Arsine gas is passed through diethyldithiocarbamate in pyridine/chloroform. The red-colored complex formed is measured at 520 nm. Organoarsenicals are required for the conversion into arsenate and are reduced to arsenite before measurement, as the volatile arsines Me_xAsH_{3-8} ($x = 1 - 3$) also form colored complexes but with a different absorption than that with arsine. This method is useful for the determination of arsenic in water samples.

Quinteros et al.[11] reported the speciation of iron using bathophenanthroline complexing reagent. The authors reported the detection of Fe(II) and Fe(III) species in beans, chickpeas, and lentils. Themelis et al.[12] described a simple and rapid flow-injection (FI) method for the simultaneous spectrophotometric determination of Fe(II) and Fe(III) in pharmaceutical products. The method is based on the reaction of Fe(II) with 2,2′-dipyridyl-2-pyridylhydrazone (DPPH) in acidic medium to form a water-soluble reddish complex (λ_{max} = 535 nm). Fe(III) reacts with DPPH under flow conditions only after its online reduction by ascorbic acid (AsA). Both analytes were determined in the same run via a double-injection valve, which enabled the simultaneous injection of two sample volumes in the same carrier stream (single-line double-injection approach). The two well-defined peaks produced corresponded to total iron [Fe(II) + Fe(III)] and Fe(II). Speciation of iron into the mixture was achieved by multiple regression analysis. The calibration curves obtained were linear over the ranges 0 to 30 and 0 to 50 mgL^{-1} for Fe(II) and Fe(III), respectively, and the precision [s_r = 1% for Fe(II) and 1.5% for Fe(III)] was satisfactory. Furthermore, the authors reported that the method proved to be selective and adequately sensitive [c_L = 0.25 and 0.17 mgL^{-1} for Fe(III) and Fe(II) speciation, respectively, in mixtures]. Application of the method to the analysis of pharmaceutical samples resulted in excellent accuracy with percent mean recoveries ranging from 99 to 102% for both Fe(II) and Fe(III). The mean relative error was e_r = 1%. Wu and coworkers[13] used diphenylcarbazide and methyl orange as color-developing reagents for Cr(VI) and Cr(III) species, respectively. Determination of these color complexes was carried out by visible spectrometer. Abbas and Madrakian[14] described a simple, accurate, sensitive, and selective kinetic spectrophotometric method for selenium speciation with rapid determination of ultratrace quantities of selenium. Selenium (IV) was collected on activated carbon after reduction to elemental selenium by ascorbic acid. The collected selenium was then determined based on its accelerating effect on the oxidation reaction of methyl orange with bromate in acidic media. The total amounts of Se(IV) and Se(VI) were collected on activated carbon after their reduction by hydrazine. Se(IV), Se(VI), and total selenium could be determined by the method. Selenium in the range of 10 to 10,000 ng could be determined by the reported method. The method was used for the determination of Se(IV), Se(VI), and total selenium in natural water with satisfactory results. Taylor and van Staden[15] reviewed the speciation of vanadium using the UV-visible spectrometer technique. The authors reported various color-developing reagents for vanadium species. The detection of the developed color species was carried out using different λ_{max} values. Abbaspour et al.[16] reported and H-point standard addition method for the simultaneous speciation of Sb(III) and Sb(V) metal ion species. The method was based on the differences between rate of complexation of pyrogallol red with these metallic species at pH 2. The determination of these species was carried out by UV-visible spectrometer at 510 nm λ_{max} value. The developed method was used for the speciation of these metal ion species in the water of rivers and springs.

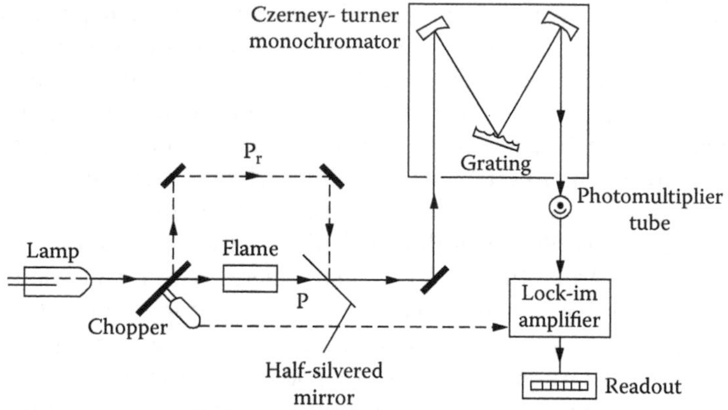

FIGURE 10.2 Atomic absorption spectrometer.

10.2 ATOMIC ABSORPTION SPECTROSCOPY

Atomic absorption spectrometry (AAS) was introduced by Alan Walsh in 1950, and since then, it has become a powerful tool in the quantitative analysis of trace elements. This method provides the total concentrations of metal ions and is almost independent of the molecular form of the metal ion in the sample. The impurities present in the sample do not affect the detection; hence, there is no need for special sample preparation in AAS. However, the complex samples containing organic pollutants need to be extracted before loading onto AAS. The absorption energy by ground-state atoms in the gaseous state forms the basis of AAS. Metal ion atoms are vaporized by flame in AAS, and atoms in gaseous form absorb the energy from the energy source (metal ion lamp) — the absorption of energy is directly proportional to the number of atoms in the flame. The absorption spectrum is recorded and used to calculate the concentration of metal ions. The basic components of AAS are energy source, chopper, nebulizer, monochromator, detector, amplifier, and recorder. The radiation source for an AAS should emit stable and intense radiation to the element to be determined, usually a resonance line of the element. Normally, hollow cathode lamps (HCLs) for different elements are used as the radiation sources. However, electrodeless discharge lamps are also available. Electrodeless discharge types of lamps provide higher energy than HCL and, hence, can improve sensitivity and detection for some elements, such as arsenic, selenium, and so forth. A schematic diagram of AAS is given in Figure 10.2.

The sensitivity, precision, and detection in AAS can be improved by coupling it with flow injection and hydride generation units. The detection limit in AAS was improved by replacing flame AAS with graphite furnace AAS. The flow-injection methodology was used by Rûika and Hansen[17] in 1975, and since then, many publications included reports of using the flow injection technique coupled with various other analytical technologies. A review of flow injection application in speciation was published by Luque de Castro.[18] Hydride atomization was suggested first by Dedina and Rubeska[19] and indicated that hydride atomization is caused by free radicals (H° and OH°) generated in the reaction zone of a hydrogen oxygen flame burning in the T-shaped quartz tube of the system. Welz and Melcher[20] reported that these radicals are formed in a reaction with oxygen at a temperature above 600°C. Furthermore, they found a significant enhancing effect of oxygen on the sensitivity of volatile hydride-forming elements at temperatures around 700 to 800°C. Parisis and Heyndrickx[21] investigated the effect of purge gases, gas transfer tubing, and reaction flasks on the sensitivity of hydride-forming elements. Hydride is usually formed in an aqueous medium using some reagents. The most important reagents are $NaBH_4$ and $SnCl_2$. Typical hydride-forming chemical reactions are as follows:

$$NaBH_4 + 3H_2O + HCl \rightarrow H_3BO_3 + NaCl + 8H^\circ \qquad (10.1)$$

$$E^{m+} + H^\circ \text{ (excess)} \rightarrow EH^\circ n + HE \text{ (excess)} \qquad (10.2)$$

where E is the analyte, and m may or may not be equal to n. These reactions generate a volatile hydride that is transported to a quartz cell by means of argon carrier gas. In the quartz cell, the hydrides are converted to gaseous metal atoms. It is believed that atomization of the hydride occurs from collision with free hydrogen radicals.[20,22] In the quartz cell, the generated analyte atoms are considered in the path of a source lamp, and a signal is generated by measuring the amount of light absorbed. The optimization of hydride generation can be achieved by controlling many factors, such as the use of fresh solutions, gas flow, cleaning of the quartz cell surface, sample volume, and so forth. Flow injection can be used in hydride generation, and a commercial assembly is available from Perkin Elmer, USA. The reaction conditions for some metal ions in a flow-injection hydride generation unit coupled to AAS are given in Table 10.1. A schematic diagram of a hydride generation unit is shown in Figure 10.3. Graphite furnace units are manufactured from Ringsdorf spectroscopic carbon (Type RW 003), containing a graphite cylinder with a transverse hole as the atomizing chamber and a hole orthogonal to the chamber as the sample injection port. The surface area of the atomization chamber may be changed by cutting a screw through the surface. The ends of the cylinder are recessed to accommodate the flat ends of the cylindrical furnace support electrodes.[23]

Normally, speciation of a particular metal ion is carried out by determining the total concentration and the concentration of one oxidation state of that metal ion. For example, arsenic speciation is carried out by the determination of As(III) or As(V) and the total arsenic, which is determined by the oxidation of As(III) to As(V) or the reduction of As(V) to As(III) depending on the method used. Prior oxidation of As(III) to As(V) by potassium dichromate or sodium thiosulfate was reported.[5–7,24,25] Speciation can also be carried out by reducing one form of metal ion into another. For example, a method was developed for the speciation

TABLE 10.1
Reaction Conditions of Hydride Formation for Some Elements with Flow Injection Coupled to AAS Assembly

Metal Ion	Reaction Conditions
As(III)	in 10% (v/v) HCl, carrier solution 10% (v/v) HCl, reducing agent 0.2% $NaBHO_4$ in 0.05% NaOH
Bi(III)	in 10% (v/v) HCl, carrier solution 10% (v/v) HCl, reducing agent 0.2% $NaBHO_4$ in 0.05% NaOH
Hg(I)	in slightly acidified, carrier solution 3% (v/v) HCl, reducing agent solution 1.1% $SnCl_2$ in 3% HCl
Hg(II)	in slightly acidified, carrier solution 3% (v/v) HCl, reducing agent solution 0.2% $NaBHO_4$ in 0.05% NaOH
Sb(III)	in 10% (v/v) HCl, carrier solution 10% (v/v) HCl, reducing agent 0.2% $NaBHO_4$ in 0.05% NaOH
Se(IV)	in 10% (v/v) HCl, carrier solution 3% (v/v) HCl, reducing agent 0.2% $NaBHO_4$ in 0.05% NaOH
Sn(II) or Sn(IV) in saturated	Carrier solution saturated boric acid (approximately boric acid with 1% (v/v) HCl 50 g/L) in 1% (v/v) HCl, reducing agent 0.4% $NaBHO_4$ in 0.05% NaOH
Te(IV)	in 10% (v/v) HCl, carrier solution 10% (v/v) HCl, reducing agent 0.2% $NaBHO_4$ in 0.05% NaOH

Source: Chakraborti, D., Adams, F., Vanmol, W., and Irgolic, K.J., *Anal. Chim. Acta*, 196, 23, 1987. With permission.

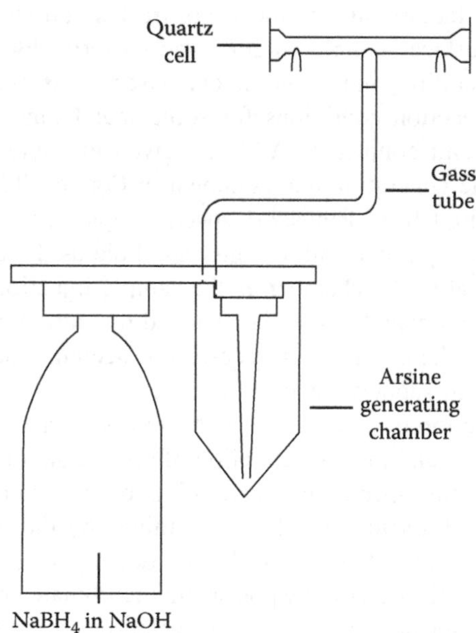

FIGURE 10.3 Hydride generation unit.

of arsenic by reducing arsenate into arsenite by L-cysteine[26–28] and mercaptoacetic acid.[27] These prereductants were used with flow injection hydride generation coupled with a homemade in-house nondispersive AAS device.

Arsenite can be reduced to arsine using sodium borohydride, which boils at 55°C and can be easily separated and measured through an appropriate detector. Arsenite is separated from arsenate by reduction at a controlled pH using sodium borohydride. In buffered aqueous solutions, only arsenate is reduced to arsenite, whereas both arsenite and arsenate are converted to arsine in solutions with pH < 1. Thus, arsenite and arsenate can be determined sequentially; alternatively, arsenite can be determined in an aliquot of the sample and the sum of both in another aliquot.[29–31] Detection limits for these methods depend on the detector used and are in the range of nanogram (ng) to microgram (μg) of arsenic. A nondispersive atomic absorption spectrometer with hydride-forming elements was developed. Recently, a method was developed for the determination of trace amounts of arsenic(III) and total arsenic with L-cysteine as prereductant using flow injection hydride generation coupled with an in-house-made nondispersive AAS device.[26] Generally, organoarsenic is not determined by the hydride generator method (as arsine is formed at pH 1 or lower).

Rahman et al.[32] reported the speciation of arsenic, and the authors have, so far, analyzed 34,000 and 101,934 hand-tubed well-water samples from Bangladesh and West Bengal, respectively, by FI-HG-AAS. From this speciation, 56% and 52% samples were found to have arsenic levels above 10 μg/L and 37% and 25% had arsenic levels above 50 μg/L from Bangladesh and West Bengal, respectively. Yalcin and Le[33] speciated arsenic species by using flow injection with hydride generation atomic fluorescence spectrometry (FI-HG-AFS) and hydride generation atomic absorption spectrometry (FI-HG-AAS). A detection limit of 0.05 μg/L arsenic in a water sample was achieved using HG-AFS. An application of the method was demonstrated at a drinking water treatment facility. As(III) and As(V) species were determined in water at various stages of treatment. Schwenzer et al.[34] carried out arsenic speciation by hydride generation atomic absorption spectrometry (HG-AAS). Most of the arsenics in the anoxic NaCl-type waters were reduced into the arsenite forms. Sakamoto et al.[35] described a method for the determination of total arsenic by hydride generation–atomic absorption spectrophotometry using a mixed acid as a pretreatment. The authors investigated the temperature and time of decomposition using inorganic, organic arsenic and environmental standard samples pretreated with nitric–perchloric–sulfuric mixed acid. By using a mixed acid as a pretreatment agent at 220°C, the decomposition time could be shortened, and the blank value of arsenic from the reagents used was reduced. The mixed nitric–perchloric–sulfuric acid was also found to be effective as a pretreatment agent for organic arsenic compounds in which a dimethylated compound, sodium cacodylate, or a biological sample, is known to be one of the undecomposables. The present approach was proven to be satisfactory as a pretreatment for the quantitative analysis of trace amounts of total arsenic in liquid or solid environmental samples, such as geothermal water, sediments, and biological samples.

Many reports were published for arsenic speciation using graphite furnace AAS methods.[36–39] A microprocessed-based flow injection analysis system (FIAS) hydride generator system is available for fast, accurate, and easy experimentation.[40] Benramdane et al.[41] investigated the distribution of arsenic species in human organs following fatal acute intoxication by arsenic trioxide. The collected autopsy samples of most organs were ground and dried, and the total arsenic was measured by electrothermal atomic absorption spectrometry (ET-AAS). Arsenic species — arsenite [As(III)], arsenate [As(V)], and its metabolites (monomethylarsonic acid [MMA] and dimethylarsinic acid [DMA]) were quantified by ET-AAS. The speciation of elements in graphite furnace also requires the preoxidation or reduction of metal ion prior to introduction to the furnace.

Nielsen and Hansen[42] described a volume-based flow injection (FI) procedure for the determination and speciation of trace inorganic arsenic, As(III) and As(V), via HG-AAS of As(III). The determination of total arsenic is obtained by the online reduction of As(V) to As(III) by means of 0.50% (w/v) ascorbic acid and 1% (w/v) potassium iodide in 4 *M* HCl. The combined sample and reduction solution is initially heated by flowing it through a knotted reactor immersed in a heated, thermostated oil bath at 140°C and, subsequently, for cooling the reaction medium, a knotted reactor immersed in a water bath at 10°C. By using the same volume-based FI-HG-AAS system without the heating and cooling reactors and employing mild hydrochloric acid conditions, As(V) is not converted to arsine, thereby allowing for the selective determination of As(III). The injected sample volume was 100 μg/L, while the total sample consumption per assay was 1.33 mL, and the sampling frequency was 180 samples per hour. The detection limit for the online reduction procedure was 37 ng/L. The relative standard deviation (RSD) was 1.1% ($n = 10$) at 5 μg/L by calibrating with As(III) standards, while it was 33 ng/L by calibrating with As(V) standards. For the selective determination of As(III), the detection limit was 111 ng/L, and the RSD was 0.7% ($n = 10$) at 5 μg/L. Both procedures were tolerant to potential interferents. Thus, without impairing the assay, interferents such as Cu, Co, Ni, and Se could, at a As(V) level of μg/L, be tolerated at a weight excess of 2000, 30,000, 200, and 200 times, respectively. The assay of a certified drinking water sample by means of multiple standard addition (five levels, each three replicates) was 9.09 ± 0.05 μg/L. Pyell et al.[43] described a flow-injection system for the determination of inorganic arsenic [As(III)/As(V)] and selenium species [Se(IV)/ Se(VI)] by electrochemical hydride generation, cryogenic trapping, and atomic absorption spectrometry. A simple and robust electrochemical flow-through cell with fibrous carbon as the cathodic material was developed for the speciation of arsenic. A cold-trap system makes it possible to eliminate interference from methylated arsenic species. Without prereduction, the system is selective to As(III) and Se(IV). The selectivity obtained with fibrous carbon as the cathode material is compared to the selectivity obtained with a second electrochemical flow-through cell using a lead foil as the cathode. Cabon and Cabon[44] speciated As(III), As(V), monomethylarsenic, dimethylarsenic, and unknown organic compounds using a continuous flow-injection hydride generation technique coupled to atomic absorption spectrometry (AAS).

After hydride generation and collection in a graphite tube coated with iridium, arsenic was determined by AAS. By selecting different experimental hydride generation conditions, it was possible to determine As(III), total arsenic, hydride reactive arsenic, and, by the difference, nonhydride-reactive arsenic. The method was applied to seawater sampled at a Mediterranean site and at an Atlantic coastal site. Sakamoto et al.[35] described a hydride generation atomic absorption spectrophotometry method with a flow-injection method for the speciation of arsenic metal ion. This approach proved to be satisfactory for the quantitative analysis of trace amounts of total arsenic in liquid or solid environmental samples, such as geothermal water, sediments, and biological samples. Shemirani and Rajabi[45] developed a simple and sensitive method for the speciation of chromium [Cr(III) and Cr(VI)] in water by ET-AAS. The procedure was based on the selective absorption of Cr(III) onto a cellulose microcolumn (pH 11, 0.5 *M* NaCl). Total chromium was subsequently determined after appropriate reduction of Cr(VI) to Cr(III). Recoveries of more than 97% were obtained with a concentration factor of 100. The relative standard deviations (n = 10) at the 40 ng/L level for chromium(III) and chromium(VI) were 2.3% and 1.8%, respectively. The detection limits were 1.8 and 5.1 ng/L for chromium(III) and chromium(VI), respectively. No interference effects were observed from other investigated species, and the method was successfully applied to natural water samples.

Atomic emission spectroscopy (AES) works on the emission of radiations from metal ions. A sample is excited by absorbing thermal or electrical energy, and the radiation emitted by the excited sample is used to determine the concentration of metal ion. An AES instrument contains an excited source, electrodes, sample handling apparatus, monochromator, detector, and recorder. Normally, flames and direct current arc are used as excitation sources. AES can also be coupled with hydride generation, graphite furnace, flow injection, and other units for speciation purposes. Flame photometry is based on the measurement of the intensity of the light emitted when a metal is introduced into a flame. The intensity of the wavelength color is directly proportional to the concentration of metal ion. This method may also be used qualitatively in nature, as different metal ions have different flame colors. Normally, this method is useful for the detection of group IA and IIA of the Periodic Table, but sometimes, copper, manganese, and iron metal ions can be detected by flame photometry. This technique may be used for speciation purposes by determining the concentrations (total concentration and the concentration of a specific oxidation state) of a particular metal ion.

10.3 INDUCTIVELY COUPLED PLASMA SPECTROMETRY

Inductively coupled plasma spectrometry (ICP) is a promising emission spectroscopic technique, and commercial ICP systems became available in 1974. In this technique, the carrier gas (argon) is heated at a high temperature (9000 to 10,000K), and a plasma is developed in which the excitation of atomic electrons takes place easily and precisely; hence, ICP is known as a better technique than

AAS. The metal ion enters into the plasma as an aerosol, and the droplets are dried and dissolved, and the matrix is decomposed in the plasma. In the high-temperature region of the plasma, atomic and ionic species, in various energy states, are formed. The emission lines can then be exploited for analytical purposes. This technique can be coupled with AAS or MS. ICP-MS is superior with respect to detection limits, multielemental capabilities, and wide linear dynamic ranges. Quadrupole mass filters are the most common mass analyzers, because they are inexpensive. Double-focusing magnetic/electrostatic sector instruments and time-of-flight mass analyzers are also used.[46,47] ICP-MS can be coupled to a hydride generation unit for the speciation of certain metal ions. A schematic diagram for ICP discharge is shown in Figure 10.4. Figure 10.5 describes the nebulizer, spray chamber, and desolvation parts in an ICP setup.

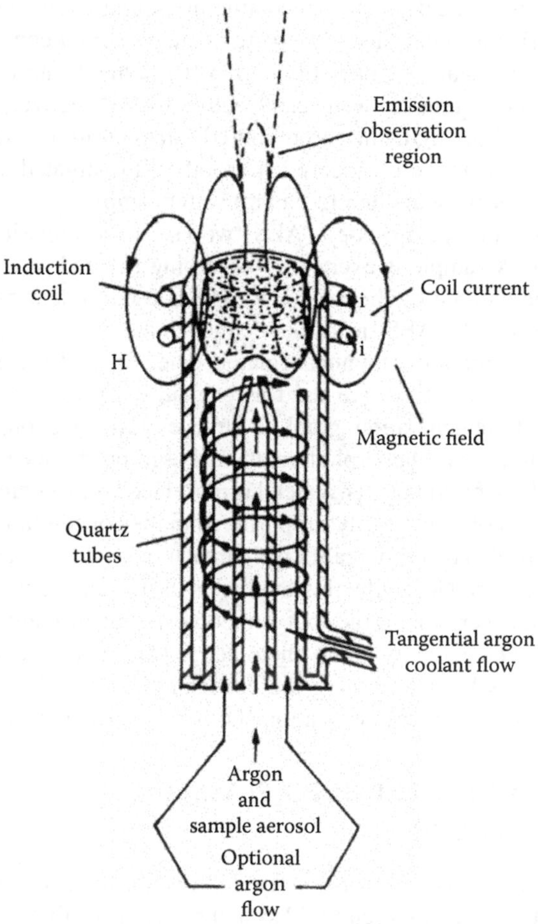

FIGURE 10.4 Inductively coupled plasma spectrometer.

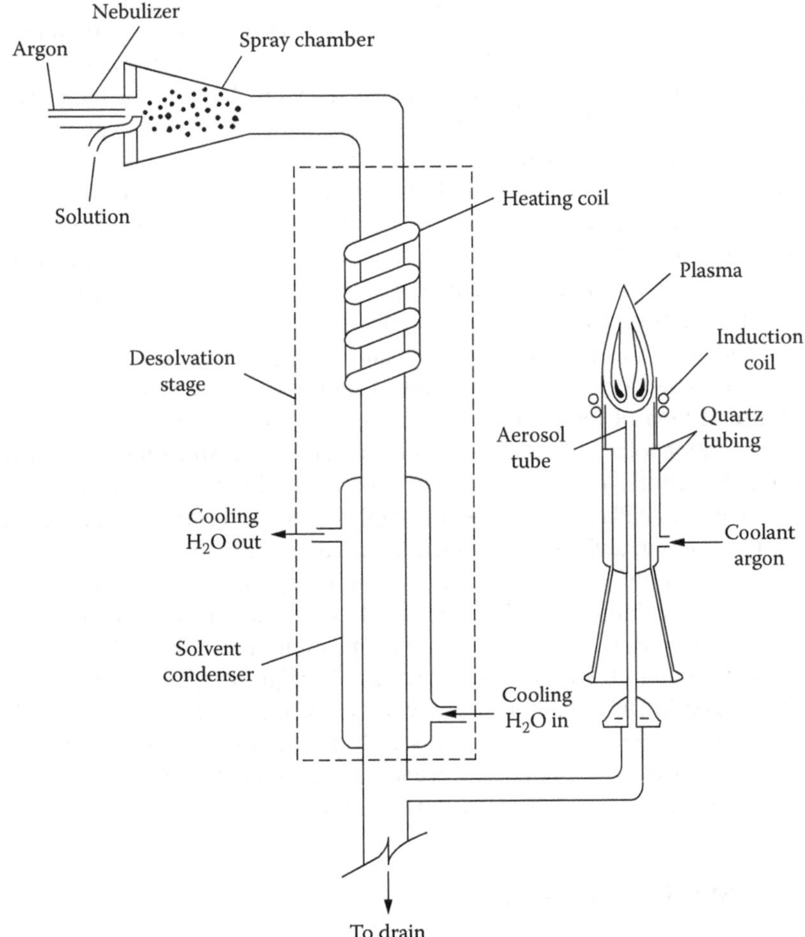

FIGURE 10.5 Nebulizer, spray chamber, and desolvation parts of an inductively coupled plasma spectrometer.

The determination of arsenic by ICP-AES after reduction to arsine by sodium tetrahydroborate was reported with a detection limit below 0.1 μg/L.[48] This high sensitivity was used in combination with the selectivity for arsenic species obtained by varying the composition of the reducing medium. Several authors reported that As(III) is reduced to AsH_3 in near-neutral solutions, whereas As(V) is reduced only in solutions of lower pH.[49] Methylarsonic acid and similar organoarsenic compounds can also be reduced to volatile substituted arsines. Yan et al.[50] reported a method for the determination of ultra-trace amounts of As(III) and As(V) in water by flow injection online sorption preconcentration and separation coupled with ICP-MS using a knotted reactor (KR). The determination of As(III) was achieved by selective formation of the As(III)-pyrrolidine dithiocarbamate complex over a sample acidity range of 0.01 to 0.7 *M*/L HNO_3, its adsorption onto the inner walls of the KR made from 150-cm-long, 0.5-mm-i.d. PTFE tubing, elution with 1 *M*/L HNO_3, and detection by ICP-MS. Total inorganic arsenic was determined after prereduction of As(V) to As(III) in a 1% (m/v) L-cysteine-0.03 *M*/L HNO_3 media. The concentration of As(V) was calculated by the difference (the total inorganic arsenic and As(III)). The presence of organoarsenic species such as monomethylarsonate and dimethylarsinate in water samples has no effect on the results of As(III) and As(V). The detection limits of 0.021 μg/L for As(III) and 0.029 μg/L for total inorganic arsenic were achieved. The precision for 14 replicate determinations of 1 μg/L As(III) was 2.8% (RSD) with drift correction and 3.9% (RSD) without drift correction. Yu et al.[51] presented a novel and simple method for inorganic antimony speciation based on selective solid-phase extraction (SPE), separation of antimony(III), and highly sensitive ICP-MS detection of total antimony and antimony(V) in the aqueous phase of the sample. Nonpolar SPE cartridges, such as an isolute silica-based octyl (C_8) cartridge, selectively retained the Sb(III) complex with ammonium pyrrolidine dithiocarbamate (APDC), while the uncomplexed Sb(V) remained as a free species in the solution and passed through the cartridge. Sb(III) concentration was calculated as the difference between total antimony and Sb(V) concentrations, with the detection limit of 1 ngL^{-1}. Factors affecting the separation and detection of antimony species were investigated. Acidification of samples led to partial or complete retention of Sb(V) on C_8 cartridge. Foreign ions tending to complex with Sb(III) or APDC did not interfere with the retention behavior of the Sb(III)-APDC complex. This method was successfully applied to the antimony speciation of various types of water samples. Recently, Waddell et al.[52] reviewed the status of elemental speciation using the ICP-MS technique.

10.4 MASS SPECTROMETRY

In MS, the species under investigation is bombarded with a beam of electrons, which produce the ionic fragments of the original species. The resulting assortment of charged particles is then separated according to their masses and charge ratio, and the spectrum produced is called a mass spectrum, which is a record of information regarding various masses produced and their relative abundance. As such, MS is not capable of speciating metal ions; however, it augments the sensitivity of detection when coupled with HG-AAS or HG-ICP instruments. Various components of a mass spectrometer include the sample handling unit, ion source, electrostatic accelerating unit, analyzer, ion collector, detector, and recorder. An outline of the main components of MS is shown in Figure 10.6, while Figure 10.7 presents a schematic representation of a mass spectrometer.

During the last decade, MS has been coupled to ICP, resulting in a delicate research tool intended for the well-trained scientist only, in a more robust and well-established analytical technique for trace and ultra-trace element determination, with a few thousand instruments in use worldwide. Despite this immense

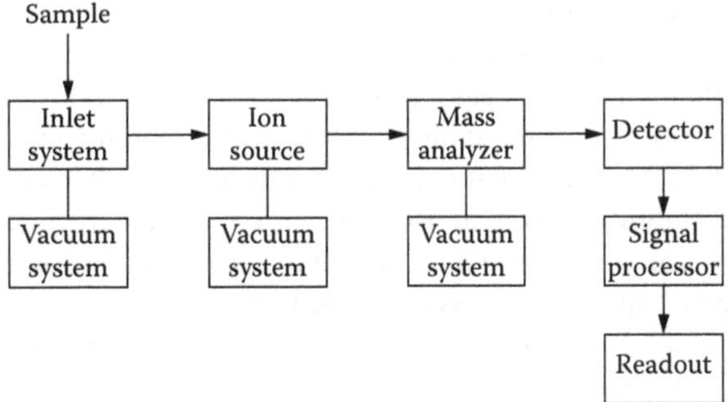

FIGURE 10.6 Outline of the main components of mass spectrometry.

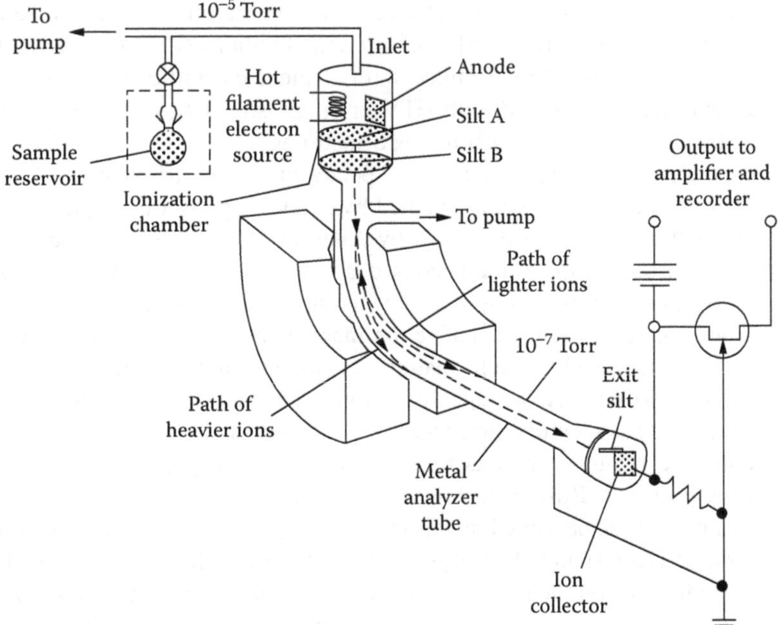

FIGURE 10.7 A mass spectrometer.

success, it should be realized that in its standard configuration, that is, equipped with a pneumatic nebulizer for sample introduction and with a quadrupole filter, ICP-MS also shows a number of important limitations and disadvantages. These limitations and disadvantages are as follows: the occurrence of spectral interference may hamper accurate trace element determination, solid samples have to be taken into solution prior to analysis, and no information on the chemical form in which an element appears can be obtained. Self-evidently, efforts have been and still are made to overcome the aforementioned limitations to the largest possible extent. The application of a double-focusing sector field mass spectrometer in

ICP-MS instrumentation offers a higher mass resolution, such that spectral overlap can be avoided to a great extent. Additionally, in a sector field instrument, photons are efficiently eliminated from the ion beam, resulting in very low background intensities, making it also well suited for extreme trace analysis. The combination of ICP as an ion source and a quadrupole filter operated in a so-called alternate stability region, an ion trap, or a Fourier transform ion cyclotron resonance mass spectrometer, allows for higher mass resolution to be obtained. With modern quadrupole-based instruments, important types of spectral interference can be avoided by working under cool plasma conditions or by applying a collision cell. The use of electrothermal vaporization (ETV) or, especially, laser ablation (LA) for sample introduction permits the direct analysis of solid samples with sufficient accuracy for many purposes. The application range of LA-ICP-MS has become very wide, and the introduction of UV lasers led to improved spatial resolution. Solid-sampling ETV-ICP-MS, on the other hand, can be used for some specific applications only, but accurate calibration is more straightforward than with LA-ICP-MS. Limited multielement capabilities resulting from the transient signals observed with ETV or single-shot LA, can be avoided by the use of a time-of-flight (TOF) ICP-MS instrument.

The positive and negative ion mass spectra of arsenic trioxide (As_2O_3) and arsenic pentaoxide (As_2O_5) were obtained by single-step laser desorption/ionization time-of-flight mass spectrometry by Allen et al.[53] Pulsed UV radiation at 266 nm was used for the simultaneous desorption and ionization of the solid sample. High-mass cluster ions that are unique to the oxidation state of each oxide sample appeared in the negative ion mass spectra. As_2O_3 produces $As_3O_5^-$, while As_2O_5 yields $As_3O_8^-$. The formation of unique negative cluster ions presents the possibility to conduct arsenic oxidation state speciation by laser desorption/ionization mass spectrometry. The ability of time-of-flight mass spectrometry to examine the relative amounts of each arsenic oxide present in a series of mixtures is discussed. Rosenberg[54] reviewed the application of mass spectrometry based on atmospheric pressure ionization techniques (atmospheric pressure chemical ionization [APCI] and electrospray ionization [ESI]) for speciation analysis. This review focuses on the increasing interest of atmospheric pressure ionization and ESI techniques, which have been used for qualitative and quantitative speciation analysis in past years. The main contents of this research are the recent applications, including redox species, use of ESI-MS for structural elucidation of metal complexes, and characterization and quantification of small organometallic species (metallothioneins and phytochelatins) with relevance to the environment, health, and food.

10.5 X-RAY SPECTROSCOPY

X-rays are generated by the bombardment of a metal target with a beam of highly energetic electrons or by exposure of a substance to a primary beam of x-rays in order to generate a secondary beam of x-rays or by employment of a radioactive source whose decay results in x-ray emission. The most important x-ray techniques are absorption, fluorescence, and diffraction. Mainly, the absorption mode is used for metal ion speciation. Kim et al.[55] reported the speciation of mercury using x-ray absorption spectroscopy (XAS). This technique has the additional benefits of being nondestructive to the sample, being element-specific, being relatively sensitive at low concentrations, and requiring minimal sample preparation. In this study, Hg L_{III}-edge extended x-ray absorption fine structure (EXAFS) spectra were collected for several mercury mine tailings (calcines) in the California Coast range. Speciation data (mercury phases present and relative abundance) were obtained by comparing the spectra from heterogeneous, roasted (calcined) mine tailings samples with a spectral database of mercury minerals and sorbed mercury complexes. Speciation analyses were also conducted on known mixtures of pure mercury minerals in order to assess the quantitative accuracy of the technique. While some calcine samples consisted exclusively of mercuric sulfide, others contained additional, more soluble mercury phases, indicating a greater potential for the release of mercury into solution. Also, a correlation was observed between samples from hot spring mercury deposits, in which chloride levels are elevated, and the presence of mercury-chloride species as detected by the speciation analysis. The speciation results demonstrated the ability of XAS to identify multiple mercury phases in a heterogeneous sample, with a quantitative accuracy of ±25% for the mercury-containing phases considered. Use of this technique, in conjunction with standard microanalytical techniques, such as x-ray diffraction and electron probe microanalysis, may be beneficial in the prioritization and remediation of mercury-contaminated mine sites. Shaffer et al.[56] used x-ray absorption near-edge structure (XANES) spectroscopy for chromium speciation in soil samples. Similarly, Cutler et al.[57] XANES and extended x-ray absorption fine structure (EXAFS) spectroscopic methods for uranium speciation. Shoji and coworkers[58] used x-ray absorption to determine the valency states of arsenic, chromium, and zinc in fine particulate matter separated from coal fly ash samples. Parsons et al.[59] reviewed metal ion speciation using x-ray absorption spectroscopy.

Dupont et al.[60] used transmission electron microscopy (TEM) coupled with energy-dispersive x-ray spectroscopy (EDXS) to determine the speciation of copper and chromium metal ions on the constitutive moieties of the lignocellulosic substrate. The EDXS analysis showed that metal ions are preferentially sorbed onto microstructures rich in lignin moieties. Energy electron loss spectroscopy (EELS) was used to determine the oxidation states of chromium in association with lignocellulosic moieties. Howe et al.[61] localized chromium and determined the oxidation states and possible complexation mode of chromium in intact plant tissue (*Trifolium brachycalycinum*) by means of XANES, synchrotron XRF microprobe spectroscopy, and EPR spectroscopy. Paktunc et al.[62] used x-ray absorption spectroscopy for the characterization and speciation of arsenic in the Ketza River mine tailings deposited. The tailings are mainly composed of iron oxyhydroxides, quartz, calcite, dolomite, muscovite, ferric arsenates, calcium-iron arsenates, arsenopyrite, and pyrite. Orescanin et al.[63] developed and reported a fast, sensitive, and selective EDXRF method for chromium speciation [Cr(III) and Cr(VI)] in environmental and industrial samples via preconcentration with ammonium pyrrolidine dithiocarbamate (ADPC). The authors studied the effect of pH and the presence of organic matter, carbonate species, vanadium, manganese, and iron on speciation. The speciation behavior was found to be affected at different pHs and by the presence of other moieties.

10.6 MISCELLANEOUS SPECTROSCOPIC METHODS

In addition to the above-discussed spectroscopic methods, new trends emerged in literature for metal ion speciation using new, advanced spectroscopic methods. These spectroscopic techniques include electrospray, fluorescence, and Raman. These methods are useful for the speciation of organometallic compounds such as peptides, proteins, metal complexes, and so forth.[2] Therefore, these methods are used to ascertain the bindings of metal ions with biomolecules in the living systems. These are also very useful methods for determining the stoichiometry of the chemical reactions of different metal ionic species. Some examples of the application of these techniques for metal ion speciation are discussed below.

Moulin et al.[64] used ES-MS for uranium speciation. Free uranyl (UO_2^{2+}), the first hydroxo complex (UO_2OH^+), and the oligomeric species $(UO_2)_3(OH)_5^+$ were observed as a function of pH. The influence of various parameters, such as the cone voltage, temperature, and gas flow rate, was discussed. The results were compared with those obtained by time-resolved laser-induced fluorescence and were in agreement. Furthermore, the same group[65] speciated the complexes of europium and strontium metal ions with diethylenetriaminepentaacetic acid (DTPA) by electrospray ionization–mass spectrometry (ESI-MS) and time-resolved laser-induced fluorescence (TRLIF). Variation in pH allowed us to readily investigate whether different species (protonated, hydrolyzed, etc.) existed in the pH range from 2 to 9 and to evaluate the stoichiometry and conditional stability constant for the Eu-DTPA complex. Johnston et al.[66] used Raman spectroscopy to investigate the speciation of aluminum metal ion in an aqueous system over a wide range of solution compositions and hydration. The authors used a ternary phase diagram to correlate the observed changes in the spectra with the composition of the solution and with the dimerization of aluminate that occurred at elevated aluminate concentrations (>1.5 *M*). The activity of water and the molar concentrations of the monomeric and dimeric aluminate species were estimated using the relative intensities of the Al-O stretching bands from the Raman spectra.

10.7 CONCLUSION

Metal ion speciation by spectroscopic methods, particularly by UV-visible spectroscopy, may be a useful and affordable technique for developing and underdeveloped countries. But sometimes, the speciation of metal ions by spectroscopic methods requires the conversion of one form of metal ion into another, which requires chemical conversion. The oxidation or reduction of one oxidation state

TABLE 10.2
Speciation of Metal Ions Using Spectroscopic Techniques

Metal Ion	Techniques	Ref.
Aluminum	Visible using hydroquinone and ferron	67
Chromium	Visible	68
	Solid-phase extraction and graphite furnace AAS	69
Iron	Visible using bathophenanthroline complexing reagent	11
	2,2′-Dipyridyl-2-pyridylhydrazone	12
	1,10-Phenanthroline	70
	Solvent extraction and flame AAS	71
Chromium	Visible using diphenylcarbazide and methyl orange	13
Aluminum	ETV-AAS	72
Aluminum	ETV-AAS	73
Aluminum	ETV-AAS	74
Arsenic	AAS	75
Arsenic	Electrothermal AAS	76
Arsenic	HG-AAS	6–8, 24, 25
Arsenic	HG-AFS	77
Arsenic	Visible (molybdenum blue method)	5
As(III), As(V), DMA, and MMA	SPE-HG-AFS	78
As(III) and As(V)	SPE-ICP-AES	79
Cadmium	ETV-AAS	80
Chromium	Selective adsorption on melamine-urea-formaldehyde resin, AAS	81
Chromium	Selective absorption on a cellulose, AAS	68
Copper	ETV-AAS	82
Iron	AAS	83
Platinum	AAS	84
Selenium	AAS	85
Selenium	AAS	86
Tin	HG-AAS	87
Tin	HG-AAS	88
Arsenic	ICP-AES	41
Arsenic	HG-ICP	48, 49
Antimony	SPE-ICP	51
Chromium	ICP-AES	89
Mercury	ICP-MS	90
Arsenic	XAS	57
Mercury	XAS	54

TABLE 10.2 (Continued)
Speciation of Metal Ions Using Spectroscopic Techniques

Metal Ion	Techniques	Ref.
Chromium	XANES	55, 57
Chromium	XRF	91
Uranium	XANES	56
Zinc	XAS	57
Vanadium	EPR	92
Manganese	EPR	93

Note: AAS: atomic absorption spectrometer, AES: atomic emission spectrometer, AFS: atomic fluorescence spectroscopy, EPR: electron paramagnetic resonance, ETV: electrothermal vaporization, HG: hydride generation, ICP: inductively coupled plasma, ETV: electrothermal vaporization, MS: mass spectrometer, SPE: solid-phase extraction, XAS: x-ray absorption spectroscopy, XANES: x-ray absorption near edge structure, and XRF: x-ray fluorescence.

to another is a tedious and time-consuming job. Moreover, the process sometimes introduces errors due to the chemical reactivity of two or more ionic forms of a particular metal ion simultaneously. Hydride generation is not a universal speciation method, because at times, hydrides of all the oxidation forms are formed, which creates large rates of confusion and error. Therefore, these techniques could not achieve prominence in routine laboratory analysis and were replaced by chromatographic and capillary electrophoretic methods. Moreover, some of these techniques, such as AAS, MS, and ICP, were used as the best element-specific detectors in chromatographic and capillary electrophoretic technologies. The future of these techniques for metal ion speciation is not good, but in the future, spectroscopic instruments will be the only choice for metal ion speciation in the chromatographic and electrophoretic modalities.

REFERENCES

1. Cornelis R, Caruso J, Crew H, Heumann K (Eds.), *Hand Book of Elemental Speciation: Techniques and Methodology*, John Wiley & Sons, Chichester (2003).
2. Ray SJ, Andrade F, Gamez G, McClenathan, Rogers D, Schilling G, Wetzel W, Hieftje GM, *J. Chromatogr. A*, 1050: 3 (2004).
3. Melamed D, *Anal. Chim. Acta*, 532: 1 (2005).
4. Pyrzynska K, Wierzbicki T, *Talanta*, 64: 823 (2004).
5. Frankenberger Jr., WT (Ed.), *Environmental Chemistry of Arsenic*, Marcel Dekker, New York (2002).
6. Bogdanova VI, *Microkim. Acta*, II: 317 (1984).
7. Matsubara C, Yamamoto Y, Takamura K, *Analyst*, 112: 1257 (1987).
8. Nasu T, Kan R, *Analyst*. 113: 1683 (1988).

9. Tamari Y, Yamamoto N, Tsuji H, Kasuka Y, *Anal. Sci.*, 5: 481 (1989).
10. Palanivelu K, Balasubramaniam N, Ramakrishna TV, *Talanta*, 39: 555 (1992).
11. Quinteros A, Farre R, Lagarda MJ, *Food Chem.*, 75: 365 (2001).
12. Themelis DG, Tzanavaras PD, Kika FS, Sofoniou MC, *Fresenius' J. Anal. Chem.*, 371: 364 (2001).
13. Wu H, Wang ZP, Chen GS, *Fenxi Shiyanshi*, 20: 65 (2001).
14. Abbas A, Madrakian T, *Talanta*, 58: 311 (2002).
15. Taylor MJC, van Staden JF, *Analyst*, 119: 1263 (1994).
16. Abbaspour A, Najafi M, Kamyabi MA, *Anal. Chim. Acta*, 505: 301 (2004).
17. Rûika J, Hansen EH, *Anal. Chem. Acta*, 78: 145 (1975).
18. Luque de Castro MD, *Talanta*, 33: 45 (1986).
19. Dedina J, Rubeska I, *Spectrochem. Acta Part B*, 35: 119 (1980).
20. Welz B, Melcher M, *Anal. Chim. Acta*, 131: 17 (1981).
21. Parisis NE, Heyndrickx A, *Analyst*, 111: 281 (1986).
22. Dedina J, *Progress Anal. Spec.*, 11: 251 (1988).
23. Bahreyni-Toosi MH, Dawson JB, *Analyst*, 108: 225 (1983).
24. Tamari Y, Yamamoto N, Tsuji H, Kasuka Y, *Anal. Sci.*, 5: 481 (1989).
25. Martinez A, Mrraler-Rubro A, Luisa Cervera M, de la Guarelia M, *J. Anal. At. Spectrom.*, 16: 762 (2001).
26. Yin X, Hoffmann E, Ludke C, *Fresenius' J. Anal. Chem.*, 355: 324 (1996).
27. Carrero P, Malave A, Burguera JL, Burguera M, Rondon C, *Anal. Chim. Acta*, 438: 195 (2001).
28. Portman JE, Riley JP, *Anal. Chim. Acta*, 31: 509 (1964).
29. Arbaz-Zaver MH, Howard AG, *Analyst*, 105: 338 (1980).
30. Aggett J, Hayashi Y, *Analyst*, 112: 277 (1987).
31. Aggett J, Boyes G, *Analyst*, 114:, 1159 (1989).
32. Rahman MM, Chowdhury UK, Mukherjee SC, Mondal BK, Paul K, Lodh D, Biswas BK, Chanda CR, Basu GK, Saha KC, Roy S, Das R, Palit SK, Quamruzzaman Q, Chakraborti D, *J. Toxicol. Clin. Toxicol.*, 39: 683 (2001).
33. Yalcin S, Le XC, *J. Environ. Monit.*, 3: 81 (2001).
34. Schwenzer SP, Tommaseo CE, Kersten M, Kirnbauer T, *Fresenius' J. Anal. Chem.*, 371: 927 (2001).
35. Sakamoto H, Susa Y, Ishiyama H, Tomiyasu T, Anazawa K, *Anal. Sci.*, 17: 1067 (2001).
36. Chakraborti D, Jonghe de W, Adams F, *Anal. Chim. Acta*, 120: 121 (1980).
37. Chakraborti D, Adams F, Irgolic KJ, *Fresenius' J. Anal. Chem.*, 323: 340 (1986).
38. Chakraborti D, Adams F, Vanmol W, Irgolic KJ, *Anal. Chim. Acta*, 196: 23 (1987).
39. Chung CH, Iwamoto E, Yamamoto M, Yamamoto Y, *Spectrochim. Acta*, 39B: 459 (1984).
40. Howard AG, Arbaz-Zavar MH, *Analyst*, 106: 213 (1981).
41. Benramdane L, Accominotti M, Fanton L, Malicier D, Vallon JJ, *Clin. Chem.*, 45: 301 (1999).
42. Nielsen S, Hansen EH, *Anal. Chim. Acta*, 343: 5 (2002).
43. Pyell U, Dworschak A, Nitschke F, Neidhart B, *Fresenius' J. Anal. Chem.*, 363: 495 (1999).
44. Cabon JY, Cabon N, *Fresenius' J. Anal. Chem.*, 368: 484 (2000).
45. Shemirani F, Rajabi M, *Fresenius' J. Anal. Chem.*, 371: 1037 (2001).
46. Jarvis KE, Gray AL, Houk RS, *Handbook of Inductively Coupled Plasma Mass Spectrometry*, Glasgow, Blackie (1992).

47. Connor GO, Evans EH, in *Inductively Coupled Plasma Spectrometry and Its Applications*, Hill SJ (Ed.), Sheffield Academic Press, Sheffield (1999).
48. Pahlavanpour B, Michael T, Laurance T, *Analyst*, 106: 467 (1980).
49. Howard AG, Arbaz-Zavar MH, *Analyst*, 105: 338 (1980).
50. Yan XP, Kerrich R, Hendry MJ, *Anal. Chem.*, 70: 4736 (1998).
51. Yu C, Cai Q, Guo ZX, Yang Z, Khoo SB, *Analyst*, 127: 1380 (2002).
52. Waddell R, Lewis C, Hang W, Hassell C, Majidi V, *Appl. Spectrosc. Rev.*, 40: 33 (2005).
53. Allen TM, Bezabeh DZ, Smith CH, McCauley EM, Jones AD, Chang DP, Kennedy IM, Kelly PB, *Anal. Chem.*, 68: 4052 (1996).
54. Rosenberg E, *J. Chromatogr. A*, 1000: 841 (2003).
55. Kim CS, Brown GE Jr., Rytuba JJ, *Sci. Total Environ.*, 261: 157 (2000).
56. Shaffer RE, Cross JO, Roze-Pehrsson SL, Elam WT, *Anal. Chim. Acta*, 442: 295 (2001).
57. Cutler JN, Jiang DT, Remple G, *Can. J. Anal. Sci. Spectrosc.*, 46: 130 (2001).
58. Shoji T, Huggins FF, Huffman GP, *Energy Fuel*, 16: 325 (2000).
59. Parsons JG, Aldrich MV, Gardea-Torresdey JC, *Appl. Spectrosc. Rev.*, 37: 187 (2002).
60. Dupont L, Bouanda J, Ghanbaja J, Dumonceau J, Aplincourt M, *J. Colloid Interface Sci.*, 279: 418 (2004).
61. Howe JA, Loeppert RH, DeRose VJ, Hunter DB, Bertsch PM, *Environ. Sci. Technol.*, 37: 4091 (2003).
62. Paktunc D, Foster A, Laflamme G, *Environ. Sci. Technol.*, 37: 2067 (2003).
63. Orescanin V, Mikelic L, Lulic S, Rubcic M, *Anal. Chim. Acta*, 527: 125 (2004).
64. Moulin C, Charron N, Plancque G, Virelizier H, *Appl. Spectrosc.*, 54: June (2000).
65. Moulin C, Amekraz B, Steiner V, Plancque G, Ansoborlo E, *Appl. Spectrosc.*, 57: 1151 (2003).
66. Johnston CT, Agnew SF, Schoonover JR, Kenney JW III, Page B, Osborn J, Corbin R, *Environ. Sci. Technol.*, 36: 2451 (2002).
67. Hodges SC, *J. Soil Sci. Soc. Am.*, 51: 57 (1987).
68. Lynch TP, Kernoghan NJ, Wilson JN, *Analyst*, 109: 839 (1984).
69. Hu G, Deming RL, *Anal. Chim. Acta*, 535: 237 (2005).
70. Faizullah AT, Townshend A, *Anal. Chim. Acta*, 167: 225 (1985).
71. Yaman M, Kaya G, *Anal. Chim. Acta*, 540: 77 (2005).
72. Campbell PGC, Bisson M, Bougie R, Tessier A, Villeneuve JP, *Anal. Chem.*, 55: 2246 (1983).
73. Keirsse A, Smeyers-Verbeeke J, Verbeelen D, Massart D, *Anal. Chim. Acta*, 196: 103 (1987).
74. Blanco-Gonzalez E, Perez-Parajon J, Garcia-Alonso JI, Sanz-Medel A, *J. Anal. At. Spectrom.*, 4: 175 (1989).
75. Van Cleuvenbergen RJA, Van Mol WE, Adams FJ, *J. Anal. At. Spectrom.*, 3: 169 (1988).
76. Shemirani F, Baghdadi M, Ramezani M, *Talanta*, 65: 882 (2005).
77. Ming-Dong L, Yan XP, *Talanta*, 65: 627 (2005).
78. Cava-Montesino P, Nilles K, Cervera ML, de la Guardina M, *Talanta*, 66: 895 (2005).
79. Jitmanee K, Oshima M, Motomizu S, *Talanta*, 66: 529 (2005).
80. Olayinka KO, Haswell SJ, Grzeskowiak R, J. *Anal. At. Spectrom.*, 4: 171 (1989).
81. Demirata B, *Mikrochim. Acta*, 136: 143 (2001).

82. Gardiner PE, Ottaway JM, Fell GS, Burns RR, *Anal. Chim. Acta*, 124: 281 (1981).
83. Cox JA, Al-Shakshir S, *Anal. Lett.*, 21: 1757 (1988).
84. Dominici C, Alimonti A, Caroli S, Petrucci F, Castello MA, *Clin. Chim. Acta*, 158: 207 (1986).
85. Muangnoicharoen S, Chiou KY, Manuel OK, *Talanta*, 35: 679 (1988).
86. Apte SC, Howard AG, *J. Anal. At. Spectrom.*, 1: 379 (1986).
87. Donard OFX, Rapsomanikis S, Weber JH, *Anal. Chem.*, 58: 772 (1986).
88. Han JS, Weber JH, *Anal. Chem.*, 60: 316 (1988).
89. Cox AG, Cook IG, McLeod CW, *Analyst*, 110: 331 (1985).
90. Bushee DS, *Analyst*, 113: 1167 (1988).
91. Arber JM, Urch DS, West NG, *Analyst*, 113: 779 (1988).
92. Jakusch T, Jin W, Yang L, Kiss T, Crans DC, *J. Inorg. Biochem.*, 95: 1 (2003).
93. Chiswell B, Mokhtar MB, *Talanta*, 34: 307 (1987).

11 Speciation of Metal Ions by Electrochemical and Radiochemical Methods

Electrochemical methods are used for metal ion speciation due to their unique advantages. These techniques are unique, as they are based on interfacial phenomena. Electrochemical methods follow the low-energy excitation principle and, hence, are species selective in nature rather than element selective. These techniques require very small sample volume and work at low limits of detection. The working principles of electrochemical methods are based on the measurement of electrical signals associated with molecular or interfacial properties of the chemical species. The development of an ion-selective electrode in the 1970s may be considered as the start of the boon for the selectivity of electrochemical methods. In spite of all these salient features, these techniques suffer from certain drawbacks. The main limitations are the poor selectivity in real-world samples and the susceptibility of the electrode surface to fouling by surface-active materials into unknown matrices.[1,2] Moreover, as in the case of spectrometric analysis, the conversion of one oxidation form into another is required for electrochemical methods, or the speciation is carried out by determining the concentrations of a particular oxidation state and total element, respectively, in case of the presence of only two oxidation states of the elements. Some reviews discuss metal ion speciation by electrochemical methods.[3–5] Radiochemical methods are also being used for metal ion speciation due to their high sensitivity, selectivity, and low limit of detection. These methods can also be used to study the transformation of one oxidation state into another during chemical and biological processes by using radioactive metal ions. Radiochemical methods used for metal ion speciation are isotopic dilution and activation analysis. As in the case of electrochemical techniques, radiochemical methods also have certain limitations. Serious drawbacks of these methods include their narrow range of applications — they are essentially used for radioactive or labile metal ions. Moreover, some metal ions lose their oxidation states during labeling experiments. This chapter discusses state-of-the-art metal ion speciation analysis by electro- and radiochemical methods.

11.1 ELECTROCHEMICAL METHODS

Electrochemical techniques are classified on the basis of their working principles. The International Union for Pure and Applied Chemistry (IUPAC) classified and defined their names.[6,7] The different classes of electrochemical techniques are electrolytic, potentiometric, conductometric, polarographic, amperometric, and coulometric methods. According to another classification, the electrochemical methods can be classified into two groups: potentiometry and voltammetry. In potentiometry, measurements are made of the potential existing between an ion-selective electrode and a reference electrode, but the detection limit of this technique is poor. On the contrary, the detection limit of voltammetry is good in comparison to that of potentiometry, so the latter technique can be used precisely for metal ion speciation.[8,9] Direct-current polarography is able to detect different oxidation states of many metal ions. In addition to the above classifications, the electrochemical techniques used for metal ion speciation studies may be categorized again into two groups — dynamic (active) and passive, depending on whether the process of measurement forces concentration (activity) changes at the electrode interface. Voltammetry (polarography and cyclic, pulse, stripping, voltammetric methods) belongs to the former class, while potentiometry is an example of the passive technique. Although there are many electrochemical methods available, this chapter will focus only on those electrochemical techniques that are used for metal ion speciation. The details of the working principles and outline of the electrochemical techniques are beyond the scope of this book, and the reader should consult some general textbooks on this issue.[10–12] The electrochemical methods are powerful tools for the speciation of many elements and have been applied to about 30 metals. The electrochemical methods appear to provide the best opportunities to experimentally model the bioavailability of elements and their complexes with organic and inorganic ligands. These techniques require less handling of samples, and samples are in contact with fewer potential sources of contamination. The electrochemical methods are useful for the speciation of metals, nonmetals, and organic compounds. The most important metal ions studied for their speciation are silver, arsenic, bismuth, cadmium, cobalt, chromium, copper, euradium, iron, gallium, indium, manganese, molybdenum, nickel, lead, antimony, selenium, tin, tellurium, uranium, vanadium, ytterbium, tungsten, and zinc.[3] The redox potential of an electrode can be varied accurately, precisely, and continuously over a wide potential range, and the study of the kinetics of metal complex dissociation at an electrode is supported by well-established theories.[13–15] Electrochemical techniques are based on the measurement of redox and half-wave potentials and labile/inert discrimination. The most commonly used electrodes in electrochemical techniques are hanging mercury drop electrode (HMDE), thin mercury film (on glassy carbon) electrode (TMFE), dropping (or static) mercury electrode (DME), jet stream mercury film electrode (JSMFE), streaming mercury electrode (SME), carbon fiber electrode (CFE), ion-selective electrode (ISE), and some chemically modified electrodes. Some important electrochemical techniques used for speciation are discussed below.

11.1.1 Polarography

Polarography was developed by J. Heyrovsky in 1922, and the current versus voltage curves are called polarograms. One oxidation state is electrochemically active, while the other is inactive under the applied specific potential range. Total metal amounts can be determined after chemical treatment of the sample. For example, As(III) is the electrochemically active species, and As(V), which is electrochemically inactive, can be determined by its reduction to As(III) species. Polarography has been used to speciate many other metal ions, such as iron, chromium, tellurium, tin, manganese, antimony, arsenic selenium, vanadium, eurobium, and uranium.[3] The static mercury drop electrode (SMDE) and dropping mercury electrodes are commonly used in polarography. Normally, polarography is not a useful speciation technique for natural water, as the concentration of trace elements in natural waters are in the range of 10^{-8} to 10^{-10} *M*, while the detection limit of polarography is 10^{-7} *M*. Pseudo-polarography, a pseudo-polarogram is a plot of anodic stripping voltammetry (ASV) stripping peak current versus deposition potential, has been used for metal ion speciation. Brown and Kowalski[13] reported arsenic, cadmium, and lead speciation in natural water samples using pseudo-polarography. Nawar et al.[16] reported copper speciation using polarography. The direct-current polarographic behavior of Cu(II) and Cu(I) in the presence of 0.15 *M* triethanolamine and 30% (v/v) methanol containing two reduction wavelengths was studied. The first wave for the reduction of Cu(II) to Cu(I) was found to be reversible, while the second wave corresponding to the reduction of Cu(I) to Cu(0) was irreversible. The first wave for Cu(II) was used to determine the concentration of Cu(II) in a mixture of Cu(II) and Cu(I). Cu(I) was calculated by the difference between total copper concentration and Cu(II) species. The serious limitation of polarography in speciation is the incapability to detect all oxidation states of a particular metal ion simultaneously. Ion-selective electrodes can detect all oxidation states of a particular metal ion, but the sensitivity of these electrodes, especially in natural water samples, is poor.

11.1.2 Stripping Voltammetry

Anodic stripping voltammetry (ASV) is the most widely used electrochemical technique for metal ion speciation in water samples. The basic principle of this technique is similar to polarography. The most commonly used ASV technique is square wave stripping at a glassy carbon thin mercury film electrode (TMFE). The concentration of metal ions up to 10^{-11} *M* can be detected by this technique, and it can be used for metal ion speciation without extra sample preparation. Dissolved air and oxygen are the interferents in this technique, and an inert atmosphere, by passing pure nitrogen, should be generated during the experiment.

Rasul et al.[17] reported the development of an inexpensive ASV technique for the speciation of arsenic in groundwater. The measurements are validated by atomic absorption, atomic emission, and other techniques. Vancara et al.[18] presented a new stripping method for the speciation of arsenic in water samples with a gold-film-plated carbon paste electrode for use in constant current stripping analysis (CCSA). In the novelized procedure, the differentiation between As(III) and chemically prereduced As(V), the effect of Cu(II) on the response of arsenic, and the stability of sample solutions were studied in detail. Compared to the voltammetric approach, the method utilizing CCSA offers a more rapid procedure with improved analytical characteristics such as reproducibility, selectivity over the Cu(II) ions, or lower detection limit (3 ppb for As(III) and 0.5 ppb for As(V), respectively). Schwartz et al.[19] used adsorptive stripping voltammetry for the speciation of butyltin species in the surface water from tap water and river water in Trier, Germany. The detection limits of these species were in the range of 0.5 to 5.0 µg/L. An improvement in the limits of detection (10 to 20 times) was achieved by using preconcentration through solid-phase extraction. Huang and Dasgupta[20] used a field-deployed instrument for the speciation of arsenic into potable water. The developed method was based on a gold electrode in 4.5 *M* hydrochloric acid. The concentration of As(V) was calculated by the difference of total and As(III) arsenic concentrations.

Sauve et al.[21] described cadmium speciation using differential pulse anodic stripping voltammetry (DPASV), assuming that DPASV is sensitive to easily dissociated inorganic ion pairs and free Cd(II) while excluding organic complexes. The authors used the developed method for the determination of Cd(II) and Cd(0) species in soil samples collected from orchard, urban, forest, and agriculture fields. Bobrowski and coworkers[22] used catalytic adsorptive stripping voltammetry (CAdSV) for chromium speciation. The authors used adsorptive preconcentration of Cr(III)-diethylenetriamine-*N,N′,N″,N,N′*-pentaacetic acid (DTPA) complex and utilized the catalytic reaction in the presence of nitrate. The CAdSV enabled the oxidation-state speciation study of chromium content by direct determination of Cr(VI) in the presence of predominant Cr(III). The variation of CAdSV peak current with time for Cr(VI) and Cr(III) is shown in Figure 11.1. The cathodic stripping voltammetry (CSV) was developed by van den Berg,[23] and it involves the cathodic stripping of an insoluble film of the mercury salt of the analyte. Widespread application in metal speciation was not found. However, reports are available for the speciation of arsenic and selenium metal ions in their higher oxidation states.[3] Some metal ions are sensitive toward pH, ionic strength, temperature, and so forth, and the optimization of the experimental condition is an important aspect. For example, Sb(III) and Sb(V) behave in different ways at various pHs. Figure 11.2 indicates the behavior of these species as a function of acidity. Sb(III) can be determined in 0.2 *M* HCl, while total antimony can be detected in the pH range of 6 to 8 *M* HCl. The concentration of Sb(V) can be calculated by the difference of total concentration and the concentration of Sb(III) species.[24]

11.1.3 Differential Pulse Voltammetry

The speciation of metal ions by differential pulse voltammetry (DPV) can be achieved with shorter time periods. The basic working principle is similar to that of polarography. Liu et al.[25] developed a sensitive and simple DPV method for

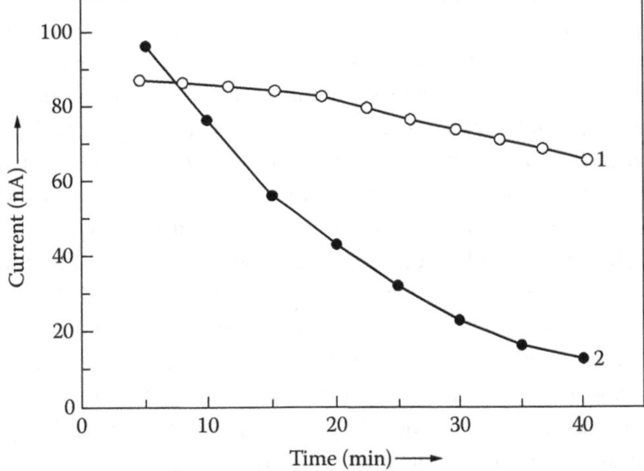

FIGURE 11.1 Variation of CAdSV peak current with time for (1) 5 n*M* Cr(VI) and (2) 50 n*M* Cr(III) metal ion species. (From Bobrowski, A., Bas, B., Dominik, J., Niewiara, E., Szalinska, E., Vignati, D., and Zarebski, J., *Talanta*, 63: 1003 (2004). With permission.)

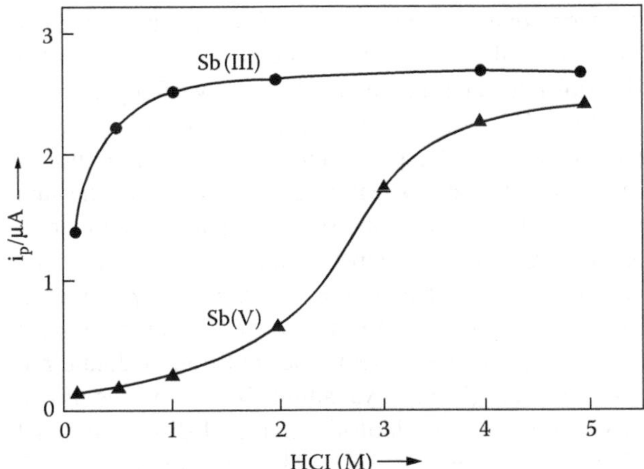

FIGURE 11.2 Effect of HCl concentration on ASV peak height of Sb(III) and Sb(V) metal ion species. (From Batley, G.E., and Florence, T.M., *J. Electroanal. Chem.*, 55: 23 (1974). With permission.)

the speciation of aluminum in natural waters using pyrocatechol violet chemically modified electrodes (PCV-CMEs). The speciation was based on the selective determination of different aluminum forms under two pH conditions. This method was used for aluminum analysis on natural water samples from different regions of China. The analyzed species were labile monomeric aluminum, nonlabile monomeric aluminum, total monomeric aluminum, acid-reactive aluminum, and acid-soluble aluminum. The authors compared the results with those obtained by

Driscoll's 8-hydroxyquinoline extraction ion-exchange method, and these were in good agreement. Furthermore, the same group[26] reported a fast, sensitive, and inexpensive DPV procedure for the speciation of Al(III) in natural waters and biological fluids using six catechols (L-dopa, dopamine, epinephrine, norepinephrine, caffeic acid, and *o*-benzenediol) as electroactive ligands. The speciation analysis was based on the measurement of the complexation capacity, namely, the different affinities of Al(III) for catechols and organic ligands under two pH conditions. The labile monomeric aluminum fraction (mainly inorganic aluminum) was determined at pH 4.6, while the total monomeric aluminum fraction was determined at pH 8.5. The developed method was applied to the speciation analysis of aluminum in natural water samples, and the results agreed with those obtained by ultrafiltration and dialysis and Driscoll's methods. The authors also explored the speciation analysis of aluminum in biological fluids. Oomen et al.[27] used a lead ion-selective electrode (Pb-ISE) and the DPASV techniques for lead speciation in artificial human intestinal fluid (chyme).

11.1.4 Potentiometric Stripping Analysis

Jagner[28] developed potentiometric stripping analysis. This technique uses the same initial steps as in ASV, that is, metal ion is deposited on TMFE at a controlled potential. This technique is different from ASV in the sense that a chemical oxidant in solution is allowed to diffuse metal ions to the electrode to oxidize the deposited metal ions, and the potential of the working electrode is followed as a function of time. Few reports are available on metal ion speciation using this mode of electrochemical technique. Munoz and Palmero[29] presented a review article on arsenic speciation using stripping potentiometry. The authors discussed the advantages of stripping potentiometry in comparison to other electrochemical methods. A good compilation on experimental optimization in a real sample is presented. Jagner[30] studied arsenic speciation using this method. Perry and Shafran[31] reported the systematic formation of aluminum polycations at room temperature by using automated potentiometric titration. Jakusch et al.[32] studied the chemistry of vanadium(IV) and vanadium(V) complexes of two ligands, 2,6-pyridinedicarboxylic acid (dipicolinic acid, H_2dipic) and 4-hydroxy-2,6-pyridinedicarboxylic acid (H_2dipic-OH). The speciation of two 2,6-pyridinedicarboxylic acids with vanadium(IV) and vanadium(V) ions was determined by pH-potentiometry at ionic strength and temperature, 0.2 *M* (KCl) and 298K, respectively. The speciation of some metal ions by different electrochemical methods is summarized in Table 11.1.

11.2 RADIOCHEMICAL METHODS

Among various radiochemical methods,[50,51] isotopic dilution, activation analysis, and radiometric analysis were used for metal ion speciation. These methods are applicable for metal ion speciation with good accuracy and precision. In activation analysis, labeled metal ion species are detected by measuring their radioac-

TABLE 11.1
Speciation of Metal Ions Using Electrochemical and Radiochemical Methods

Metal Ion	Sample Matrix	Techniques	Ref.
	Electrochemical Methods		
As(III), As(V), and total As	Seawater	CCSA	33
As(III), As(V), and total As	Seawater and urine	CCSA	34
As(III), As(V), and total As	River water	CCSA	35
As(III), As(V), and total As	Synthetic water	CCCSP	36
As(III), As(V), and total As	Tap water, surface water, and waste water	CPS	37
As(III), As(V), and total As	Water and rice	PSA	38, 39
As(III) and As(V)	—	PSA	28
Arsenic, cadmium, and lead	Water and algae	Polarography	17, 40
Arsenic	Groundwater	ASV	17
Arsenic	—	ASV	18
Arsenic	Soil	ASP	21
Antimony	—	Polarography/ASV	24
Copper	Borate glass	Polarography	16
Chromium	—	Polarography/ASV	41
Iron		Polarography/ASV	42
Lead	Human digestive fluid (Chyme)	DPASV	27
Manganese	Seawater	ASV	43
Manganese	Sea and estuarine water	Polarography	44
Selenium	—	Polarography/ASV	45
Tellurium	—	Polarography/ASV	46
Tin	—	Polarography/ASV	47
Uranium	—	Polarography/ASV	48
Vanadium	—	Polarography/ASV	49
	Radiochemical Methods		
Arsenic	Human breast milk	ID-MS	61
Chromium	Aerosol particles	ID-MS	59
Radionuclides of natural uranium, thorium, plutonium, and americium	Sea sediment	SEM	70
^{239}Plutonium and ^{240}Plutonium	Seawater	-Spectrometer	62
Selenium	Water and serum	ID-ICP-MS	58

Note: ASV: anodic stripping voltammetry, DPASV: differential pulse anodic stripping voltammetry, CCSA: constant-current stripping analysis, CCCSP: constant-current coulometric stripping analysis, CPS: cathodic stripping analysis, and PSA: potentiometric stripping analysis.

tivities. In the isotopic dilution method, pure but radioactive forms of metal ion to be detected are mixed with sample in known amounts. After equilibration, a fraction of the metal ion of interest is isolated and analyzed based on the activity of this isolated fraction. Radiometric analysis involves the use of radioactive reagent to completely separate the metal ion from the bulk of the sample, and then the activity of the isolated portion is measured. Alternatively, metal ion may be titrated with a radioactive reagent, and an end point is then established by activity measurement.[52]

11.2.1 Isotopic Dilution

Isotopic dilution analysis (IDA) was applied to inorganic mass spectrometry for metal ion speciation.[53] The utility and importance of IDA lies in the fact that it provides accuracy and precision.[54] And, with its use, the problems associated with instrumental drift and matrix effects during mass spectrometric detection can be overcome. Furthermore, IDA relies on the measurement of isotope ratios, not external calibration. A prerequisite for IDA is that the target metal ion should have more than one stable isotope — as this may be the case of the majority of the elements, most can be investigated using this method. Two stable isotopes of the target metal ion are chosen, which should ideally have a large difference in natural abundance. The isotopically enriched spike should have the isotope of lowest natural abundance enriched to as high an abundance as possible (the spiked isotope), with the lower-abundance isotope heavily depleted (the reference isotope). The isotope amount ratio in the sample is measured, after spiking and equilibration, and entered into the isotope dilution equation:

$$C_x = [(C_sW_sM_x)/(W_xM_s)] \times [(A_s - RB_s)/(RB_x - A_x)] \qquad (11.1)$$

where

- C_x is the concentration of metal ion in the sample
- C_s is the concentration of metal ion in the spiked solution
- W_s is the mass of spiked metal ion
- M_x is the molar mass of metal ion in the sample
- W_x is the mass of the sample
- M_s is the molar mass of metal ion in the spiked solution
- A_s is the abundance of reference isotope in the spiked solution
- B_s is the abundance of spiked isotope in the spiked solution
- B_x is the abundance of spiked isotope in the sample
- A_x is the abundance of reference isotope in the sample
- R is the reference and spiked isotope amount ratio in the sample after spiking

Cornelis[55] presented a review on metal ion speciation using radiochemical methods. Clough et al.[53] reviewed the speciation of metal ions using the isotopic dilution technique coupled with inductively coupled plasma–mass spectrometry (ICP-MS). To achieve maximum accuracy and precision in this method, sample preparation, selection of the most appropriate isotopic internal standard, characterization of the isotopically enriched analog, addition of the isotopically enriched analog, instrumental analysis, calculation of the results, and estimation of uncertainty are the most important parameters.[56] IDA has been used for the speciation of various elements in soils, sediments, natural and artificial water, gases, and reference materials.[53] Some cases of metal ion speciation by IDA were discussed herein. Lambertsson et al.[57] speciated ^{198}Hg, ^{200}Hg, $CH_3{}^{198}Hg(I)$, and $CH_3{}^{201}Hg(I)$ isotopes using IDA. Turner and coworkers[58] studied the speciation of selenium species by IDA in water and serum samples. The isotopic amount ratios reported were in the range of 0.06 to 0.08. Nusko and Heumann[59] used an isotope dilution–mass spectrometric (ID-MS) method for Cr(III) and Cr(VI) speciation in aerosol particles using extractive separation and thermal ionization ID-MS. Cr(III) and Cr(VI) spike species, enriched in ^{53}Cr, were applied for the isotope dilution step. After leaching filter-collected aerosol samples by an alkaline solution, species separation was carried out by extraction with a liquid anion exchanger in methyl isobutyl ketone. Cr(VI) in the organic phase was reextracted into an ammonia solution, and chromium was then isolated from both fractions of species by electrodeposition. The detection limits of 30 pg/m^3 for Cr(III) and of 8 pg/m^3 for Cr(VI) were achieved in atmospheric aerosols for volumes of air samples of about 120 m^3. These low detection limits allowed for the determination of chromium species in continental aerosol particles, depending on different seasons. The Cr(III)/Cr(VI) ratio was always found to be about 0.3, whereas dust from soil erosion, which is probably the primary source of chromium in the atmosphere, showed higher ratios. The accuracy of the method was demonstrated in two interlaboratory comparisons of Cr(VI) determinations in welding dust samples. Furthermore, the authors reported that the ID-MS method contributed to the certification of a corresponding standard reference material organized by the Standard Reference Bureau of the European Union. Chromium speciation, including the determination of elemental chromium Cr(0), was carried out in aerosols of different welding processes for stainless steel. These analyses showed distinct differences in the distribution of chromium species in the welding process and can be used as an exact calibration method for routine methods in this important monitoring of workplaces. Rodriguez-Gonzalez et al.[60] used different spike solutions for the determination of butyltin compounds by isotope dilution analysis. The developed method was used for the determination of butyltin compounds in PACS-2 certified reference material. Additionally, those spike solutions were evaluated during the course of an interlaboratory exercise organized by the National Research Council of Canada and the Laboratory of the Government Chemist (UK) in order to quantify tributyltin (TBT) in a pilot sediment. The aim of this study was to evaluate the capabilities of ID-MS to reduce uncertainty in the certification of reference materials for the speciation of organotin compounds. All participants were supplied with a 17Sn-enriched TBT solution from the Laboratory of the Government Chemist (UK). The authors also performed the analysis of the pilot sediment using a 119Sn-enriched spike (mixed mono-, di-, and tributyltin) and a 118Sn–119Sn double spike. The use of these additional spike solutions not only allowed for the determination of monobutyltin and dibutyltin in the pilot sediment but also for the evaluation and correction of possible extraction-derived rearrangement reactions. Moser-Veillon et al.[61] used stable isotope tracer methodology for the determination of selenite and seleno-methionine in lactating, nonlactating, and never-pregnant women.

11.2.2 Activation Analysis

In activation analysis, radioactivity is produced in metal ions by bombarding them with neutrons, photons, or charged particles. The radioactivity produced is directly proportional to the concentration of metal ions, which is measured by sophisticated devices. The chemical speciation for certain elements can be facilitated to a great extent by incorporating a single radioisotope into the system and measuring the radiation of the isolated species.[55] This radio spiking can be applied to *in vitro* and *in vivo* experiments. The radionuclides are, however, also present as anthropogenic contaminants from various nuclear fission activities. The radio tracer should be added under such conditions that it behaves in the same way as the isotope it represents. It should possess an adequate radioactive half-life, preferably by a γ-emitter because of the ease of detection. Radiotracer labeling is now widely used to study the speciation of many elements. Determination of the location of a radio isotracer in a cell by autoradiography proved to be impractical because of the lack of resolution. Hirose and Aoyama[62] reported the speciation of ^{239}Pu and ^{240}Pu using α-spectrometer in surface water from the U.S. Western North Pacific region. Chai et al.[63] discussed the speciation of trace elements in biological and environmental samples by nuclear analytical techniques, mainly molecular activation analysis, position-sensitive spectrometry with a variety of exciting sources, and synchronous-radiation-based analytical techniques (although radioisotope or enriched stable isotope-based speciation techniques are also used). The authors also discussed the merits and drawbacks of the nuclear analytical techniques, as there were reagent blanks, contamination, and artifacts. Chat et al.[64–67] used neutron activation analysis for the speciation of arsenic, technetium, and eptunium metal ions. The speciation of some metal ions using radiochemical techniques is summarized in Table 11.1.

11.3 MISCELLANEOUS METHODS

The most commonly used methods for metal ion speciation were discussed in Chapter 2 through Chapter 11. Of course, these techniques are the only choices for metal ion speciation, but they are associated with certain limitations that were discussed in the individual chapters. The growing scientific knowledge is compelling scientists to explore new approaches to metal ion speciation. Therefore, some new trends are emerging in the area of metal ion speciation. Biosensors and chemical sensors are gaining importance in this area. However, only a few reports are available on this issue. Many metal ions in their different oxidation states were detected using various biosensors.[68] Recently, Pons et al.[69] reported an optical fiber reflectance sensor coupled to a multisyringe flow injection system for the determination and speciation of iron in fresh- and seawater samples. The authors reported 0.001 μg as the detection limit with 2.2% standard deviation. Furthermore, the applicability of this system to real-world samples was advocated by the authors.

11.4 CONCLUSION

Electrochemical methods are good detection techniques for metal ions, but the speciation of metal ions required the conversion of one form into another. This chemical conversion is a tedious, time-consuming job that, sometimes, involves adding errors due to the good chemical reactivity of ionic forms of a particular metal ion. The radiochemical methods are also useful for the speciation of radionuclides at trace levels, but these are not universal for speciation purposes, as they cannot be used for nonradioactive metal ions. Sometimes, the targeted metal ion is destroyed or changed oxidation states, which create the rates of error. Briefly, electrochemical and radiochemical techniques could not be routinely used for laboratory analysis; however, these techniques can be used as good detectors in chromatographic and capillary electrophoretic technologies.

REFERENCES

1. Bond AM, Modern *Polarographic Methods in Analytical Chemistry*, Marcel Dekker, New York (1980).
2. Templeton DM, Ariese F, Cornelis R, Danielsson LG, Muntau H, van Leeuwen HP, Lobinski R, *Pure Appl. Chem.*, 72: 1453 (2000).
3. Florence TM, *Analyst*, 111: 489 (1986).
4. Williams G, D'Silva C, *Analyst*, 119: 2337 (1994).
5. Taillefert M, Luther GW, Nuzzio DB, *Electroanalysis*, 12: 401 (2000).
6. IUPAC Commission on Electroanalytical Chemistry, *Pure Appl. Chem.*, 45: 81 (1976).
7. Fogg AG, Wang J, *Pure Appl. Chem.*, 71: 891 (1999).
8. Rajeshwar K, Ibanez JG, *Environmental Electrochemistry: Fundamentals and Application in Pollution Abatement*, Academic Press, New York (1997).
9. Van Loon JC, Barefoot RR, *Analyst*, 117: 563 (1992).
10. Galus Z, *Fundamentals of Electrochemical Analysis*, Ellis Horwood, Chichester (1976).
11. Bockris JON, Reddy AKN, *Modern Electrochemistry*, 2nd ed., Plenum Press, New York (1998).
12. Bard AJ, Faulkner LR, *Electrochemical Methods: Fundamental and Applications*, 2nd ed., John Wiley & Sons, New York (2001).
13. Brown SD, Kowalski BR, *Anal. Chem.*, 51: 2133 (1979).
14. Van Leeuwan HP, *J. Electroanal. Chem.*, 99: 93 (1979).
15. Buffle J, *J. Electroanal. Chem.*, 125: 273 (1979).

16. Nawar N, El-Askalany ME, El-Defrawy MM, *Analyst*, 120: 2505 (1995).
17. Rasul SB, Munir AKM, Hossain ZA, Khan AH, Alauddin M, Hussam A, *Talanta*, 58: 33 (2002).
18. Vancara I, Vytas K, Bobrowski A, Kalcher K, *Talanta*, 58: 45 (2002).
19. Schwarz J, Henze G, Thomas FG, *Fresenius' J. Anal. Chem.*, 352: 479 (1995).
20. Huang H, Dasgupta PK, *Anal. Chim. Acta*, 380: 27 (1999).
21. Sauve S, Norvell WA, McBride M, Hendershot W, *Environ. Sci. & Technol.*, 34: 291 (2000).
22. Bobrowski A, Bas B, Dominik J, Niewiara E, Szalinska E, Vignati D, Zarebski J, *Talanta*, 63: 1003 (2004).
23. van den Berg CMG, *Analyst*, 117: 589 (1992).
24. Batley GE, Florence TM, *J. Electroanal. Chem.*, 55: 23 (1974).
25. Liu J, Wang X, Chen G, Gan N, Bi S, *Analyst*, 126: 1404 (2001).
26. Liu J, Bi S, Yang L, Gu X, Ma P, Gan N, Wang X, Long X, Zhang F, *Analyst*, 127: 1657 (2002).
27. Oomen AG, Tolls J, Sips AJ, Van den Hoop MA, *Arch. Environ. Contam. Toxicol.*, 44: 107 (2003).
28. Jagner D, Josefson M, Westerlund S, *Anal. Chem.*, 53: 2144 (1981).
29. Munoz E, Palmero S, *Talanta*, 65: 613 (2005).
30. Jagner D, *Analyst*, 107: 593 (1982).
31. Perry CC, Shafran KL, *J. Inorg. Biochem.*, 87: 115 (2001).
32. Jakusch T, Jin W, Yang L, Kiss T, Crans DC, *J. Inorg. Biochem.*, 95: 1 (2003).
33. Hua C, Jagner D, Renman L, *Anal. Chim. Acta*, 201: 263 (1987).
34. Huiliang H, Jagner D, Renman L, *Anal. Chim. Acta*, 207: 37 (1988).
35. Svancara I, Vytras K, Boborowski A, Kalcher K, *Talanta*, 58: 45 (2002).
36. Jagner D, Wang Y, Ma F, *Electroanalysis*, 8: 862 (1996).
37. Jurica L, Manova A, Dzurov J, Beinrohr E, Broekaert JAC, *Fresenius' J. Anal. Chem.*, 366: 260 (2000).
38. Adeloju SB, Young TM, Jagner D, Batley GE, *Anal. Chim. Acta*, 381: 207 (1999).
39. Wang YQ, Cui LJ, *Guangdong Weiliang Yuansu Kexue*, 6: 59 (1999); *Chem. Abstr.*, 130: 349281 (1999).
40. Morrison GMP, Florence TM, *Anal. Chim. Acta*, 209: 97 (1988).
41. Batley GE, Matousek JP, *Anal. Chem.*, 52: 1570 (1980).
42. Leon LE, Sawyer DT, *Anal. Chem.*, 53: 706 (1981).
43. O'Halleran RJ, *Anal. Chim. Acta*, 140: 51 (1982).
44. Colombini MP, Fuoco R, *Talanta*, 30: 901 (1983).
45. Hamilton TW, Ellis J, Florence TM, *Anal. Chim. Acta*, 110: 87 (1979).
46. Batley GE, Florence TM, *J. Electroanal. Chem.*, 61: 205 (1975).
47. Phillips SL, Shain I, *Anal. Chem.*, 34: 262 (1962).
48. Kubota H, *Anal. Chem.*, 32: 610 (1960).
49. Heyrovsky J, Kuta J, *Principles of Polarography*, Academic Press, New York (1966).
50. Friedlander G, Kennedy JW, Macias ES, Miller JM, *Nuclear and Radiochemistry*, 3rd ed., Wiley Interscience, New York (1981).
51. Brune D, Forkman B, Persson B, *Nuclear Analytical Chemistry*, Chartwell-Bratt (1984).
52. Geary W, in *Radiochemical Methods, Analytical Chemistry by Open Learning*, James AM (Ed.), John Wiley & Sons, Chichester (1986).

53. Clough R, Truscatt J, Belt ST, Evans EH, Fairman B, Catterick T, *Appl. Spectrosc. Rev.*, 38: 101 (2003).
54. Hill S, Pitts L, Fisher A, *Trends Anal. Chem.*, 19: 120 (2000).
55. Cornelis R, *Analyst,* 117: 583 (1992).
56. Sargent M, Harte R, Harrington C (Eds.), *Guidelines for Achieving High Accuracy in Isotope Dilution Mass Spectrometry*, Royal Society of Chemistry, Cambridge (2002).
57. Lambertsson L, Lundberg E, Nilsson M, Frech W, *J. Anal. At. Spectrom.*, 16: 1296 (2001).
58. Turner J, Hill S, Evans EH, Fairman B, Wolf Brich C, *J. Anal. At. Spectrom.*, 15: 743 (2000).
59. Nusko R, Heumann KG, *Fresenius' J. Anal. Chem.*, 357: 1050 (1997).
60. Rodriguez-Gonzalez P, Encinar JR, Garcia Alonso JI, Sanz-Medel A., *Analyst,* 128: 447 (2003).
61. Moser-Veillon PB, Mangels AR, Patterson KY, Veillon C, *Analyst,* 117: 559 (1992).
62. Hirose K, Aoyama M, *Anal. Bioanal. Chem.*, 372: 418 (2002).
63. Chai Z, Mao X, Hu Z, Zhang Z, Chen C, Feng W, Hu S, Ouyang H, *Anal. Bioanal. Chem.*, 372: 407 (2002).
64. Chatt A, *Studies on Speciation of Technetium and Neptunium in Groundwater, 31st Annual Conference of the Spectroscopy Society of Canada*, St. Jovite, PQ, October 1–3 (1984).
65. Chatt A, *Neutron Activation Analysis for Speciation of Trace Elements*, Ghana Atomic Energy Commission, Legon-Accra, Ghana, November 27 (2001).
66. Chatt A, *Simultaneous Multielement Speciation Analysis Using Neutron Activation, 85th Canadian Society for Chemistry Conference & Exhibition*, Vancouver, BC, June 1–5 (2002).
67. Chatt A, *Speciation Neutron Activation Analysis for Arsenic in Marine Fish, 14th Radiochemical Conference*, Marianske Lazne, Czech Republic, April 14–19 (2002).
68. Bontidean I, Csöregi E, in Cornelis R, Caruso J, Crews H, Heumann K, *Handbook of Elemental Speciation: Techniques and Methodology*, John Wiley & Sons, Chichester, UK (2003).
69. Pons C, Forteza R, Cerda V, *Anal. Chim. Acta*, 528: 197 (2005).
70. Desideri D, Meli MA, Roselli C, Testa C, Degetto S, *J. Radio. Anal. Nucl. Chem.*, 248: 727 (2001).

12 Perspectives and Legislation of Metal Ion Speciation

The distribution and toxicities of metal ion species were discussed in Chapter 2 and Chapter 3. However, metal ions occur naturally in rocks, soils, water, foods, and so on, in both useful and harmful forms and concentrations. Serious health problems and diseases can be due to high concentrations of metal ions. Many metals are important from environmental and health points of view, especially in the study of human diseases (pathology), because of their potential toxic effects on human organs. Due to different toxicities of a metal ion species, the International Union for Pure and Applied Chemistry (IUPAC) referred to speciation as the chemical form or compound in which an element occurs in a living system or in the environment.[1] Furthermore, this concept reveals the quantitative distribution of an element. The toxicological effect of an administered agent is correlated with the chemical speciation prevailing rather than with the total dose. Scientists and the public have entirely different interpretations of the word "safe," and the former class is reliant on chemical speciation. Theoretically, "safe" implies "protection from harm" and "free from danger." However, in practice, safety is firmly linked to risk and a compromised state, which is sought to involve the risk, cost, and benefits. Basically, scientific risk assessment and communication of the risks determined to be due to different metal ion species are of great concern to the public.[2] The public is encouraged by the press and scientific debates to involve politicians in the legislation and funding of projects dealing with chemical and environmental toxicities. In view of the toxicities associated with different metal ion species, their distribution and control in the environment are essential and urgent in the present developing world. Therefore, legislations are required to control the distribution and toxicities of metal ion species.

12.1 THE FUTURE OF SPECIATION

Speciation techniques were discussed in detail in Chapter 5 through Chapter 11. An analytical comparison of these chapters indicates the popularity of chromatographic methods followed by capillary electrophoresis spectroscopic and other techniques. Among various chromatographic modalities, ion chromatography is mainly intended for metal ion separation and speciation. Therefore, this modality is being used worldwide for metal ion speciation in different biological and environmental samples. At present, these techniques are capable of providing speciation data for legislation purposes. But, still, the speciation of some metal ions requires costly instrumentation, particularly for the use of element-selective detectors, such as atomic absorption spectroscopy (AAS), inductively coupled plasma (ICP) spectrometry, and mass spectrometry (MS). which are not often feasible for conducting speciation analyses in developing and underdeveloped countries. Therefore, the role of analytical techniques is crucial in distributing speciation technologies worldwide. The techniques should involve inexpensive paraphernalia that can be adapted to the needs of every nation, worldwide. And, as reproducibility in metal analyses is one of the serious and challenging issues for scientists, advancements are required in analytical techniques in terms of least interference and low limits of detection in unknown matrices. The techniques should be developed with good selectivity, sensitivity, and reproducible results. Eaton and Abou-Shakra[3] presented views of future developments in metal ion speciation techniques.

The future development of analytical techniques is a broad topic and cannot be described in detail in this chapter. However, some important points that may be used to make the speciation technique popular are briefly discussed. Applications of speciation methodologies may be increased by developing inexpensive and highly selective sample preparation devices and mobile and stationary phases and detectors. Moreover, improvements in the linking of chromatographic and capillary electrophoretic techniques could render them ideal techniques for metal ion speciation. Method validation is one of the required and important processes needed to optimize metal ion speciation. Control of the chromatograms, data acquisition and processing, and multiple combination of the instrument may be a boon in metal ion speciation. Briefly, growth in the advancement of metal ion speciation techniques is increasing continuously, and we hope that metal ion speciation methodologies will be selective, sensitive, reproducible, and inexpensive in the near future. It is estimated that the prospectives of metal ion speciation are bright and will attract scientists and regulatory authorities for the betterment of human beings.

12.2 RISK ASSESSMENT AND LEGISLATION

Before formulating certain rules and regulations, discussion of the determination of risk assessment due to metal ion species is very important. Risk assessment should take into account the consideration of socioeconomic aspects as well as the balancing of risk against benefits. The main aim of risk assessment is to decide and fix the maximum value of metal ion species at which there should be no harm to the ecosystem. It is worth mentioning here that such a fixed value is not possible, as any arbitrary value may be toxic to one or more species, while it may be nontoxic to other species in the ecosystem. Therefore, the minimum value of any metallic species should be correlated to the particular species. In

1995, the Organization for Economic Corporation and Development (OECD) recommended a test for metal ions and other metallic compounds.[4] Some methods were used to assess the toxicities of metal ions, and these are briefly described herein.

Risk assessment due to metal ions may be estimated by using the uncertainty factors approach, the statistical extrapolation method, and the homeostasis (adaptation) approach. In the first approach, an uncertainty factor is applied to EC_{50}, and this factor is chosen to take into account the difference between laboratory and field populations, the difference between acute and chronic exposures, statistical uncertainties, and inter- and intraspecies variations. The precautionary nature of the uncertainty factor approach leads to the calculation of a predicted no-affect concentration (PNEC) value, which is much lower than the true value. This lower value may be due to the fact that the concentration of the studied metal ion is higher in the laboratory than in the field.[5] This approach was given approval at the international level in 1992 by the OECD.[6] The description of this method was found in a U.S. Environmental Protection Agency (US EPA) assessment scheme[7] and a European Commission (EC) technical guidance document on risk assessment.[8] In the statistical extrapolation method developed by Koojiman,[9] calculations of the hazardous concentrations at the lower boundary of concentrations may be harmful to a specific living species. An assessment factor (*T*) is calculated and applied to the geometric mean of the available LC_{50} values.[10] Normally, the assessment factor is inversely proportional to the number of species growing in an ecosystem. Other authors later modified this method for its applications under varied environmental conditions.[11,12] The homeostasis method deals with the adaptation generated in the organisms due to the presence of toxic metal ions around them. The organisms regulate their uptake of metal ions from natural background levels in such a way that their internal concentration remains constant (homeostasis). These aspects were considered by van Assche et al.[13]

12.3 METAL ION SPECIATION AND LEGISLATION

There are few regulations controlling the use and concentrations of metal ions in the environment globally. Certain international agencies, such as the Food and Agricultural Organization of the United Nations (FAO), the World Health Organization (WHO), the US EPA, and the European Commission, formulated the maximum permissible limits of metal ions in drinking water and different foodstuffs. Some other nation-based agencies, individual in different countries, also control the distribution and permissible limits of metal ions. But it is a matter of great thought and concern that none of the agencies came forward to deal with the toxicities and permissible limits of different metal ion species of a particular metal ion — speciation. Because of the wide distribution and toxicities of different species of various metal ions, it is crucial to design certain rules and regulations related to the use of such types of metal ions and permissible limits. Of course, no regulatory authorities put forward such attempts; however, some scientific reports are available that deal with the issue of metal ion speciation legislation. Some courses on medical geology and toxic trace elements are conducted by the U.S. Armed Forces Institute of Pathology (AFIP) in collaboration with the U.S. Geological Survey (USGS), the International Union of Geological Sciences (IUGS), the Commission on Geological Sciences and Environmental Planning (CGSEP), the United Nations Educational, Scientific and Cultural Organization (UNESCO), the International Working Group on Medical Geology (IWGMG), the International Association for Medical Geology (IAMG), and the International Council of Scientific Unions (ICSU) to address the topics and research opportunities on medical geology, environmental toxicology, geochemistry, and the health impacts of metal ions, trace elements, and other metal compounds. With particular emphasis on toxic metals and metal species, the courses were designed for environmental scientists, toxicologists, chemical analysts, environmental health care professionals, industrial hygienists, geologists, ecologists, hydrologists, regulatory authorities (working for the protection of the environment), and medical professionals.

12.4 INTERNATIONAL SCENARIO OF LEGISLATION

As discussed above, many legislations have been put forward regarding the toxicities and permissible concentrations of metal ions, but only a few regulations deal with the toxicities and concentrations of specific species of metal ions. Some background papers presented by the FAO, the WHO, and the Joint Expert Committee of Food Additives and Contaminants (JECFA) agencies indicated information regarding the chemical nature as well as the uptake and fate of the various species. An international organization dealing with food contaminants was established at Codex Alimentarius under the umbrella of FAO and WHO. The Codex Alimentarius Commission (CAC) controls the amounts of contaminants, food additives, and toxins in foodstuffs.[14,15] Since 1962, this commission has been formulating specific horizontal or general standards for food products.[14] The European Union also formulated certain standards regarding the maximum level of various contaminants in foodstuffs.[16,17] The Codex Alimentarius Commission developed a position paper indicating the maximum permissible level of arsenic metal ion. The maximum permissible limit of inorganic arsenic was designed and ascertained by the Joint Expert Committee on Food Additives and Contaminants as 15 μg/kg body weight. The permissible concentration of organic arsenic was not fixed due to the lack of toxicological data. In the United States, the US EPA fixed certain parameters for releasing industrial effluents containing metal ions into the environment.[18] The U.S. Occupational Safety and Health Administration (OSHA) is also looking for permissible exposure limits of metal ions. As per this organization, the different maximum permissible limits for various chromic acid, chromates, chromium (III), and chromium metal[19] are fixed. In addition, regulatory authorities in the United States and the United Kingdom designed the different maximum permissible limits for various nickel compounds, that is, nickel carbonyl and nickel subsulfide.[20]

12.5 CONCLUSION

It is estimated that about 75,000 chemicals are in use everyday, with different species. Some are also being inhaled or are in direct contact of human beings; hence, speciation is an integral part of the specification criteria, because only certain defined chemical species are permitted as sources of the essential elements. In this way, in the lack of legislation governing the speciation concept, people are consuming toxic metal ion species, which results in some serious side effects and ailments. The seriousness of the problem of speciation was previously discussed in Chapter 3. However, it is noteworthy that the lack of awareness about speciation may be a serious and challenging problem to the world community in the near future due to the geometrical contamination of our environment.

Little attention has been paid to regulating metal ion speciation at the international level. However, some international legislations concerning trace elements in the environment are under development. But, again, it is a matter of great thought and concern that little attention about the speciation concept is being given during their preparation. The reason behind the absence of metal ion speciation is the lack of metal ion species, toxicological data, and awareness among communities. However, in the present situation, the analytical methodologies are sufficient and capable of providing the data to the regulatory authorities, but it still needs advancement. In view of these facts, it is important to determine the toxicities of all metallic species in order to facilitate decision making for the speciation regulators. The toxicological data should arouse social awareness. The analytical techniques should be developed in such a way that metal ions and their speciation states, which can determine metal ions and their speciation states in the real sample with high selectivity, sensitivity, and reproducibility.

The idea of speciation is extremely difficult to grasp and to accept by non-scientific communities; hence, we must do far more to contribute toward public understanding of our research by having a better understanding of their concerns and by spending more time communicating. A social reconstruction of science is required, accepting that modern technology may frighten many laypersons. Briefly, there is a great need for a rethink of risk, safety, and communication associated with metal ion speciation at international levels. Politicians and regulatory authorities should come forward to stipulate the regulations and legislations to control the speciation of metal ions in our environment. Let us hope for fast actions on this issue.

REFERENCES

1. Templeton DM, Ariese F, Cornelis R, Danielsson LG, Muntau H, van Leeuwen HP, Lobiniski R, *Pure Appl. Chem.*, 72: 1453 (2000).
2. Williams DR, *What is Safe*? The Royal Society of Chemistry, Cambridge (1998).
3. Eaton AN, Abou-Shakra FR, in *Handbook of Elemental Speciation: Techniques and Methodology*, Cornelis R, Caruso J, Crew H, Heumann K (Eds.), John Wiley & Sons, p. 461, Chichester (2003).

4. Illing HPA, *Toxicity and Risk — Context, Principles and Practice*, Taylor & Francis, Basingstoke (2001).
5. ECETOC, *Environmental Hazard Assessment of Substances*, ECETOC, Brussels (1993).
6. OECD, Report of the OECD Workshop on the Extrapolation of Laboratory Aquatic Data to the Real Environment, OECD, Paris (1992).
7. US EPA, Estimating Concern Levels for Concentrations of Chemicals Substances in the Environment, US EPA, Washington, DC (1984).
8. European Commission, Technical Guidance Document in Support of Commission Directive 93/67/EEC on Risk Assessment for New Notified Substances and Commission Regulation (EC), No. 1488/94 on Risk Assessment for Existing Substances, Office for Official Publications of the European Communities, Luxembourg (1996).
9. Koojiman SALM, *Water Res.*, 21: 269 (1987).
10. ACGIH Worldwide, *TLVs® and BEIs® — Threshold Limit Values for Chemical Substances and Physical Agents and Biological Exposure Indices*, ACGIH Worldwide, Cincinnati, OH (2001).
11. van Straalen NM, Denneman CAJ, *Ecotoxicol. Environ. Safety*, 18: 241 (1989).
12. Aldenberg T, Slob W, *Confidence Limits for Hazardous Concentration Based on Logistically Distributed NOEC Toxicity Data*, RIVM, Bilthoven (1991).
13. van Assche F, van Tilborg WJM, Waeterschoot H, *Environmental Risk Assessment for Essential Elements Case Study, Zinc*, International Workshop on Risk Assessment of Metals and Their Inorganic Compounds, Angers, November 13–15, 1996. The International Council on Metals and the Environment, Ottawa, Canada (1996).
14. Codex Alimentarius, General Standards for Contaminants and Toxins in Food, CODEX-STAN 193-1995 (Rev, 1-1997), Joint FAO/WHO Food Standards Programme, FAO, Rome (2000).
15. Berg T, Development of the Codex Standards for Contaminants and Toxins in Food, in Rees N, Waston D (Eds.), *International Standards for Food Safety*, Aspen Publishers, Gaithersburg, MD (2000).
16. European Council Regulation EC 315/93, February 8, 1993, on Community Procedures for Contaminants in Food, p. 1 (1993).
17. European Commission Regulation No. 466/2001, March 8, 2001, for Setting Maximum Levels for Certain Contaminants in Foodstuffs, p. 1. The Subsequent Amendments and EC Directive 2001/22/EC, March 8, 2001, Lying Down Sampling and Analytical Methods for the Official Control of Lead, Cadmium and Mercury Levels in the Foodstuffs, p. 14 (2001).
18. Effluent Limitations Guidelines and New Source Performance Standards for the Metal Products and Machinery Point Source Category, US EPA *Federal Register*: May 13, 2003 (Volume 68, Number 92), Rules and Regulations, p. 25685 (2003).
19. Darrie G, The Importance of Chromium in Occupational Health, in Ebdon L, Pitts L, Cornelis R, Crew H, Donard OXF, Quevauvilliter P (Eds.), *Trace Element Speciation for Environment, Food and Health*, The Royal Society of Chemistry, Cambridge, p. 315 (2001).
20. Williams SP, Occupational Health and Speciation Using Nickel and Nickel Compounds, in Ebdon L, Pitts L, Cornelis R, Crew H, Donard OXF, Quevauvilliter P (Eds.), *Trace Element Speciation for Environment, Food and Health*, The Royal Society of Chemistry, Cambridge, p. 297 (2001).

Index

A

B

C

D

E

F

G

H

I

T

U

V

X